Polymer Nanocomposites for Energy Applications

Polymer Nanocomposites for Energy Applications

T. Daniel Thangadurai
Manjubaashini Nandhakumar
Sabu Thomas
Ange Nzihou

WILEY-VCH

Authors

Prof. T. Daniel Thangadurai
Department of Chemistry
KPR Institute of Engineering and Technology
Coimbatore 641407
Tamilnadu
India

Dr. Manjubaashini Nandhakumar
University of Madras
National Centre for Nanoscience & Nanotechnology
Chennai
India

Prof. Sabu Thomas
Mahatma Gandhi University
Centre for Nanoscience & Nanotechnology
Kerala
India

Prof. Ange Nzihou
Université de Toulouse
IMT Mines Albi
RAPSODEE CNRS UMR 5302
Campus Jarlard
F.81013 Albi Cedex 09
France

Princeton University
School of Engineering and Applied Science
Princeton
NJ, USA

and

Princeton University
Andlinger Center for Energy and the Environment
Princeton
NJ, USA

Cover Image: © Vink Fan/Shutterstock.com

Library of Congress Card No.: applied for

British Library Cataloguing-in-Publication Data
A catalogue record for this book is available from the British Library.

Bibliographic information published by the Deutsche Nationalbibliothek
The Deutsche Nationalbibliothek lists this publication in the Deutsche Nationalbibliografie; detailed bibliographic data are available on the Internet at <http://dnb.d-nb.de>.

Print ISBN: 978-3-527-35048-3
ePDF ISBN: 978-3-527-83853-0
ePub ISBN: 978-3-527-83854-7
oBook ISBN: 978-3-527-83855-4

Typesetting Straive, Chennai, India
Printing and Binding CPI Group (UK) Ltd, Croydon, CR0 4YY

C120382_250822

Contents

1

Origin of Polymer Materials

1.1 History of Polymers

A polymer is an organic compound, natural or synthetic, with a high molecular weight made up of repetitive structural units. Large-size chains are formed from the covalent union of various monomer units [1]. A single-polymer molecule may consist of hundreds to a million monomers and may have a linear, branched, or network structure. Polymeric solid can be thought of as a material that contains many chemically bonded parts or units which themselves are bonded together to form a solid. The word polymer means "many parts." Two industrially important polymeric materials are plastics and elastomers. Plastics are a large and varied group of synthetic materials that are processed by forming or molding into a shape. There are many types of plastics, such as polyethylene (PE) and nylon [2]. Elastomers or rubbers can be elastically deformed a large amount when a force is applied to them and can return to their original shape when the force is released. Polymers have many properties that make them attractive to use in certain conditions. Many polymers (i) are less dense than metals or ceramics, (ii) resist atmospheric and other forms of corrosion, (iii) offer good compatibility with human tissue, and (iv) will exhibit excellent resistance to the conduction of electrical current [3]. Covalent bonds hold the atoms in the polymer molecules together and secondary bonds then hold groups of polymer chains together to form the polymeric material [4]. Copolymers are polymers composed of two or more different types of monomers.

Many examples of synthetic polymers have been mentioned; polyesters or nylons are used more than others like the one which is used for medical applications for organs, degradable sutures, etc. [5]. Based on economic and application considerations, plastic materials can be divided into a commodity and engineering plastics. In the first group, PE, polypropylene (PP), and polyvinyl chloride (PVC) are considered, and in the second, polycarbonate (PC), polyether ether ketone (PEEK), polyimide (PI), etc., are considered. Fibers, natural, artificial (modified natural), and synthetic are characterized by high aspect ratio, high strength and modulus, and other properties depending on their applications. Elastomers exhibit the ability to stretch and retract rapidly [6]. The commercial development of this poly-fiber (PF)

Polymer Nanocomposites for Energy Applications, First Edition.
T. Daniel Thangadurai, Manjubaashini Nandhakumar, Sabu Thomas, and Ange Nzihou.

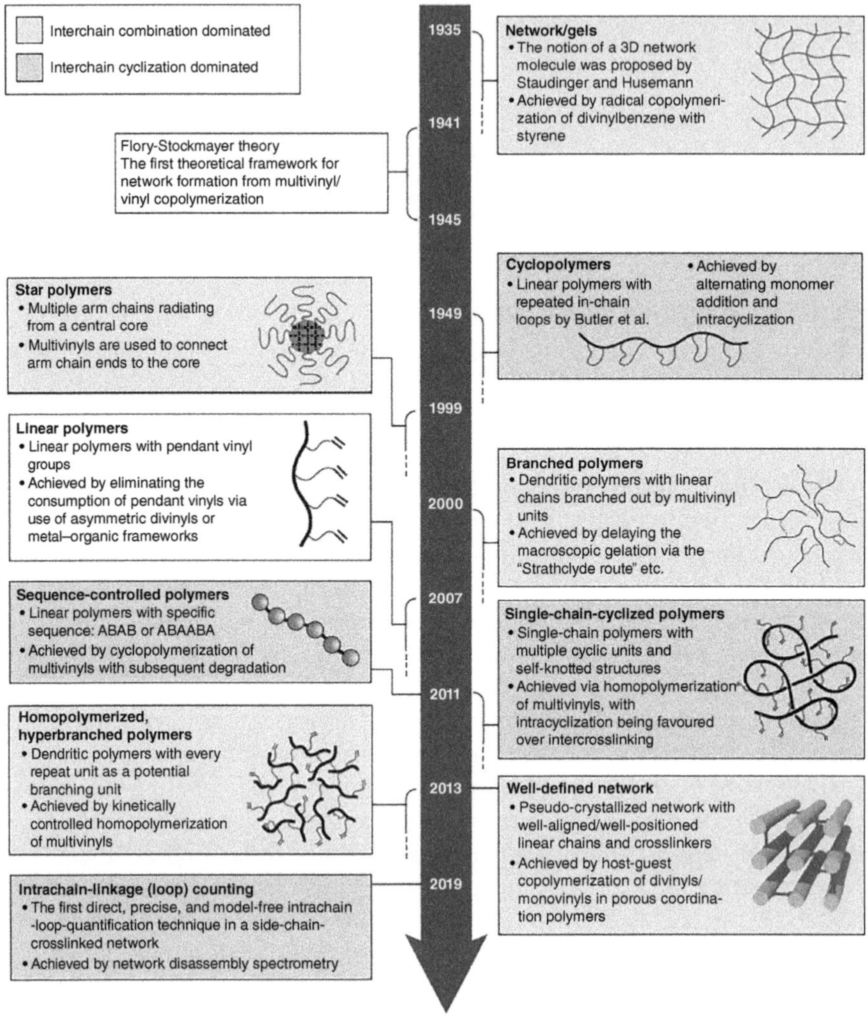

Figure 1.1 Timeline of architecturally complex polymers. Source: Gao et al. [8]/Springer Nature.

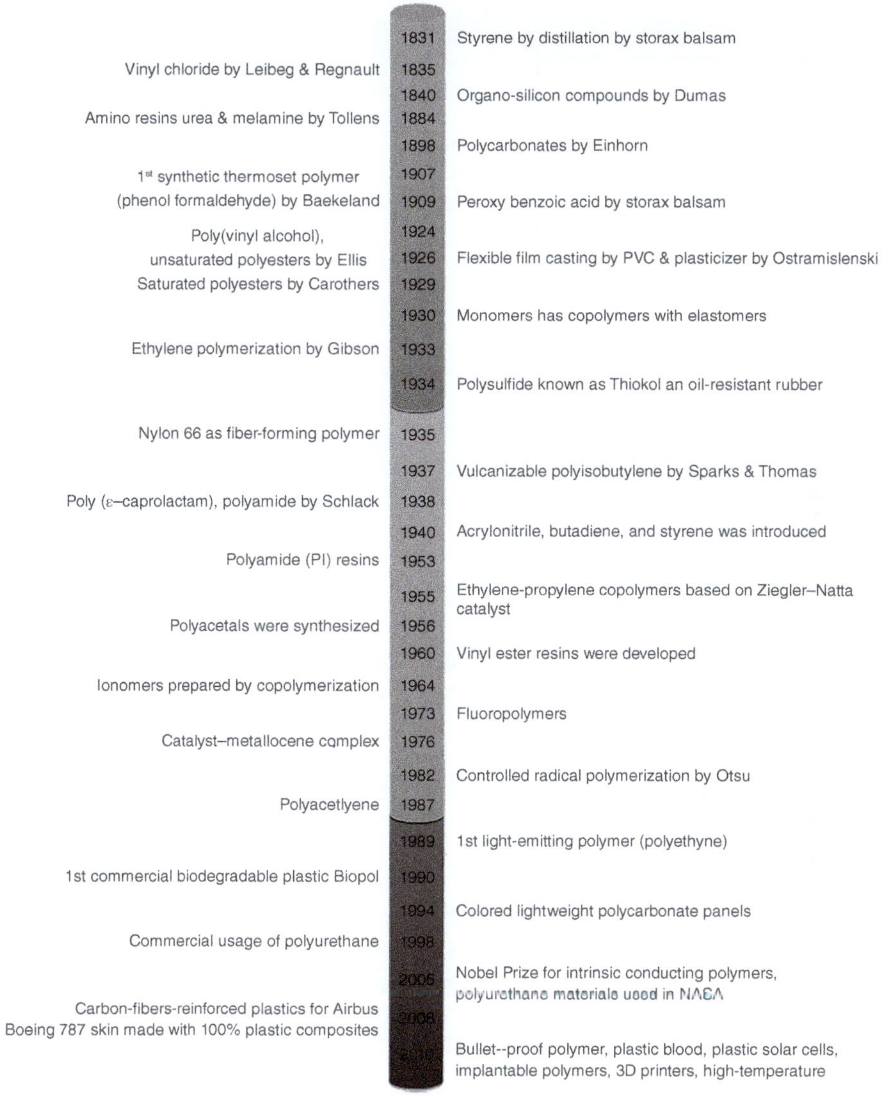

Figure 1.2 History of growth of polymer materials.

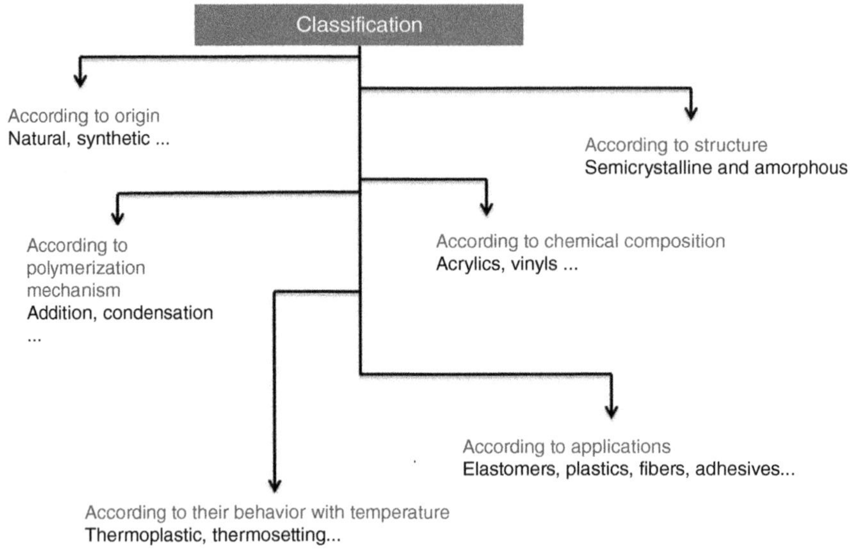

Figure 1.3 Classification of polymers.

material is considered to be the beginning of the truly synthetic plastic era and the plastic industry, although cellulose nitrate (semisynthetic) had been known and in use for some time [7] (Figures 1.1–1.3).

1.1.1 Examples of Polymers

Polyethylene PE: most popular plastic	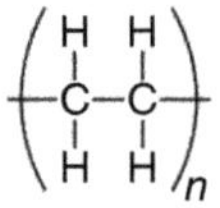
Polypropylene PP: used as plastic and as fiber	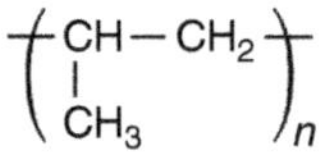
Polystyrene PS: economic and resistant.Styrofoam™: foam of PS	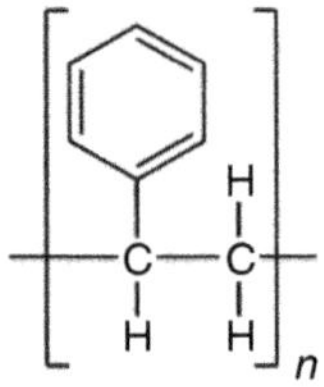
Polyvinylidene fluoride PVDF: high electric and fire resistance	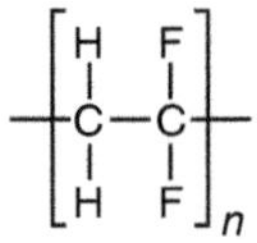

Polymethyl methacrylate PMMA: transparent plastic substitute for crystal	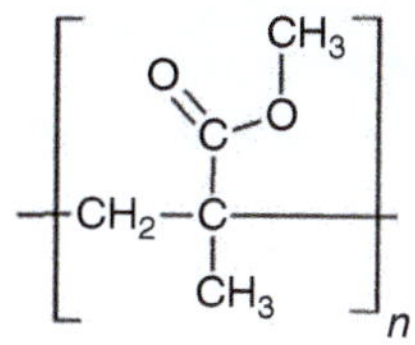
Nylon (polyamides): used as fibers	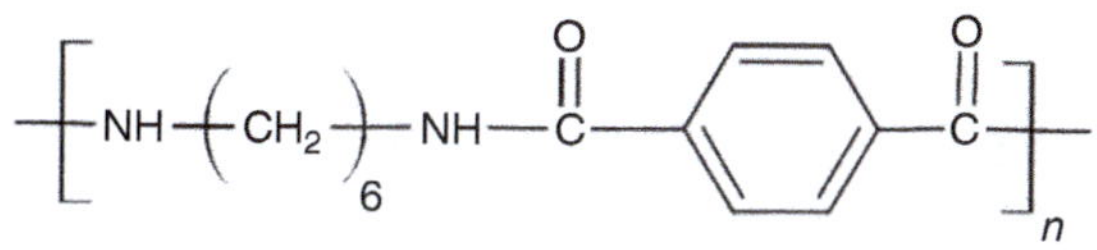

1.2 Types of Polymers

1.2.1 Based on Applications

Synthetic polymers can be classified into different types of materials:

Elastomers: Materials with very low modulus of elasticity and high extensibility.
Plastics: Polymers in which, when a sufficiently intense force is applied, they irreversibly deform.
Fibers: Present a high modulus of elasticity and low extensibility.
Coatings: Substances, normally liquid, that adheres to the surface of other materials.
Adhesives: Substances that combine a high adhesion and a high cohesion.

Polymer plasticscan be divided into two classes, thermoplastics and thermosetting plastics, depending on how they are structurally and chemically bonded. Thermoplastic polymers comprise the four most important commodity materials, such as PE, PP, polystyrene, and PVC. The term “thermoplastic” indicates that these materials melt on heating and may be processed by a variety of molding and extrusion techniques. Alternately, “thermosetting” polymers cannot be melted or remitted. Thermosetting polymers include alkyds, amino and phenolic resins, epoxies, polyurethanes, and unsaturated polyesters (Figure 1.4).

Figure 1.4 Types of synthetic polymers. Source: Stratasys Ltd., U.S. Department of Agriculture (USDA)/Wikimedia commons/Public Domain, Gigira/Shutterstock.

1.2.2 Classification Based on Temperature Effect

1.2.2.1 Thermosetting Polymers

Most polymers can be broadly classified as either thermoplastics or thermosets. The fundamental physical difference between these two is the bonding between molecular chains – thermoplastics have only secondary bonds between chains, while thermosets also have primary bonds between chains. The names are not only associated with the chemical structure of each but their general thermal and processing characteristics as well, since this basic structural difference greatly impacts material properties. Thermosetting polymers chemically decompose when heated, instead of flowing which is due to a crosslinked structure. Thermoplastic polymers can be melted or molded, while thermosetting polymers cannot be melted or molded in the general sense of the term. Thermoplastic and thermosetting polymers are sometimes identified by other names such as "linear" and "crosslinked," respectively. It should be noted that the term linear here applies to molecular structure and not to mechanical characteristics. The polymer can be a hard and stiff glass-like solid, a soft and flexible elastomeric rubber, or a viscous liquid depending only on working temperature as compared to two reference temperatures identified as the glasstransition temperature (T_g) and the melt temperature (T_m). All thermoplastic materials may exist in one of these three phases upon changes in the working temperature, while thermosetting polymers generally exist only in the first two phases [9]. The T_g and T_m for different polymers range from well below to well above ambient, and therefore a particular polymer may be glassy, elastomeric, or liquid at room temperature depending only on its chemical composition.

1.2.2.2 Thermoplastic Polymers

Thermoplastic polymers may be either amorphous or crystalline. Crystallinity is important to point out the degree of crystallinity by standards for crystalline metals, ceramics, and other materials [10]. That is, polymers are rarely over 50% crystalline. Crystalline polymers are often more dense than amorphous polymers due to the closer packing of their long-chain molecules and, in general, the following properties are enhanced: hardness, friction and wear, less creep or time-dependent behavior, corrosion resistance, and/or resistance to environmental stress cracking.

1.3 Properties of Polymers

A polymer is an organic material and the backbone of every organic material is a chain of carbon atoms. The carbon atom has four electrons in the outer shell. Each of these valence electrons can form a covalent bond to another carbon atom or a foreign atom. The key to the polymer structure is that two-carbon atom scan has up to three common bonds and still bond with other atoms. The elements found most frequently in polymers and their valence numbers are H, F, Cl, Br, and I with one-valence electron, O and S with two-valence electrons, N with three-valence electrons, and C and Si with four-valence electrons [11].

The ability for molecules to form long chains is vital to producing polymers. Consider the material polyethylene, which is made from ethane gas, C_2H_6. Ethane gas has two carbon atoms in the chain, and each of the two carbon atoms shares two valence electrons with the other. If two molecules of ethane are brought together, one of the carbon bonds in each molecule can be broken and the two molecules can be joined with carbon to carbon bond. After the two mersare joined, there are still two free valence electrons at each end of the chain for joining other mers or polymer chains [12]. The process can continue linking more mers and polymers together until it is stopped by the addition of another chemical that fills the available bond at each end of the molecule. This is called a linear polymer and is a building block for thermoplastic polymers.

1.3.1 Molecular Weight

Macromolecular molecular weight:

$$M_n = M_o X_n$$

where

M_n = number-average molecular weight
M_o = monomer molecular weight
X_n = degree of polymerization (average number of monomer units in a chain).
Mean molecular weight–definition of the various molecular weights

$$M_n = \sum M_i x_i$$

$$M_w = \sum M_i w_i$$

where

M_n = number-average molecular weight
M_w = weight-average molecular weight
M_i = mean molecular weight in size range i
x_i = fraction in number of molecules in range i
w_i = fraction in weight of molecules in range i.

1.3.2 Structural Aspects

The polymer chain is often shown in two dimensions, but it should be noted that it has a three-dimensional structure. Each bond is at 109° to the next and, therefore, the carbon backbone extends through space like a twisted chain of Tinker Toys. When stress is applied, these chains stretch and the elongation of polymers can be thousands of times greater than it is in crystalline structures. The length of the polymer chain is very important. As the number of carbon atoms in the chain is increased to beyond several hundred, the material will pass through the liquid state and become a waxy solid. When the number of carbon atoms in the chain is over a thousand, the solid material PE, with its characteristics of strength, flexibility, and toughness, is obtained. The state change occurs because as the length of the molecules increases, the total binding forces between molecules also increase. It should also be noted that the molecules are not generally straight but are a tangled mass. Thermoplastic materials, such as PE, can be pictured as a mass of intertwined worms randomly thrown into a pail. The binding forces are the result of van der Waals forces between molecules and mechanical entanglement between the chains. When thermoplastics are heated, there is more molecular movement and the bonds between molecules can be easily broken. This is why thermoplastic materials can be remelted. There is another group of polymers in which a single large network instead of many molecules is formed during polymerization. Since polymerization is initially accomplished by heating the raw materials and bringing them together, this group is called thermosetting polymers or plastics. For this type of network structure to form, the mers must have more than two places for bonding to occur; otherwise, only a linear structure is possible. These chains form jointed structures and rings and may fold back and forth to take on a partially crystalline structure. Since these materials essentially comprise one giant molecule, there is no movement between molecules once the mass has been set. Thermosetting polymers are more rigid and generally have higher strength than thermoplastic polymers. Also, since there is no opportunity for motion between molecules in a thermosetting polymer, they will not become plastic when heated (Figure 1.5).

The physical properties of a polymer depend not only on the type of monomers that make up the polymer but also on the stereochemical arrangements of the atoms. In a linear asymmetric polymer chain, the pendant groups can either be arranged into orderly configurations or they can be completely random. The steric order is called tacticity. If all chiral centers have the same configuration, the arrangement of the side groups is called *isotactic*, and if every other chiral center has the same

Figure 1.5 Representation of different molecular structures of the polymer.

Table 1.1 Melting and glass transition temperatures for some of the more common polymeric materials.

Polymer	T_g (atactic)	T_g (isotactic)	T_g (syndiotactic)
Poly(methyl acrylate)	281	272	299
Poly(ethyl acrylate)	249	253	263
Poly(methyl methacrylate)	378	319	433
Poly(n-butyl methacrylate)	293	249	361
Poly(isopropyl acrylate)	267	264	285
Poly(methyl α-chloroacrylate)	416	353	452
Poly(isopropyl α-chloroacrylate)	363	321	392

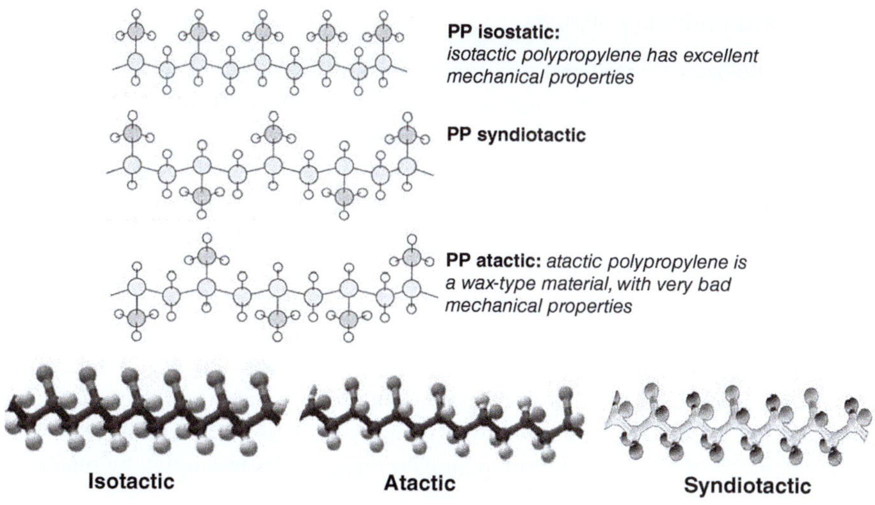

Figure 1.6 Different types of arrangements in polymeric chains.

arrangement, it is called *syndiotactic*, whereas a random arrangement of the side groups is called *atactic* or *heterotactic* (Table 1.1 and Figure 1.6).

An example of tacticity is PP.

1.3.3 Copolymers in Sequence

A copolymer is a polymer that is made up of two or more monomer species. Many commercially important polymers are copolymers, e.g. polyethylene-vinyl acetate (PEVA), nitrile rubber, and acrylonitrile butadiene styrene (ABS). The process in which a copolymer is formed from multiple species of monomers is known as copolymerization [13]. It is often used to improve or modify certain properties of plastics (Figure 1.7).

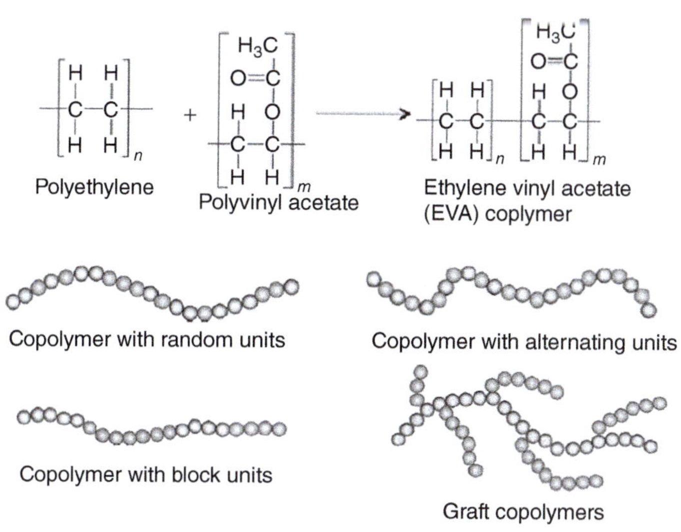

Figure 1.7 Types of copolymer sequences.

1.3.4 Crystallinity of Polymers

1.3.4.1 Solid-State Crystallinity

Crystallinity defines the degree of long-range order in a material and strongly affects its properties. The more crystalline the polymer, the more regularly aligned its chains. Increasing the degree of crystallinity increases hardness and density which is illustrated in poly(ethene). High-density polyethylene (HDPE) is composed of linear chains with little branching. Molecules pack closely together, leading to a high degree of order. This makes it stiff and dense, and it is used for milk bottles and drainpipes. The numerous short branches in low-density polyethylene (LDPE) interfere with the close packing of molecules, so they cannot form an ordered structure. The lower density and stiffness make it suitable for use in films such as plastic carrier bags and food wrapping. Often, polymers are semicrystalline, existing somewhere on a scale between amorphous and crystalline [14]. This usually consists of small crystalline regions (crystallites) surrounded by regions of an amorphous polymer.

1.3.4.2 Factors Favoring Crystallinity

In general, factors causing polymers to be more ordered and regular tend to lead to a higher degree of crystallinity:

a. *Fewer short branches*: allowing molecules to pack closely together
b. *Higher degree of stereoregularity*: syndiotactic and isotactic polymers are more ordered than atactic polymers
c. *More regular copolymer configuration*: having the same effect as stereoregularity [15].

1.3.5 Morphology of the Polymeric Crystals

Morphology refers here to the size and shape of crystals and crystal aggregates. For example, PE is the model semicrystalline polymer. The polymerization methods available during the decades have made it possible to make PE of different crystallinity and morphologies. The high-pressure process first developed in the 1930s yielded branched PE with ~50% crystallinity. Low-pressure processes, developed in the 1950s utilizing metal–organic chemistry, yielded linear PE (~75% crystallinity). Low- and medium-density PE was made by the low-pressure technique replacing part of the ethylene with higher 1-alkenes. The metallocene technology was put into commercial use in the 1990s, and it provides PE with a narrow molar mass distribution and a uniform distribution of comonomer units [16]. The recent availability of monodisperse *n*-alkanes with several hundred carbon atoms using a preparation method has provided new insight into several important aspects of PE morphology. PE is used in large quantities and for many different applications. The properties of PE are controlled by morphology (Figures 1.8 and 1.9).

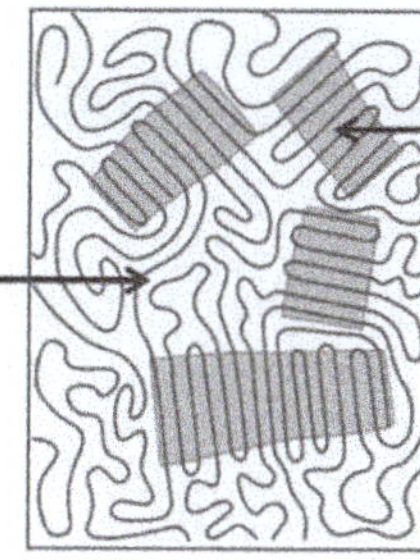

Figure 1.8 Polymeric monocrystals.

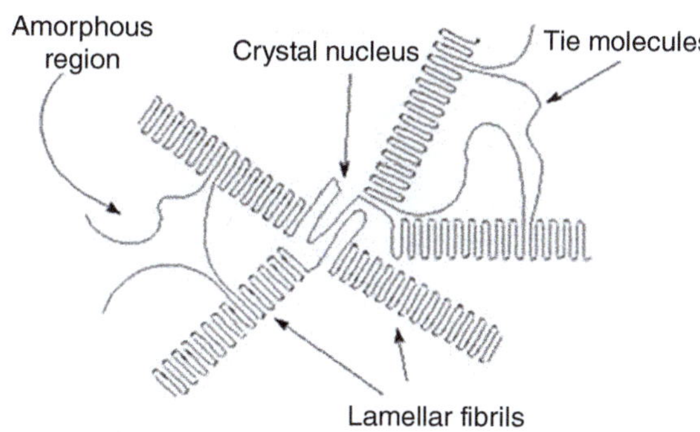

Figure 1.9 Polymer crystalline spherulite.

1.3.5.1 Solid-State Thermal Transitions

Crystalline solid–solid transitions are often observed in single-component systems. By changing temperature (or pressure), a crystalline solid can be transformed into another crystalline solid without entering an isotropic liquid phase. These transitions result in material polymorphs. In most cases, crystalline solid–solid transitions are first-order transitions that undergo discontinuous changes in volume, *enthalpy*, and entropy due to crystal packing changes. The magnitude of these changes is usually small compared to the changes occurring from crystalline solid–liquid transitions. Although this transition requires symmetry breaking in structure to qualify as a first order transition, the positional changes of the molecules to transform from one structure to the next must occur cooperatively, and the displacements cannot be too large [17] (Figures 1.10–1.12).

1.3.6 Mechanical Behavior

Polymers exhibit a wide range of stress–strain behaviors as shown in Figure 1.13. The brittle polymer elastically deforms and fractures before deforming plastically. The 100% amorphous curve is a plastic polymer and is similar to curves for many metals. Its behavior begins in the linear elastic deformation region. As the curve transitions from the elastic to plastic deformation typically there is peak stress. For polymer materials, this peak stress is identified as yield stress. As the material is

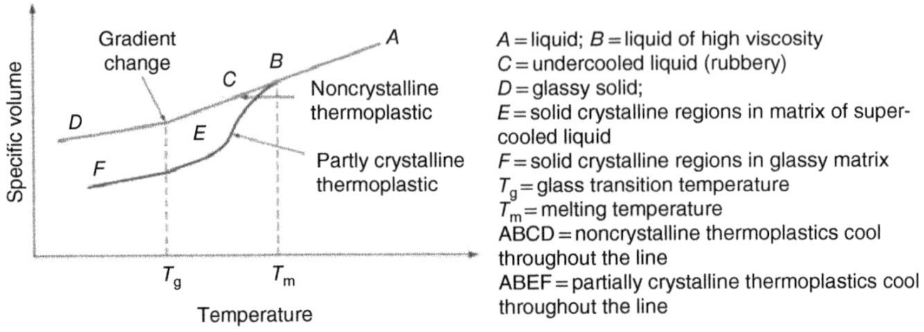

Figure 1.10 Cooling curves for thermoplastic polymers. Source: Rosler et al. [18], John Wiley and Sons, Inc.

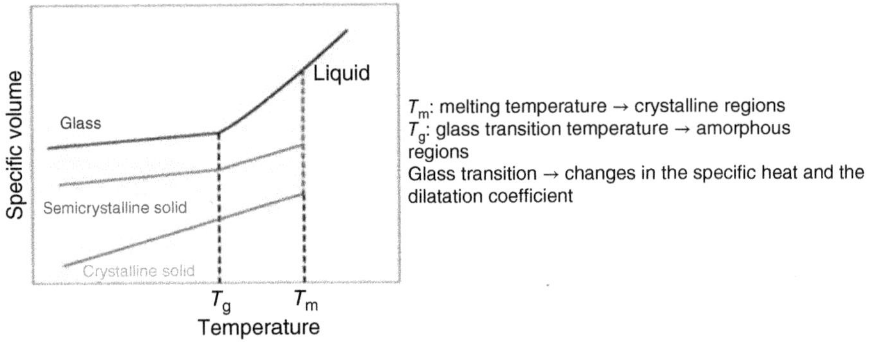

Figure 1.11 Specific volume vs. temperature upon cooling from the liquid melt for polymers with a different structure. Source: Rosler et al. [18], John Wiley and Sons, Inc.

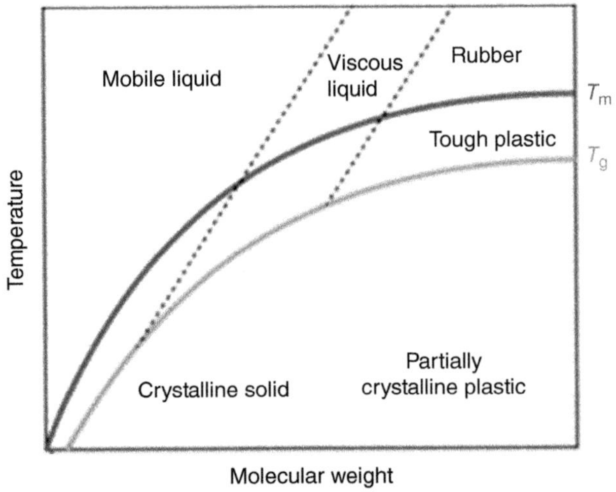

Figure 1.12 Influence of molecular weight. Source: Reprinted with permission from Rosler et al. [18].

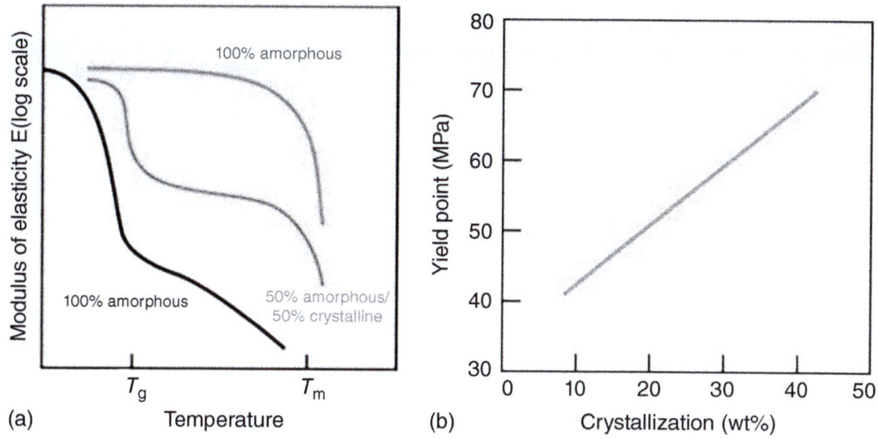

Figure 1.13 (a) Mechanical behaviors of thermoplastics: amorphous, crystalline, and 50/50 amorphous/crystalline, (b) effect of the crystallinity in the modulus of elasticity.

pulled further, fracture occurs. The stress value when the fracture occurs is defined as the tensile strength for polymer materials. The tensile strength can be greater than, equal to, or less than the yield strength. These materials exhibit rubber-like elasticity and will return to their original shape and form unless they are extended to the point of fracture [19].

1.3.7 Polymer Rheology and Processing

Measurements of the rheological properties offer a fast and reliable way to determine molecular weight distribution and long-chain branching, which in combination with the processing conditions, have a decisive influence on the end use product properties [18]. Shear viscosity, elongation viscosity, normal stress differences, stress relaxation, and some other measures and rheological phenomena, of relevance to polymer processing, are discussed in this section. The most widely used polymer processing technologies are extrusion and injection molding.

Polymer processing consists of melting or dissolving a polymeric material, shaping, and solidifying. The choice of whether a melt or solution technique is used depends on (i) chemical stability and (ii) the ability to melt or dissolve [20]. For example, PE is generally melt-processed, since it is stable in the melt but only soluble with some difficulty. Polyacrylonitrile and its copolymers, in composition, are soluble but infusible, so that they are always processed in solution.

1.3.7.1 Polymer Processing Techniques

The following polymer processing techniques are generally employed: (i) Mixing (Polymer additives, Mixing mechanics, Mixing devices), (ii) Extrusion (Extrusion process, Single-screw extruder, Single-screw extruder, Extrusion dies), (iii) Molding (Injection molding, Compression molding, Blow molding, Rotational molding), (iv) Calendering (Process, Arrangements of rolls), and (v) Coating (Fluid-coating process) [21].

The term rheology was first coined by Professor Eugene Bingham of Lafayette College, Indiana, in the 1920s, which originates from the Greek word "rheos" meaning everything flows upon time interval. Thus, rheologyis defined as the science of studying the flow and deformation of matter induced by applied shear forces. From the scope of rheology, all forms of shear behaviors, including the flow of ideal viscous liquids and the deformation of classical elastic solids, can be described by their response to external stresses [22]. These rheological characteristics are highly dependent on the deformation process concerning time. Polymer rheology in the liquid state aims at understanding the complex flow behavior of these materials to model and optimize processing operations. The complexity of the materials also couples with that of the *processing flow* conditions. It suffices to recall that the most used machinery is the extruder where polymers follow helicoidal trajectories before being pumped into a dye and then shaped in many different ways. Rheological studies reduce the flow complexity to a set of basic simple flows because complex flows can be considered as a combination of simpler ones.

The melt rheological properties of filled polymers are sensitive to the structure, concentration, particle size, shape, and surface characteristics of the fillers; rheology offers original means to assess the state of the dispersion in nanocomposites and to investigate the influence of flow conditions upon nanofiller dispersion itself [23].

A better understanding of the rheological properties of polymers is very important for determining the preferred industrial-scale processing conditions, as well as for achieving the desirable physical/mechanical properties in the finished products [24]. Rheometry is the measuring equipment used to assess these rheological properties [25]. This section gives a general idea of current measuring systems: (i) capillary rheometer, (ii) couette (concentric cylinder) rheometer, and (iii) cone-and-plate rheometer.

1.3.7.2 Rheology of Nanocomposites

Rheology is the deformation and material flow study. It is dedicated to the study of viscoelastic materials that include both liquid and solid properties. The rheological behavior of the reinforced polymers depends on many parameters, such as the nature of the fillers (size and shape), the concentration, and the interactions between the fillers and between the polymer and the fillers. These parameters cause not only an increase in viscosity but also particular phenomena such as the existence of a flow threshold, a thixotropic, shearing thinning, or shearing thickening behavior. In the case of nanocomposites, all the rheological characteristics of the conventional filled polymers are generally observed. However, because the size of the charges is extremely small, the surface developed with the polymer matrix is therefore very large [26]. To obtain a reproducible measurement, independent of a human factor such as the altitude of the flow cut, the gradient (of shear, stress, or strain) applied to the fluid must be known. For this, appropriate equipment must be used.

To characterize the rheological properties, various devices allow the polymer viscosity measurements. These devices are called kinematic viscometers or rheometers [27]. For the viscosity measurement, the "two-plate" model is used. It defines the space in which mechanical stress will be applied to the fluid. In this model, the soft material is trapped between two surfaces: one is fixed and the other is mobile, and moves at a constant speed. A shearing motion is then applied to the material to study its ability to flow [28].

1.3.7.3 Theory and Modeling of Nanocomposites Rheology

There are several rheological mathematical models applied to rheograms to transform them into information on fluid rheological behavior [29]. For non-Newtonian fluids, the twomost applied models are the Herschel–Bulkley model and the Bingham model.

1.3.8 Polymer Viscoelasticity

One important characteristic of polymeric materials is their viscoelastic behavior. This means that polymer is elastic because, after a strain due to the application of stress, it is capable of recovering. On the other hand, polymers are viscous because of their capability to creep after the strain.

As shown in Figure 1.14, the creep can be nonlinear, where the strain changes in a nonlinear fashion with time. Upon the release of the stress, the polymer can recover some, but usually undergoes some amount of permanent deformation.

Viscoelasticity is the property of materials that exhibit both viscous and elastic characteristics when undergoing deformation. A viscous material exhibits time-dependent behavior when stress is applied while under constant stress and deforms at a constant rate, and when the load is removed, the material has "forgotten" its original configuration, remaining in the deformed state [30]. On the other hand, an elastic material deforms instantaneously when stretched and remembers its original configuration, returning instantaneously to its original state once the stress is removed. Viscoelastic materials have elements of both of these properties and, as such, exhibit time-dependent strain showing a "fading memory." Such behavior may be linear (stress and strain are proportional) or nonlinear. Whereas elasticity is usually the result of bond stretching along with crystallographic planes in an ordered solid, viscoelasticity is the result of the diffusion of atoms or molecules inside an amorphous material.

The simplest models for the deformation behavior of an ideal material are those of Hookean linear elasticity in the solidstate and Newtonian linear viscosity in the liquid state. The endpoint of elastic deformation is either fracture or plastic flow, with

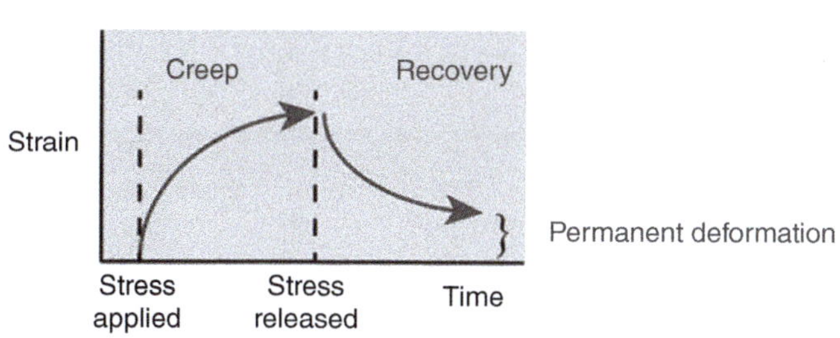

Figure 1.14 Representation of creep and recovery in viscoelastic material.

the latter taking place at constant yield stress in the ideal case. While the behavior of many real materials does approximate to these idealized models that of polymers deviates markedly from them. In particular, their solid-state deformation is time-dependent and nonlinear and so resembles some combination of elastic and viscous responses, at the same time as their melt rheology is also significantly nonlinear [31].

Polymers are characterized by the fact that their behavior under load or deformation is, to a large extent, time-dependent even at room temperature. Moreover, their response to a load or deformation will depend, in some cases, upon any previous load, deformation, or temperature history. This time dependence manifests itself in several forms: two of these are creeping, that is to say, a progressive increase in deformation under a constant load, and stress relaxation, a gradual decrease in stress under a constant deformation. Both these phenomena influence and, in many cases, limit the application of plastics for structural and load-bearing applications.

The contribution of viscous deformation was shown to result in time-dependent permanent deformation (*creep*) under instantaneous, constant stress. Similarly, under an instantaneous constant deformation or strain, viscous deformation can cause *stress relaxation*. Nearly all synthetic and natural polymers used as biomaterials, as well as biological tissues, exhibit viscoelasticity to varying degrees. Noncrystalline or semicrystalline polymers and glasses exhibit free volume and a *glass transition temperature* (T_g) due to the inability of relatively large molecules to organize into a perfectly crystalline network. Below T_g, molecular motions are restricted by time, and the material exhibits glassy or brittle elastic behavior. Near T_g, or more precisely within a temperature range surrounding T_g, the free volume allows some viscous molecular rearrangement, and the material exhibits a viscoelastic response, which becomes increasingly elastomeric (nonlinear elastic) with the increased temperature above T_g. Near the melting temperature, elastic deformation is lost and the material exhibits only viscous flow. Additional background on structure–property relationships including the effects of molecular weight, crystallinity, and crosslinking for noncrystalline and semicrystalline materials (polymers and glasses) is reported. Hydrated biological tissues can also exhibit viscoelasticity due to the permeability of the tissue to fluid flow (poroelasticity) [32].

A stress relaxation test is a simple means of investigating the viscoelasticity of a polymer. To perform this test, a fixed tensile or compressive strain is applied to a sample, and the stress, which decays over time, is monitored [33, 34]. The decrease in stress at a constant strain corresponds to a decrease in the apparent modulus of the polymer. The modulus of a viscoelastic material during a stress relaxation test is often modeled using Eq. (1.1):

$$E(t) = E_\infty + (E_0 - E_\infty)\exp(-t/\tau) \tag{1.1}$$

where (E_0) and (E_∞) are the instantaneous and long-term elastic modulus of the material, respectively, t is time, and τ is the relaxation time constant. As discussed previously, the uncertainty in the literature concerns the origin of changes in the time constant commonly observed when fibers are added to a polymer. In an isotropic solid, the shear modulus (G) and elastic modulus (E) are related by

Poisson's ratio (ν) as shown in Eq. (1.2). For an isotropic, viscoelastic material, at each point in time, the same relationship should hold, as shown in Eq. (1.3). Poisson's ratio is usually considered to be constant in this treatment:

$$E = 2G(1 + \nu) \tag{1.2}$$

$$E(t) = 2G(t)(1 + \nu) \tag{1.3}$$

Thus, Eqs. (1.1) and (1.2) can be used to obtain the time dependence of the shear modulus of a viscoelastic material, as shown in Eq. (1.4):

$$2G(t)(1 + \nu) = 2G_\infty(1 + \nu) + 2(1 + \nu)(G_0 - G_\infty)\exp(-t/\tau) \tag{1.4}$$

$$G(t) = G_\infty + (G_0 - G_\infty)\exp(-t/\tau) \tag{1.5}$$

1.4 Physicochemical Properties of Polymers

1.4.1 Polymers are Very Resistant to Chemicals

Consider all the cleaning fluids in your home that are packaged in plastic. Reading the warning labels that describe what happens when the chemical comes in contact with skin or eyes or is ingested will emphasize the need for chemical resistance in the plastic packaging. While solvents easily dissolve some plastics, other plastics provide safe, nonbreakable packages for aggressive solvents.

1.4.2 Polymers are Both Thermal and Electrical Insulators

A walk through your house will reinforce this concept, as you consider all the appliances, cords, electrical outlets, and wiring that are made or covered with polymeric materials. Thermal resistance is evident in the kitchen with pot and panhandles made of polymers, coffee pot handles, the foam core of refrigerators and freezers, insulated cups, coolers, and microwave cookware. The thermal underwear that many skiers wear is made of PP, and the fiberfill in winter jackets is acrylic and polyester [35].

1.4.3 Polymers are Very Light in Weight with Significant Degrees of Strength

Consider the range of applications, from toys to the frame structure of space stations, or from delicate nylon fiber in pantyhose to Kevlar, which is used in bulletproof vests. Some polymers float in water, while others sink. But, compared to the density of stone, concrete, steel, copper, or aluminum, all plastics are lightweight materials.

1.4.4 Polymers are Processed in Various Ways

Extrusion produces thin fibers or heavy pipes or films or food bottles. Injection molding can produce very intricate parts or large car body panels. Plastics can be molded into drums or be mixed with solvents to become adhesives or paints. Elastomers and some plastics stretch and are very flexible. Some plastics are stretched in processing to hold their shape, such as soft drink bottles. Other polymers can be foamed like polystyrene (Styrofoam™), polyurethane, and PE.

1.4.5 Polymers are Materials With a Seemingly Limitless Range of Characteristics and Colors

Polymers have many inherent properties that can be further enhanced by a wide range of additives to broaden their uses and applications. Polymers can be made to mimic cotton, silk, and wool fibers; porcelain and marble; and aluminum and zinc. Polymers can also make possible products that do not readily come from the natural world, such as clear sheets and flexible films.

1.4.6 Polymers are Usually Made of Petroleum, but not Always

Many polymers are made of repeat units derived from natural gas or coal or crude oil. But building block repeat units can sometimes be made from renewable materials such as polylactic acid from corn or cellulosics from cotton linters. Some plastics have always been made from renewable materials, such as cellulose acetate used for screwdriver handles and gift ribbons. When the building blocks can be made more economically from renewable materials than from fossil fuels, either old plastics find new raw materials or new plastics are introduced.

1.4.7 Polymers are Used to Make Items That Have no Alternatives to Other Materials

Polymers can be made into clear, waterproof films. PVC is used to make medical tubing and blood bags that extend the shelf life of blood and blood products. PVC safely delivers flammable oxygen in non-burning flexible tubing, an antithrombogenic material, such as heparin, can be incorporated into flexible PVC catheters for open-heart surgery, dialysis, and blood collection. Many medical devices rely on polymers to permit effective functioning [36].

References

1 Jensen, W.B. (2008). The origin of the polymer concept. *J. Chem. Edu.* 88: 624–625.

2 Dorel, F. (2008). Polymer history. *Des. Monomers Polym.* 11: 1–15.

3 Mark, J.E. (2007). *Physical Properties of Polymers Handbook*. LLC: Springer Science + Business Media. ISBN-13: 978-0-387-31235-4.

4 Kulkarni, V.S. (2016). Use of polymers and thickeners in semisolid and liquid formulations. In: *Essential Chemistry for Formulators of Semisolid and Liquid Dosages*, 43–69. Academic Press.

5 Ai, O., Pap, M., Ye, E., and Ly, Z. (2017). An introduction to polymers and some profiles of polymer industries in Nigeria. *J. Polym. Sci.* 1: 3–6.

6 Gowariker, V.R., Viswanathan, N.V., and Shreedhar (2005). *Polymer Science*. New Delhi, India: New Age International Publishers.

7 Chanda, M. (2006). *Introduction to Polymer Science and Chemistry*. FL, USA: CRC Press, Taylor and Francis Group.

8 Gao, Y., Zhou, D., Lyu, J. et al. (2020). Complex polymer architectures through free-radical polymerization of multi vinyl monomers. *Nat. Rev. Chem.* 4: 194–212.

9 Fried, J.R. (2014). *Polymer Science and Technology*. Prentice-Hall.

10 Brinson, H.F. and Catherine Brinson, L. (2015). *Polymer Engineering Science and Viscoelasticity*. US: Springer.

11 Van Krevelen, D.W. and TeNijenhuis, K. (2009). *Properties of Polymers Their Correlation with Chemical Structure; Their Numerical Estimation and Prediction from Additive Group Contributions*. Netherlands: Elsevier.

12 Nicholson, J.W. (2012). *The Chemistry of Polymer*. Cambridge, UK: RSC Publishing.

13 Bazylak, L.I., Zaikov, G.E., and Haghi, A.K. (2014). *Polymers and Polymeric Composites: Properties, Optimization and Applications*. Apple Academic Press.

14 Hana, C.C., Shia, W., and Jin, J. (2013). Morphology and crystallization of crystalline/amorphous polymer blends. In: *Encyclopedia of Polymers and Composites*. Berlin Heidelberg: Springer-Verlag https://doi.org/10.1007/978-3-642-37179-0_25-1.

15 Yu, C., Xie, Q., Bao, Y. et al. (2017). Crystalline and spherulitic morphology of polymers crystallized in confined systems. *Crystals* 7: 147.

16 Balani, K., Verma, V., Agarwal, A., and Narayan, R. (2015). Physical, thermal, and mechanical properties of polymers. In: *Biosurfaces: A Materials Science and Engineering Perspective. The American Ceramic Society*. Wiley.

17 Cheng, S.Z.D. (2008). Thermodynamics and kinetics of phase transitions. In: *Phase Transitions in Polymers. The Role of Metastable States*, 17–59. Elsevier Science.

18 Rosler, J., Baker, M., and Harders, H. (2007). Mechanical behavior of polymers. In: *Mechanical Behavior of Engineering Materials*. Berlin, Heidelberg: Springer.

19 Tesoro, G. (1984). *Textbook of Polymer Science*. New York: Wiley.

20 Polychronopoulos, N.D. and Vlachopoulos, J. (2019). Polymer processing and rheology. In: *Functional Polymers*, Polymers and Polymeric Composites: A Reference Series (ed. M. Jafar Mazumder, H. Sheardown and A. Al-Ahmed). Cham: Springer https://doi.org/10.1007/978-3-319-95987-0_4.

21 Ferguson, J. (1995). Application of rheology to polymer processing. In: *Rheological Fundamentals of Polymer Processing*, NATO ASI Series (Series E: Applied Sciences), vol. 302 (ed. J.A. Covas, J.F. Agassant, A.C. Diogo, et al.). Dordrecht: Springer https://doi.org/10.1007/978-94-015-8571-2_8.

22 Zhang, W., Chen, J., and Zeng, H. (2020). Polymer processing and rheology. In: *Polymer Science and Nanotechnology*, 149–178. https://doi.org/10.1016/b978-0-12-816806-6.00008-x.

23 Ianniruberto, G. (2015). *Introduction on Polymer Rheology*, Reference Module in Chemistry, Molecular Sciences, and Chemical Engineering. Elsevier https://doi.org/10.1016/b978-0-12-409547-2.11228-4.

24 Barnes, H.A., Hutton, J.F., and Walters, K. (1989). *An Introduction to Rheology*. Amsterdam: Elsevier.

25 Mezger, T.G. (2006). *The Rheology Handbook*. Vincentz Coatings Compendia.

26 Han, C. (2007). *Rheology and Processing of Polymeric Materials*. Oxford, UK: Oxford University Press.

27 Knauert, S.T., Douglas, J.F., and Starr, F.W. (2007). The effect of nanoparticle shape on polymer nanocomposite rheology and tensile strength. *Polym. Sci.* 45: 1882–1897.

28 Balmforth, N.J., Craster, R.V., Perona, P. et al. (2007). Viscoplastic dam breaks and the Bostwick consistometer. *J. Non-Newtonian Fluid Mech.* 142: 63–78.

29 Ouarhim, W., Hassani, F.Z.S.A., Qaiss, A.e.k., and Bouhfid, R. (2020). Rheology of polymer nanocomposites. In: *Rheology of Polymer Blends and Nanocomposites*, 73–96. Elsevier.

30 Seyssiecq, I., Ferrasse, J., and Roche, N. (2003). State-of-the-art: rheological characterization of wastewater treatment sludge. *Biochem. Eng. J.* 16: 41–56.

31 Roeder, R.K. (2013). Mechanical characterization of biomaterials. In: *Characterization of Biomaterials*, 49–104. Elsevier https://doi.org/10.1016/b978-0-12-415800-9.00003-6.

32 (2009). Viscoelastic behaviour of polymers. In: *Physicochemical Behavior and Supramolecular Organization of Polymers*. Dordrecht: Springer https://doi.org/10.1007/978-1-4020-9372-2_2.

33 Papanicolaou, G.C. and Zaoutsos, S.P. (2011). Viscoelastic constitutive modeling of creep and stress relaxation in polymers and polymer matrix composites. In: *Creep and Fatigue in Polymer Matrix Composites*, 3–47. Springer https://doi.org/10.1533/9780857090430.1.3.

34 Numaira, O., Kortschot, M.T., and Mohini, S. (2017). Understanding the stress relaxation behavior of polymers reinforced with short elastic fibers. *Materials* 10: 472.

35 The United States Environmental Protection Agency (2005) Municipal solid waste in the United States. EPA530-R-06-011.

36 American Chemistry Council (2005). National post-consumer plastics bottle recycling report.

2

Synthesis of Polymers

2.1 Features of the Polymerization Reactions

Polymerization occurs in varied forms like repetitive chemical bonding of individual molecules or monomers. Assorted combinations of heat, pressure, and catalysis alter the chemical bonds that hold monomers together, causing them to bond with one another. Linearly, polymers are created by chains of monomers [1]. Copolymers can be formed using two or more different monomers, and two or more polymers can be combined to produce an alloy, or blend, that displays the characteristics of each component.

The chemical process which converts molecules into polymer macromolecules (MMs) is known as polymerization. This process is the combination of different reactions that determine the feature of the product obtained depending also on the starting chemical composition. These reactions are in general: initiation, propagation, termination, and transfer. Propagation is often identified as a polymerization reaction as it is responsible for MM formation. The polymerization is a one-pot process and therefore the synthesis of MMs by successive addition of monomer units in highly distinct steps cannot be considered a polymerization. For example, proteins and nucleic acids are synthesized in this way. Synthetic polymers are normally obtained by submitting a monomer to conditions under which the separated monomer molecules react to bind each other to form a MM. However, the same monomer we can obtain as a homopolymer, MMs differing in molecular weight. Therefore, the product of the polymerization may not be formed by a simple MM, but is in general, a mixture of MMs having a different length. The MW of the obtained MM, or better its average, depends for a given monomer on reaction conditions and reaction time as well as the conversion of monomer into MMs [2]. Therefore, all polymerization reactions are characterized by the evolution with time of both conversion and MW.

Converting monomers to long-chain polymer is the final step in the polymer manufacturing sequence. Polymerization is usually highly favorable in thermodynamic terms, mainly on energetic grounds because ordering molecules into linked chains is a process where the entropy is decreased. Advances in catalysis have given a high degree of control over both structure and molecular mass so that the grades of a given polymer can be tailored for specific end users. It is possible to look at polymerization

Polymer Nanocomposites for Energy Applications, First Edition.
T. Daniel Thangadurai, Manjubaashini Nandhakumar, Sabu Thomas, and Ange Nzihou.

in at least two different ways: the nature of the catalyst used and the way the chains grow to form the final product. Polymerizations can be conducted in the gaseous, liquid, or solid state, and now in the liquid crystal state to produce highly oriented MMs [3].

A polymer is a large single chain-like molecule in which the repeating units derived from small molecules, called monomers, are bound together. The process by which monomers are transformed into a polymer is called polymerization. For example, ethylene polymerizes to form polyethylene, $nCH_2{=}CH_2 \rightarrow [-CH_2-CH_2-]_n-$. A basic understanding of polymerization processes is important not only because polymerization affects the structure, and hence properties, but also because some processing routes can convert monomers directly to a finished shape. They offer the manufacturing industry considerable benefits both in direct and indirect costs. An extra dimension to polymer structure is added by the possibilities of copolymerization, where two or more different monomers are polymerized together. In one sense, it is comparable to alloying different metals to produce an appropriate balance of properties in the final product [4]. There are two major types of polymerization methods used to convert monomers to polymers: addition and condensation polymerization:

(i) addition polymerization is known as chain, chain-growth, or chain-reaction polymerization
(ii) condensation polymerization is known as step-growth or step-reaction polymerization.

A polymerization in which a cyclic monomer yields a monomeric unit that is acyclic or contains fewer cycles than the monomer. If the monomer is polycyclic, the opening of a single ring is sufficient to classify the reaction as ring-opening polymerization (ROP) [5].

2.2 Chain Polymerization

In chain polymerization, the polymer is formed through a chain reaction in which the growth proceeds exclusively by a reaction between monomers and reactive sites on the polymer chain, with the regeneration of the reactive sites at the end of each growth step. In this kind of reaction, the polymerization is conducted by a chain reaction without the formation of any small molecules [6] (Figure 2.1).

2.3 Ring-Opening Polymerization

The ROP is another form of chain-growth polymerization in which the terminal end group of a polymer chain acts as a reactive center, where further cyclic monomers can be added by ring-opening and addition of the broken bond. Typical monomers that can be polymerized via ROP are difunctional monomers that carry two different reactive groups like one amine or alcohol and one carboxylic acid that

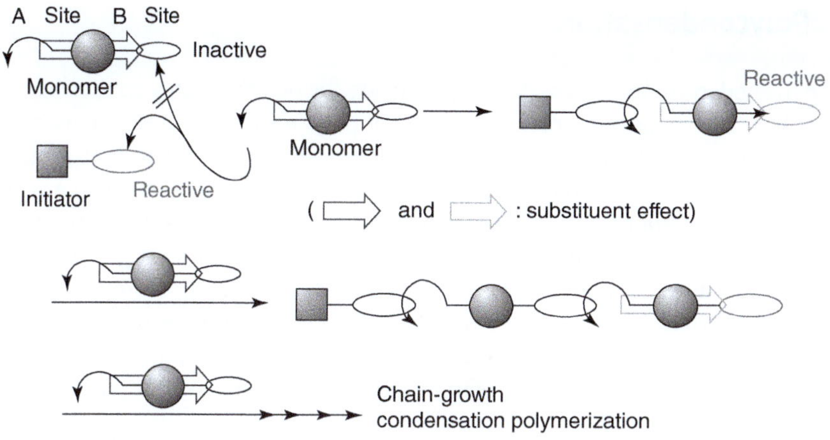

Figure 2.1 Schematic of chain-growth condensation polymerization.

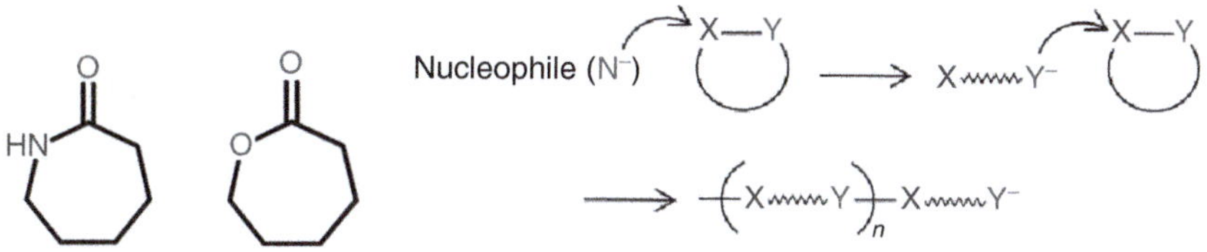

Figure 2.2 Schematic of ring-open-type polymerization.

have undergone a cyclization reaction, for example, caprolactam and caprolactone. To polymerize these moieties, one of the rings has to open before polymerization. This can be achieved, for example, by adding a small amount of a nucleophilic reagent (a Lewis base) as an initiator. This reaction is called *anionic ring-opening polymerization* (AROP) (Figure 2.2).

Cationic ring-opening polymerization (CROP) is also possible. In this case, a small amount of an electrophilic reagent (a Lewis acid) is added to the monomer to initiate polymerization. However, not all cyclic monomers containing a heteroatom undergo CROP.

The cyclic monomer undergoes ROP depending on the ring size, to be more specific, on the ring strain which is caused by bond angle distortion (angular strain), the repulsion between eclipsed hydrogen atoms (conformational strain), and non-bonding interactions between substituents attached to different parts of the ring (transannular strain). Small cyclic monomers with large ring strain such as 3-, 4-, and 5-membered rings of cyclic esters and amides polymerize readily through ROP, whereas 7- and 8-membered lactones and lactams are less reactive due to their much lower ring strain (c. 6.0 J mol^{-1}) but are still reactive enough to undergo ROP. An entropy gain due to additional degrees of rotation after ring-opening can be an additional driving force, as is the case with cyclic carbonates. Some examples of cyclic monomers that polymerize through anionic or CROP include cyclic ethers, lactones, lactams, carbonates, aziridines, and epoxides [7].

2.4 Polycondensation

In the polycondensation process, the polymers are formed through a condensation reaction between molecules of all degrees of polymerization. A condensation reaction is understood as a chemical reaction in which two functional groups interact to form a different functional group with the loss of a small molecule. The word condensation suggests a process in which two or more entities are brought together to form a "dense" entity, such as in the condensation of a gas into the liquid state; this does not imply, however, that condensation reaction products have a greater density than the reactants. When the two functional groups reacting are in the same molecule, the condensation is termed intramolecular; on the other hand, when the functional groups are in different molecules, it is termed intermolecular condensation [8].

2.5 Polyaddition

It is a polymerization reaction in which the growth of the polymer chains proceeds by addition reactions between molecules of all degrees of polymerization [9].

2.6 Step-Growth Polymerization

In this type of polymerization, only one kind of reaction is involved in the formation of a polymer and the reaction proceeds step by step. The main feature of this type of reaction is that two monomers, which bear different functionalities, can react with each other, or with a polymer of any size, through the same kind of reactions. In this case, the individual polymer molecules can grow throughout the whole process; each reaction step of a polymer molecule implies that the reactive end of a monomer or polymer encounters another species with which it can form a link [10]. The functional group at the end of a monomer is usually assumed to have the same reactivity as that on a polymer chain of any size.

Many step-growth polymerizations involve equilibrium between reactants and products, the latter comprising macromolecular species and eliminated small molecules. The course of these polymerizations and the distribution of the molecular weights are statistically controlled. High polymers cannot coexist with many monomers in a system in equilibrium. Step-growth polymerizations are reversible and also involve interchange reactions in which terminal functional groups in a growing chain react with linking units of other molecules producing changes in the molecular weight distributions [11]. There are step-growth polymerization reactions in which a small molecule is not produced (e.g. the reaction between a diol and a diisocyanate); these reactions are considered irreversible and are usually very fast, leading to high degrees of polymerization (Figure 2.3).

Aliphatic polyesters bearing pendant alkyne groups: These polyesters can be successfully prepared by step-growth polymerization of different building blocks,

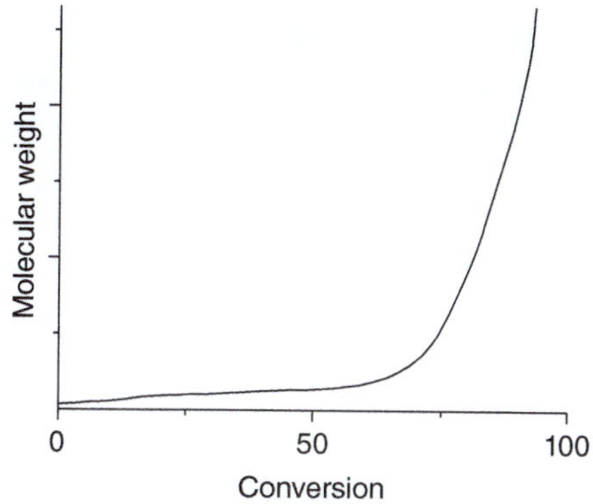

Figure 2.3 Profile of molecular weight vs. conversion in a step-growth polymerization.

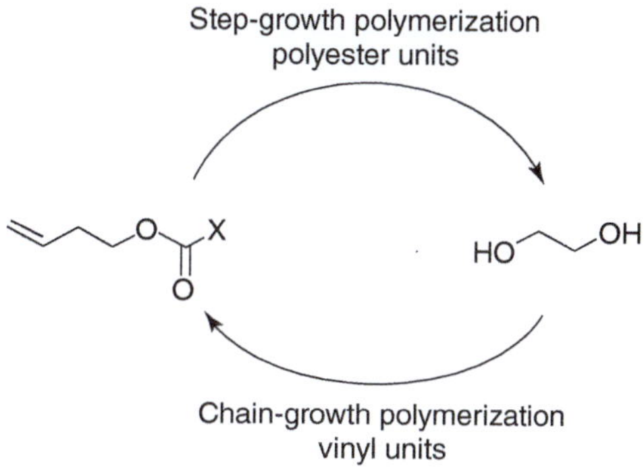

Figure 2.4 Polymerization of bifunctional monomers by step-growth and chain polymerizations.

such as adipic and succinic acid, in combination with an acetylene-based diol, examples of which are 2-methyl-2-propargyl-1,3-propanediol, 1,4-butanediol, and the most common ethylene glycol. Alkyne groups can survive the high reaction temperatures (200 °C) in the presence of a radical inhibitor [12]. Subsequently, the alkyne groups are reacted by a "click" cycloaddition reaction to obtain a functionalized polyester based on poly(ethylene succinate) and poly(butylene adipate) (Figure 2.4).

2.7 Dendrimers

The use of monomers with functionality, larger than 2, allows the formation of tridimensional structures by step-growth polymerization. In particular, the use of monomers with the general formula A_xB_y could generate several architectures such as hyper branched, dendrimers, and star polymers. Each architecture will have different physical and chemical properties. In particular, dendrimers are regularly branched polymers that can be produced by the divergent approach, or the convergent approach. In the divergent approach, the dendrimer grows from the center and spreads radially out in layers, each of them built by a stepwise reaction. In the convergent approach, the dendrimer production starts at the end with the

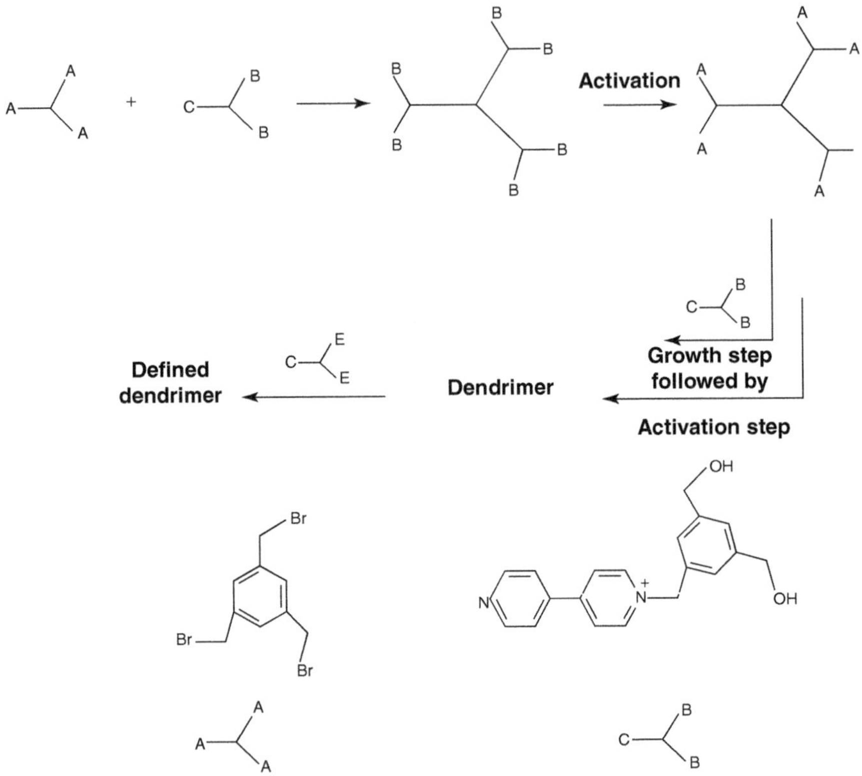

Figure 2.5 Divergent approach for the preparation of dendrimer.

so-called surface functional groups coupled to AB_y building blocks, which allow the preparation of a larger dendron, and the process is repeated to increase the size of the dendrons. In the end, the dendrons are attached to a multifunctional core molecule to produce dendrimers [13]. Examples of AB_y monomer used, for example, in polyesters, are those where A and B are –COOH and –OH groups, respectively, or in some cases their derivatives such as –COCl, $-O-(CH_2)_2-OH$, $-O-Si(CH_3)_3$, or CH_3COO^- (Figure 2.5).

Phenolic resins are a generic name given to a wide range of crosslinked polymers produced by phenol and formaldehyde. There are two types of phenolic resins, resol and novolac. The type of resin being made depends on the pH of the catalyst and the ratio of phenol to formaldehyde [14]. Novolac is made under acidic conditions, and the ratio of phenol to formaldehyde used ranges from 1.49 to 1.72, whereas resol resins are synthesized in a basic medium using a ratio of phenol to formaldehyde in the range of 1.0–0.33 (Figure 2.6).

The most conventional kinetic scheme of free radical polymerization includes initiation, propagation, and bimolecular termination reaction steps. Additional reactions such as chain transfer are introduced to improve the process description. Free radicals are highly reactive chemical species produced by the homolytic dissociation of covalent bonds [15]. Such species are produced through physical (thermos

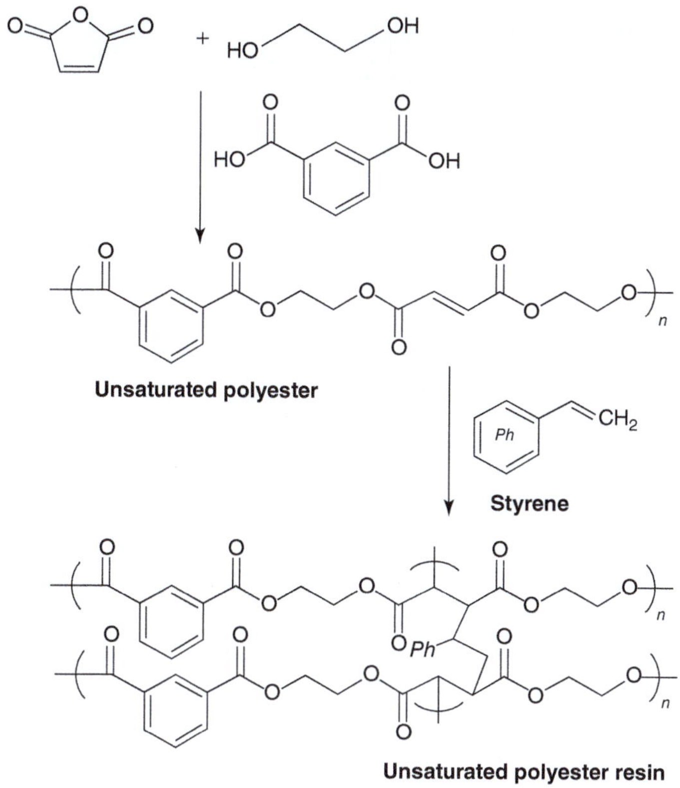

Figure 2.6 Synthesis of unsaturated polyester resin.

excitation and radiation) or chemical methods (oxidation–reduction, addition, etc.). Generally, their survival time is less than a second, except for those radicals highly stabilized by specific chemical groups; the hybridization state is sp^2.

Polymerization by coordination is one of the most versatile methods to produce a variety of polymers. Stereoregularity is one of the outstanding characteristics of coordination polymerization that relies on the use of catalytic systems based on organometallic or coordination complexes of special structures and symmetries to make highly stereospecific polymers [16].

2.8 Anionic Polymerization

A general mechanism for living anionic polymerization of a vinyl monomer is illustrated in Figure 2.7, encompassing only initiation and propagation steps; chains are terminated only by the deliberate addition of a Bronsted acid or an electrophile. Important aspects of this mechanism, and that of any living polymerization, are that one initiator generates one polymer chain and that the product after all of the monomer has been consumed is a polymer with an active anionic chain end. With this mechanism, the defining characteristics of living polymerizations concerning

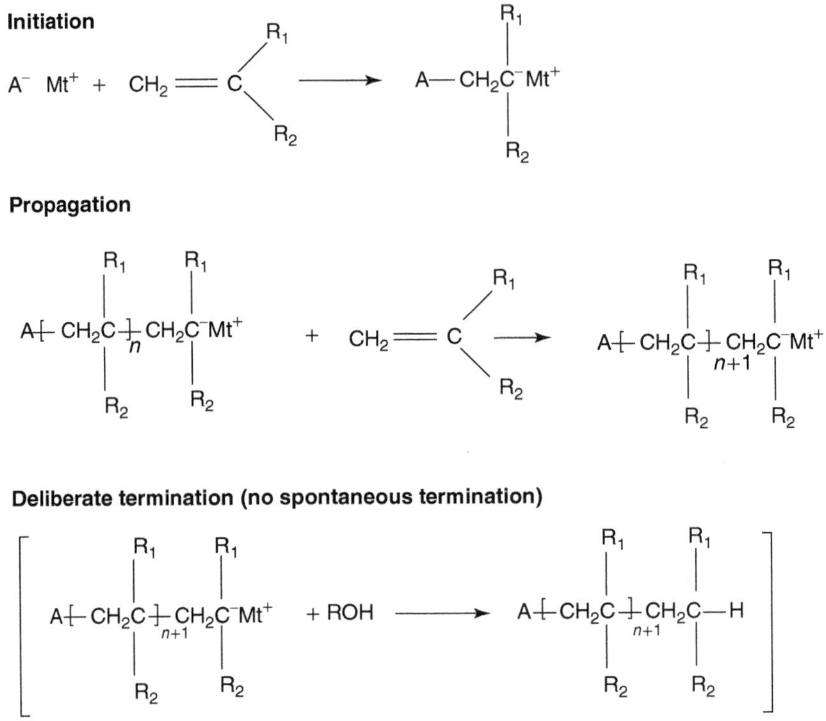

Figure 2.7 General mechanism for a living anionic polymerization.

the synthesis of a wide variety of polymers with low degrees of compositional heterogeneity can be understood [17] (Figure 2.7).

2.9 Cationic Polymerization

Cationic polymerization of vinyl monomers involves the highly reactive carbenium ions, which can only be formed when they are stabilized by the substituents on the vinyl group. Stabilization of the carbenium ion can be achieved by resonance structures, as is the case for vinyl ether and styrene monomers, or by an inductive effect, which is the stabilizing mechanism for 1,1-disubstituted alkenes [18]. Despite this stabilization, the reactivity of such carbenium ions is still very high, giving rise to the occurrence of side reactions, such as isomerization, hydride abstraction, as well as various transfer reactions. Moreover, carbocationic polymerizations are very sensitive to minor nucleophilic impurities. Nonetheless, in recent years, a range of less sensitive polymerization initiators have been developed that can be used in an aqueous environment. The counter anion in carbocationic polymerizations is of major importance to "regulate" the reactivity of the cationic species. Initial studies involved non-nucleophilic counter anions, such as BF_4^-, PF_6^-, SbF_6^-, or ClO_4^-, completely exposing the carbenium ions resulting in uncontrolled cationic polymerizations [19] (Figure 2.8).

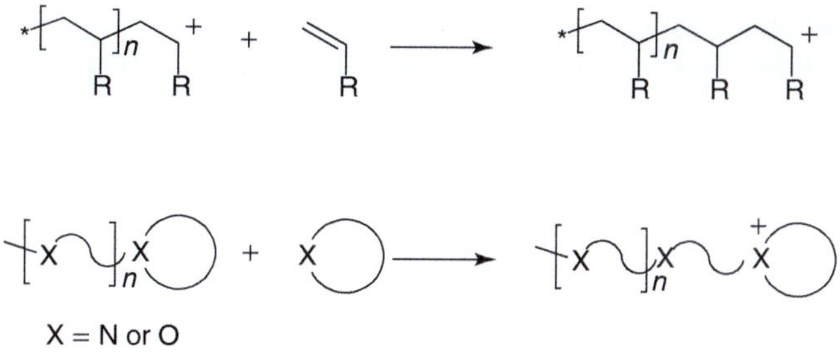

Figure 2.8 Schematic representation of the propagation step, monomer addition, in carbocationic polymerizations (top) and cationic ring-opening polymerizations (bottom).

2.10 Controlled Radical Polymerization

Several controlled radical polymerization (CRP) methods have been developed, and the three most promising are stable-free radical polymerization (SFRP), most commonly nitroxide-mediated polymerization (NMP) but may also include organometallic species transition-metal-catalyzed atom transfer radical polymerization (ATRP), and degenerative transfer with alkyl iodides, methacrylate macromonomers, and dithioesters via reversible addition-fragmentation chain transfer (RAFT) polymerization (Figure 2.9).

In the case of SFRP or ATRP, the equilibrium is pushed to the left-hand side (deactivated), forming an excess of dormant species as a result of the persistent radical effect. In all radical polymerizations, biradical termination occurs at a rate R_t, which is dependent on the concentration of radicals, $[P^*]$, where $R_t = k_t\ [P^*]$. Therefore, at the same polymerization rate (the same $[P^*]$), essentially the same

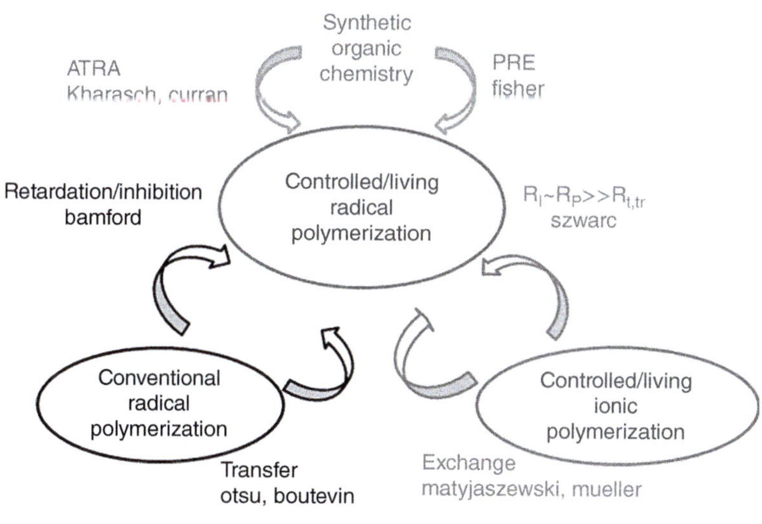

Figure 2.9 Development of CRP by the integration of advances in several fields of chemistry. Source: Matyjaszewski and Spanswick [20]/Elseiver/CC BY 4.0.

number of chains terminates regardless of being in conventional or CRP systems. However, in the conventional process, all chains are terminated, whereas in CRP, as a result of the greater number of growing chains, the terminated chains constitute a small fraction of all the chains (~1–10%). The remaining chains are dormant species, capable of reactivation, functionalization, and chain extension to form block copolymers, etc.

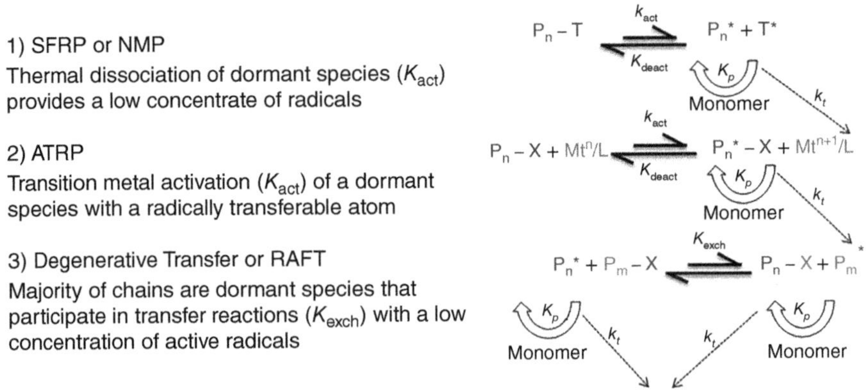

2.11 Atom Transfer Radical Polymerization (ATRP)

ATRP is currently the most widely used CRP technique. This originates from the commercial availability of all necessary ATR Preagents, including transition metal compounds and ligands used as catalysts as well as alkyl halide initiators and also the large range of monomers polymerizable by this technique under a wide range of conditions. The advantages include: (i) only catalytic amounts of transition metal complexes (commercially available) are necessary; (ii) many commercially available initiators, including multifunctional and hybrid systems; (iii) large range of monomers (except unprotected acids); (iv) easy end-functionalization; and (v) large range of polymerization temperatures.

Radical polymerization is a very important commercial process for preparing high-molecular-weight polymers because it can be used for many vinyl monomers under mild reaction conditions, requiring the absence of oxygen but tolerance to water, and large temperature ranges (−20 to 200 °C) [21]. The only disadvantage to conventional radical polymerization is the poor control of macromolecular structures including degrees of polymerization, polydispersity, end functionalities, chain architectures, and compositions. Thus, it is desirable to prepare, by radical polymerization, new well-defined block and graft copolymers, stars, combs, and networks that have not been previously prepared using ionic living polymerizations. Therefore, controlled T living radical polymerizations allow for the synthesis of new well-defined and functional materials from a larger range of monomers under simpler reaction conditions than are appropriate for ionic processes.

2.12 Reversible Addition Fragmentation Chain Transfer (RAFT)

RAFT polymerization is a reversible deactivation radical polymerization (RDRP) and one of the more versatile methods for providing living characteristics to radical polymerization [22]. The advantages of RAFT polymerization are,

- Ability to control polymerization of most monomers polymerizable by radical polymerization which includes (meth) acrylates, (meth) acrylamides, acrylonitrile, styrenes, dienes, and vinyl monomers.
- Tolerance of unprotected functionality in monomer and solvent (e.g. OH, NR_2, COOH, $CONR_2$, SO_3H), and polymerizations can be carried out in aqueous or protic media.
- Compatibility with reaction conditions (e.g. bulk, organic or aqueous solution, emulsion, mini-emulsion, suspension).
- Ease of implementation and inexpensive relative to competing technologies.

In an ideal living polymerization, all chains are initiated at the beginning of the reaction, grow at a similar rate, and survive the polymerization: there is no irreversible chain transfer or termination. If initiation is rapid concerning propagation, the molecular weight distribution is very narrow and chains can be extended by further adding monomers into the reaction [23]. In a radical polymerization, all chains cannot be simultaneously active. In RDRP, such as RAFT polymerization, these attributes are displayed in the presence of reagents that are capable of reversibly deactivating propagating radicals such that the majority of living chains are maintained in a dormant form and reaction conditions that support a rapid equilibrium between the active and dormant chains (Figure 2.10).

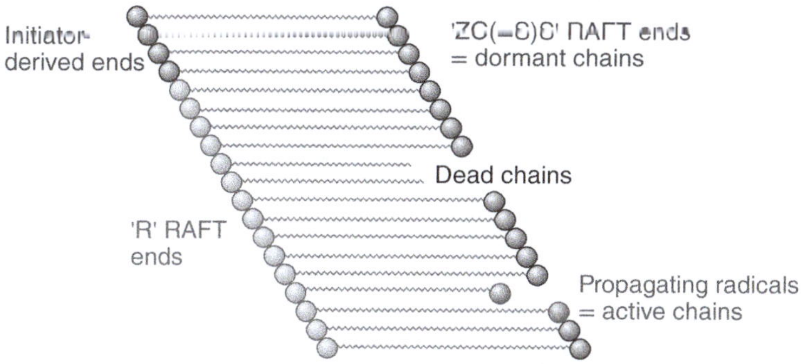

Figure 2.10 Schematic of RAFT polymerization. The number of chains of each type shown here is not in proportion to that expected for a well-designed experiment. On average, all living chains grow simultaneously and have equal chain length because the equilibration of the dormant and active chain ends is rapid concerning propagation. A RAFT agent is represented as "ZC(=S)S."

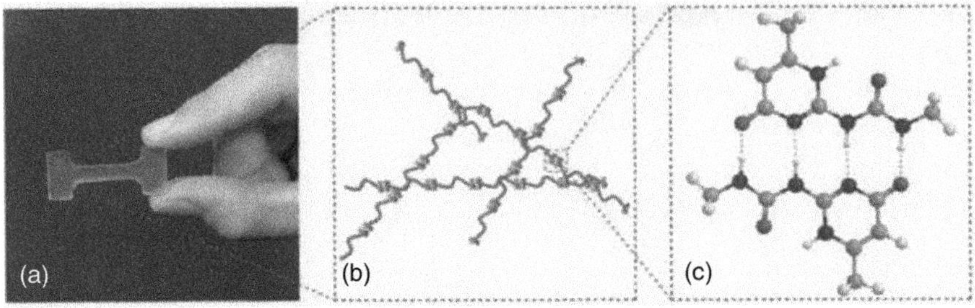

Figure 2.11 (a) Supramolecular polymeric material based on a low-molecular-weight compound equipped with two ureido-pyrimidinone (UPy) units; (b) schematic picture of the underlying polymeric network; and (c) schematic picture of the self-complementary UPy dimer. Source: Pol [25]/John Wiley & Sons.

2.13 Supramolecular Polymerization

Supramolecular polymers are defined as polymeric arrays of monomeric units that are brought together by reversible and highly directional secondary interactions, resulting in polymeric properties in dilute and concentrated solutions as well as in the bulk [24]. The directionality and strength of the supramolecular bonding are important features of systems that can be regarded as polymers and that behave according to well-established theories of polymer physics.

Supramolecular polymerizations can be classified based on three different principles: (i) the physical nature of the noncovalent force that lies at the origin of the reversible interaction (physical origin classification), (ii) the type of monomer(s) used (structural monomer classification), and (iii) the evolution of the Gibbs free energy of the polymer as a function of conversion (thermodynamical classification). In principle, a fourth classification scheme based on the dimensionality of the aggregate is possible (Figure 2.11).

2.14 Bulk Polymerization

Bulk polymerization is carried out in the absence of any solvent or dispersant and is thus the simplest in terms of formulation. It is used for most step-growth polymers and many types of chain-growth polymers [26]. In the case of chain-growth reactions, which are generally exothermic, the heat evolved may cause the reaction to become too vigorous and difficult to control unless efficient cooling coils are installed in the reaction vessel [27]. Bulk polymerizations are also difficult to stir because of the high viscosity, associated with high-molecular-weight polymers. Bulk polymerization is advantageous in the production of cast-molded products such as *polystyrene scintillators*.

Bulk polymerization has the advantages of simple operation, obvious adsorption effect, and good selectivity for target ions. However, there are some disadvantages, which are as follows: (i) the obtained polymers require to be comminuted, milled, and sieved, and the process is cumbersome and time-consuming, and yields <50%; (ii) the particle size distribution of MIPs is heterogeneous and the shape is irregular, which are disadvantageous to many chromatographic and separation applications; (iii) there is the destruction of some binding sites during grinding, leading to a great loss of adsorption capacity; and (iv) the adsorption sites in the polymers are too far from the surface so that binding capacity for the target ion is poor [28].

2.15 Solution Polymerization

Solution polymerization occurs in the existence of inert solvent and a suitable catalyst. This polymerization is used for manufacturing many polymers such as polystyrene, polyethylene, and polymethyl methacrylate for bulk polymerization and polyacrylonitrile, and polyesters such as nylon for solution polymerization [29]. Examples of solution polymerization are (i) ethylene in isooctane polymerized at 150–180 °C, 300–700 psi with chromium-silica-alumina catalyst, (ii) an aqueous solution of acrylonitrile at 80 °C with a persulfate catalyst resulting in polyacrylonitrile precipitate, and (iii) nylon rope trick-interfacial polycondensation of a solution of sebacoyl chloride in tetrachloroethane plus the aqueous solution of hexamethylene diamine yielding nylon 6/10 at the boundary (interface) of both solutions [30].

Advantages of solution polymerizations are: (i) heating and stirring are much easier than in bulk polymerization due to the solvent medium; (ii) the viscosity problem is minimized by the presence of a solvent medium; (iii) heat removal is accomplished via solvent reflux, and (iv) conversion of monomer(s) to polymer is high and typically about 80–90% [31, 32].

2.16 Suspension Polymerization

Suspension polymerization is used to make many *thermoplastic* polymers. In suspension polymerization, all reactions are carried out in relatively large *droplets* or *polymer particles* stabilized by a small amount of water-soluble gum. *Organic peroxide* initiators are used to generate radicals within the droplets. A solvent may be used to dissolve a monomer at a relatively high concentration. The main advantages of suspension polymerization over *emulsion* systems are that no *surfactants*, which are difficult to remove from the product, are used, and no ionic end groups are present, which may be unstable during processing at high temperatures [33]. Suspension polymerization methods for producing beaded polymers from acrylic and styrene-based *monomers* are very well-established processes. They are typically used in the production of *ion exchange resins* and supports solid-phase synthesis [34].

2.17 Methods for the Synthesis of Functional Polymers

The synthesis or modification of polymeric substrates to produce polymers with polar or functional groups can be carried out by several synthetic pathways. When the process starts from a previously existing nonfunctional polymer, and functional groups are introduced to the chain, the term functionalization is used [35].

2.17.1 Direct Copolymerization

In this polymerization, two types of monomers react in which one of them has a functional or pendant functional group; for instance, the copolymerization of maleic anhydride (MA) and styrene (St) generates the alternating copolymer poly(St-alt-MA). Another example is the direct copolymerization of α-olefins, polypropylene (PP), and polyethylene (PE) with functional monomers such as MA or glycidyl methacrylate (GM); this goal remains one of the challenges in synthetic polymer science. In the latter example, the difficulty lies in the fact that the majority of this kind of polyolefins is produced by Ziegler–Natta or metallocene coordination chemistry, and both exhibit intolerance to Lewis bases because of their high oxophilic nature, which leads to the deactivation of the catalysts. The most important mechanism in catalyst deactivation is the formation of very stable complexes between the catalyst and the heteroatoms (N, O, and halogens) present in the functional polar monomers. As a consequence, the post-modification of polyolefins with polar or functional groups, as opposed to the direct synthesis, offers the highest viability for the production of polyolefins with functional or reactive grafts [36].

2.17.2 End-Functionalization

The possible routes for the end-functionalization are (i) modification of the chain end of preexisting polymers, (ii) break of the growth of a polymeric chain, and (iii) break of the chain growth followed by its end-functionalization. In all these routes, it is common to perform the indicated action using agents or chemical groups that contain the desirable functional group [37].

2.17.3 Functionalization-Grafting

It consists of one of the following techniques: (i) synthesis *in situ* of grafts of homopolymers or copolymers (small or long chains) containing functional groups, starting from units located along an existing polymer backbone; (ii) reaction or interchange of labile atoms of the polymeric backbone by functional monomers or reactive species; and (iii) use of preexisting polymers possessing functional species and chain ends with chemical affinity to the backbone of the polymer [38].

There are several ways in which structures having functional chemical groups or chains of homopolymers or copolymers grafted onto a polymeric backbone can be generated. In general, there are three common methods for the synthesis of

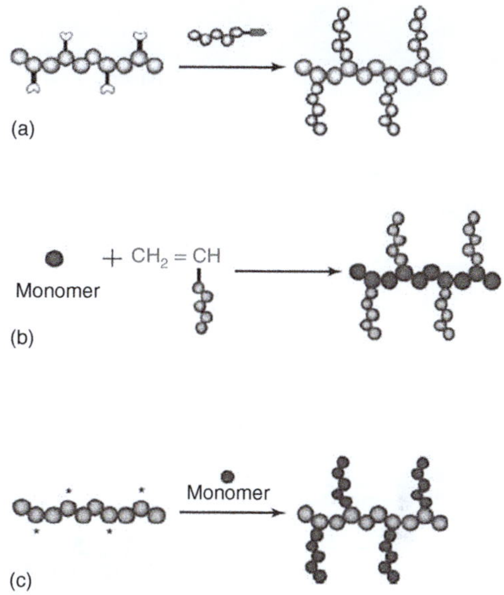

Figure 2.12 Schematic representation of (a) grafting-onto, (b) grafting-through, and (c) grafting-from methods.

this type of copolymers: "grafting-onto," "grafting-from," and "grafting-through." Figure 2.12 schematically shows the synthesis of graft copolymers. The first technique, "grafting-onto" (also called grafting-to), involves the reaction of an end-functional structure previously synthesized with a complementary functional monomeric unit present in the polymer backbone [39]. Thus, grafting-onto is widely used for the synthesis of comb polymers. The "grafting-through" method uses macromonomers (polymer chains of very low molecular weight or oligomers) containing groups at their end that is capable of carrying out polymerization, particularly vinyl groups [40]. The addition of a second monomer (comonomer) in the presence or absence (in the case of auto-initiation) of an initiator leads to the construction of several complex topologies of graft copolymers. Indeed, depending on the reactivity ratios and the distance between adjacent grafts, brushes, regular grafts, centipede, barbwire, etc. structures can be produced. Finally, "grafting-from" is the most popular method used to obtain graft polymers and requires a polymer containing reactive groups, functional species, or initiator moieties along its backbone, which in the presence of monomers, produce the growth of grafts from the surface of the substrate. Thus, the number of grafts can be controlled by the number of active sites present along the backbone [41] (Figures 2.12 and 2.13).

2.17.4 Click Chemistry in Polymerization

Click polymerization is a powerful polymerization technique for the construction of new MMs with well-defined structures and multifaceted functionalities. Click chemistry, conceptually introduced by Sharpless and co-workers in 2001, refers to a new class of organic reactions that enjoy the features of high efficiency, regioselectivity, modularity, atom economy, simple product isolation, and functional group tolerance [42].

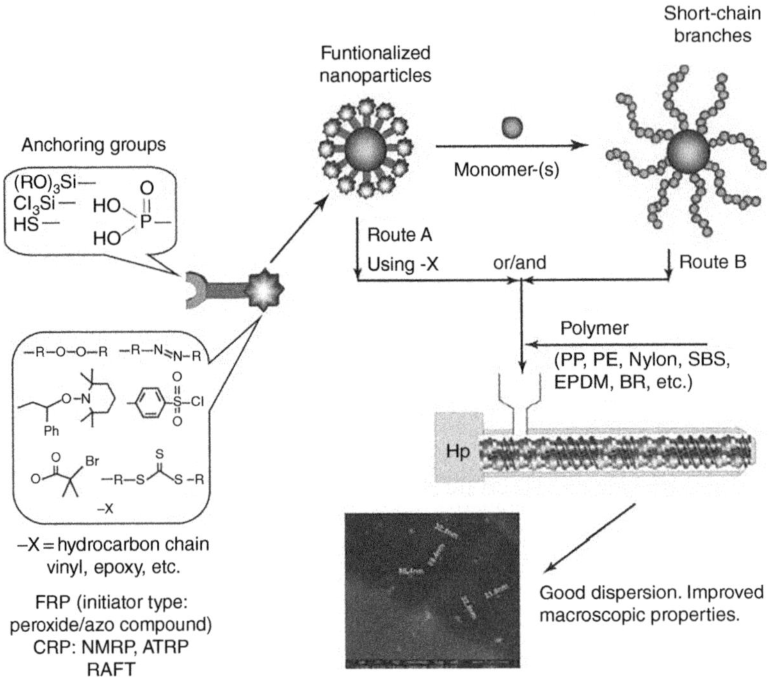

Figure 2.13 Grafting on surfaces and their dispersion in polymer matrices.

Click polymerization also possesses its unique properties, such as easy access to and facile variation of the monomers, large-scale purification, good solubility of the products, and versatile properties. Contrast to traditional polymerizations, the powerful click polymerization makes it possible to integrate units with different characters, for instance, rigidity and flexibility, topology, polarity, and charge, into one polymeric scaffold [43].

Among different click polymerization techniques, the Cu(I)-catalyzed azide and alkyne polycycloaddition have been established as a powerful and selective method for polymer synthesis. Another attractive click polymerization is the thiol-based one, which is essentially a novel active free radical polymerization [44].

2.18 Polymer Nanoparticles

Polymer nanoparticles (PNPs) can be defined as being colloidal systems, which are usually from around 5–10 nm to an upper size limit of 1000 nm, although the range generally obtained is 100–500 nm. The term "polymer nanoparticles" is a collective term used for any type of polymer nanosized particles, but specifically for polymer nanospheres and nanocapsules. Polymer nanospheres are matrix particles, that is, particles whose entire mass is solid [45]. They can further serve as carriers for other biologically active molecules which may be either adsorbed at the sphere surface or encapsulated within the particles. Here, biologically active materials include

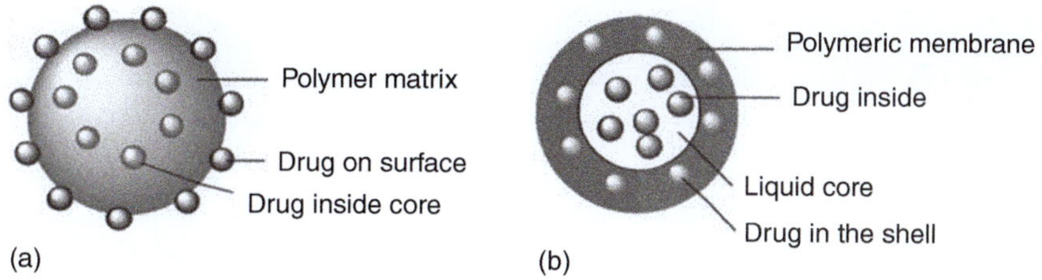

Figure 2.14 Polymer nanoparticle structures. (a) Polymer nanoparticles or polymer nanospheres and (b) polymer nanocapsules.

drugs, genes, nucleic, fluorescence, and other functional materials. Unlike polymer nanospheres, polymer nanocapsules are vesicular systems in which the bioactive agents are confined to an aqueous core and surrounded by the polymeric shell around them [46]. A schematic representation of PNPs is shown below (Figure 2.14).

The manufacturing techniques of PNPs are classified due to whether the particle formation involves a polymerization reaction or NPs are produced directly from a preformed polymer or by direct polymerization of monomers using classical polymerization or polyreactions.

2.19 Synthesis Techniques of Polymer Nanoparticle

PNPs can be conveniently prepared by methods like solvent evaporation, salting-out, dialysis, and supercritical fluid technology [47], involving the rapid expansion of a supercritical solution or rapid expansion of a supercritical solution into a liquid solvent, can be utilized for the preparation of PNP from preformed polymers. On the other hand, various emulsion polymerization techniques such as micro-, mini-, and surfactant-free, and interfacial polymerization have been utilized for the direct synthesis of PNPs [48]. Methods like surfactant-free emulsion polymerization can also be successfully applied for manufacturing surfactant-free PNPs, and they are environment-friendly. The choice of preparation method is made based on several factors such as the type of polymeric system, area of application, and size requirement. For instance, a polymeric system that is developed for application in the biomedical or environmental fields should be completely free from additives or reactants such as surfactants or traces of organic solvents. In this case, techniques like rapid expansion of supercritical solution (RESS) or rapid expansion of a supercritical solution into a liquid solvent (RESOLV) can be selected, as they do not utilize any surfactant or organic solvent during the PNPs preparation. These are just a few of the many factors that have to be considered before choosing a particular technique for the PNP preparation [49] (Figure 2.15).

2.19.1 Solvent Evaporation

Solvent evaporation was the first method developed to prepare PNPs from a preformed polymer. Although originally proposed by polymer chemists, its main

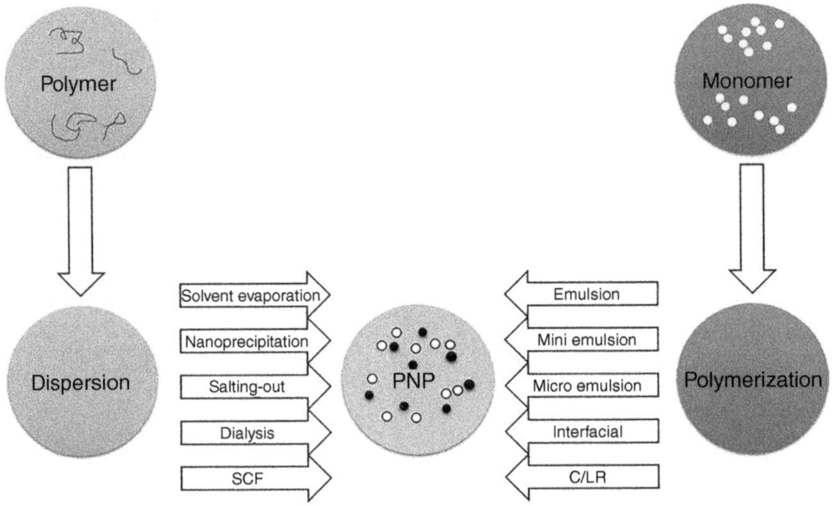

Figure 2.15 Schematic representation of various techniques for the preparation of PNPs. SCF – supercritical fluid technology, C/LR – controlled/living radical. Source: Based on Anton et al. [50].

developments are found in pharmaceutical technology where biodegradable polymers have been applied in the production of drugs. In this method, polymer solutions are prepared in volatile solvents and emulsions are formulated. In the past, dichloromethane and chloroform were widely used but are now replaced with ethyl acetate which has a better toxicological profile. The emulsion is converted into an NPs suspension on evaporation of the solvent for the polymer, which is allowed to diffuse through the continuous phase of the emulsion. Solvent evaporation is the most widely employed technique to prepare NPs of polymers in the current literature on techniques using a dispersion of preformed polymers. In the conventional methods, two main strategies are being used for the formation of emulsions: the preparation of single emulsions, e.g. oil-in-water (o/w) or double emulsions, e.g. (water-in-oil)-in-water, (w/o)/w. These methods utilize high-speed homogenization or ultrasonication, followed by evaporation of the solvent, either by continuous magnetic stirring at room temperature or under reduced pressure. Afterward, the solidified nanoparticles can be collected by ultracentrifugation and washed with distilled water to remove additives such as surfactants. Finally, the product is lyophilized. Generally, a polymer dissolved in an organic solvent forms the oil phase, whereas the aqueous phase containing the stabilizer forms the water phase [51] (Figure 2.16).

2.19.2 Salting-Out

The methods discussed previously require the use of organic solvents, which are hazardous to the environment as well as to physiological systems. To overcome this hurdle, Bindschaedler first disclosed a modified version of the emulsion process that involves a salting-out process, which avoids surfactants and chlorinated

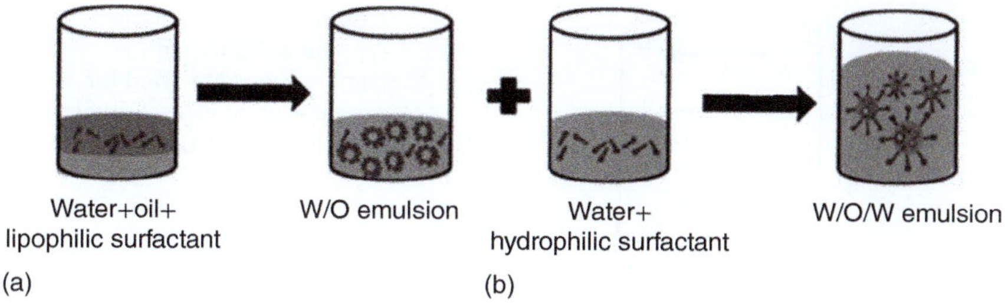

Figure 2.16 Preparation of w/o/w double emulsion. (a) High-shear emulsification and (b) low-shear emulsification.

solvents. The emulsion is formulated with a polymer-solvent which is normally totally miscible with water, i.e. acetone, and emulsification of the polymer solution in the aqueous phase is achieved, more like an *Ouzo effect*, without employing any high-shear forces, by dissolving a high concentration of salt or sucrose chosen fora strong salting-out effect in the aqueous phase [48]. Magnesium chloride, calcium chloride, and magnesium acetate are commonly used suitable electrolytes. The miscibility properties of water with other solvents are modified as these components dissolve in the water. A reverse salting-out effect, obtained by dilution of the emulsion with a large excess of water, leads to the precipitation of the polymer dissolved in the droplets of the emulsion. In fact, upon dilution, migration of the solvent for the polymer from the emulsion droplets is induced due to the reduction of the salt or sucrose concentration in the continuous phase of the emulsion [52].

2.19.3 Nanoprecipitation

The nanoprecipitation method was developed by Fessi for the preparation of PNPs. It is also called a solvent displacement method. The basic principle of this technique is based on the interfacial deposition of a polymer after displacement of a semipolar solvent, miscible with water, from a lipophilic solution. The rapid diffusion of the solvent into the nonsolvent phase results in the decrease of interfacial tension between the two phases, which increases the surface area and leads to the formation of small droplets of organic solvent. The nanoprecipitation system consists of three basic components: the polymer (synthetic, semisynthetic, or natural), the polymer-solvent, and the nonsolvent of the polymer [53]. The organic solvent (i.e. ethanol, acetone, hexane, methylene chloride, or dioxane), which is miscible in water and easy to remove by evaporation, is chosen as a polymer solvent. Due to this reason, acetone is the most frequently employed polymer solvent in this method. Sometimes, it consists of binary solvent blends, acetone with a small amount of water, and blends of acetone with ethanol and methanol. On the other hand, the nonsolvent phase consisting of a nonsolvent or a mixture of nonsolvents is supplemented with one or more naturally occurring or synthetic surfactants. Nanoprecipitation is a simple, fast, and reproducible method that is widely used for the preparation of both nanospheres and nanocapsules [54].

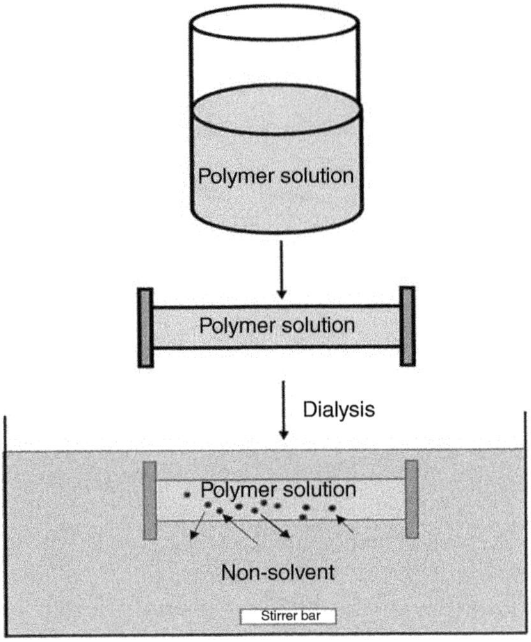

Figure 2.17 Schematic representation of an osmosis-based method for preparation of PNPs. Source: Raj et al. [55]/MDPI/CC BY 4.0.

2.19.4 Dialysis

Dialysis offers a simple and effective method for the preparation of small, narrow-distributed PNPs. A polymer is dissolved in an organic solvent and placed inside a dialysis tube with proper molecular weight cutoff. Dialysis is performed against a nonsolvent miscible with the former miscible (Figure 2.17).

The displacement of the solvent inside the membrane is followed by the progressive aggregation of the polymer due to a loss of solubility and the formation of homogeneous suspensions of NPs [56].

2.19.5 Supercritical Fluid Technology

Supercritical fluid and dense gas technology are expected to offer an interesting and effective technique of particle production, avoiding most of the drawbacks of the traditional methods. Two principal processes have been developed for the production of NPs using supercritical fluids: (i) RESS and (ii) RESOLV [57].

2.19.6 Rapid Expansion of Supercritical Solution (RESS)

In traditional RESS, the solute is dissolved in a supercritical fluid to form a solution, followed by the rapid expansion of the solution across an orifice or a capillary nozzle into the ambient air. The high degree of supersaturation, accompanied by the rapid pressure reduction in the expansion, results in homogeneous nucleation and, thereby, the formation of well-dispersed particles [58]. Results from mechanistic studies of different model solutes for the RESS process indicate that both nanometer- and micrometer-sized particles are present in the expansion jet (Figure 2.18).

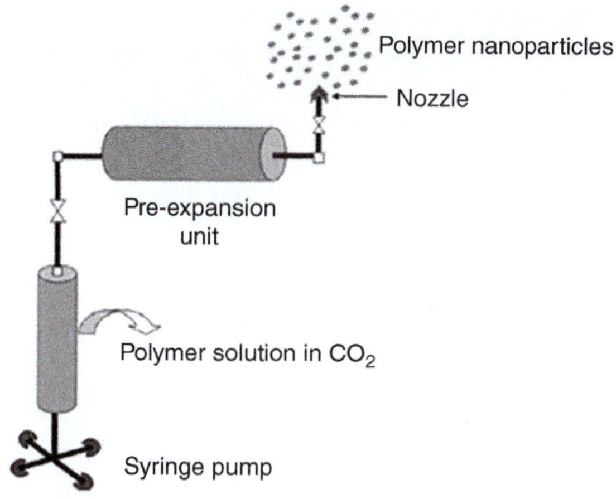

Figure 2.18 Experimental setup for the preparation of PNPs by RESS. Source: Raj et al. [55]/MDPI/CC BY 4.0.

2.19.7 Rapid Expansion of Supercritical Solution into a Liquid Solvent (RESOLV)

A simple, but significant modification to RESS involves the expansion of the supercritical solution into a liquid solvent instead of ambient air, termed as RESOLV. The liquid solvent suppresses the particle growth in the expansion jet, thus making it possible to obtain primarily nanosized particles [59] (Figure 2.19).

2.19.8 Polymerization of Monomers

The techniques discussed previously involve the production of PNPs from preformed polymers and did not involve any polymerization processes. To attain the

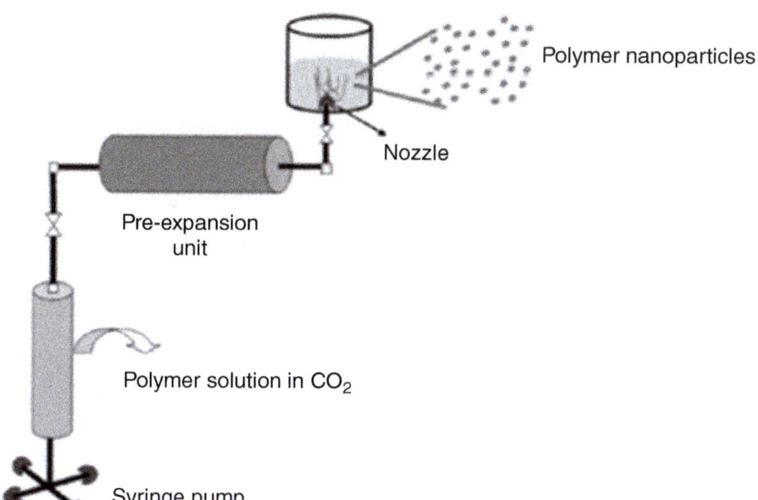

Figure 2.19 Experimental setup for the RESOLV process. Source: Raj et al. [55]/MDPI/CC BY 4.0.

desired properties for a particular application, suitable PNPs must be designed, which can be done during the polymerization of monomers. Processes for the production of PNPs through the polymerization of monomers are discussed, focusing principally on mini-, micro-, and emulsion polymerization techniques as the three major techniques currently in use [60]. Other heterophase polymerizations, such as interfacial and living/CRP, are also employed in the preparation of PNPs.

2.19.9 Emulsion Polymerization

Emulsion polymerization is the most common method used for the production of a wide range of specialty polymers. The use of water as the dispersion medium is environment friendly and also allows excellent heat dissipation during polymerization. Based on the utilization of surfactant, it can be classified as conventional and surfactant-free emulsion polymerization [61].

2.19.10 Conventional Emulsion Polymerization

Conventional emulsion polymerization accounts for the majority of the world's production by emulsion polymerization. In the conventional system, the ingredients comprise water, a monomer of low water solubility, a water-soluble initiator, and a surfactant. At the end of the reaction, PNPs are typically ~102 nm in size, each containing many polymer chains [62].

2.19.11 Surfactant-Free Emulsion Polymerization

Conventional emulsion polymerization systems utilize varying quantities of surfactants that need to be eliminated from the final product but are hard to completely remove. Removal of surfactants is a time-consuming process that increases the cost of production. Moreover, increasing environmental and energy concerns cannot be effectively addressed because of these drawbacks [63]. As an alternative, emulsion polymerization has been performed in the absence of added emulsifiers, often referred to in the literature as surfactant-free, emulsifier-free, or soap-less emulsion polymerization. This technique has received considerable attention for use as a simple, green process for PNPs production without the addition and subsequent removal of the stabilizing surfactants. Surfactant-free emulsion polymerization has emerged as a simple, green process for PNPs production without the addition and subsequent removal of stabilizing surfactants [64].

2.19.12 Mini-Emulsion Polymerization

Research publications on mini-emulsion polymerization and the development of a wide range of useful polymer materials have recently increased substantially. A typical formulation used in mini-emulsion polymerization consists of water, monomer mixture, co-stabilizer, surfactant, and initiator. The key difference between

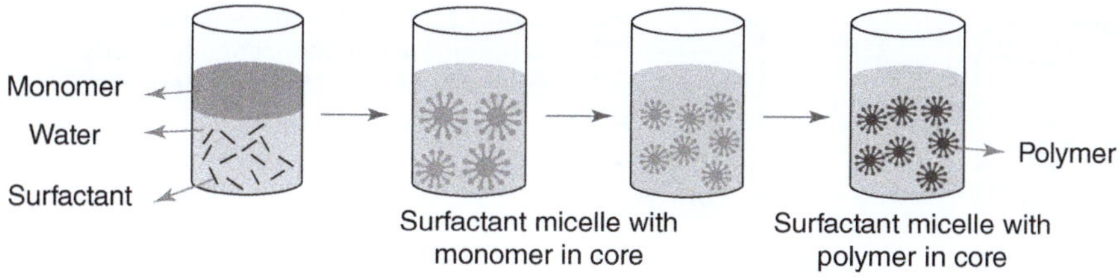

Figure 2.20 Schematic representation of mini-emulsion polymerization process.

emulsion polymerization and mini-emulsion polymerization is the utilization of a low molecular mass compound as the co-stabilizer and also the use of a high-shear device (ultrasound, etc.). Mini-emulsions are critically stabilized, require a high shear to reach a steady state, and have an interfacial tension much greater than 0 [65] (Figure 2.20).

2.19.13 Micro-Emulsion Polymerization

Micro-emulsion polymerization is a new and effective approach for preparing nano-sized polymer particles and has attracted significant attention. Although emulsion and micro-emulsion polymerization appear similar because both methods can produce colloidal polymer particles of high molar mass, they are entirely different when compared kinetically. Emulsion polymerization exhibits three reaction rate intervals, whereas only two are detected in micro-emulsion polymerization. Both particle size and the average number of chains per particle are considerably smaller in micro-emulsion polymerization [66]. In micro-emulsion polymerization, an initiator, typically water-soluble, is added to the aqueous phase of a thermodynamically stable micro-emulsion containing swollen micelles. The polymerization starts from this thermodynamically stable, spontaneously formed state and relies on high quantities of surfactant systems, which possess an interfacial tension at the oil–water interface close to zero. Furthermore, the particles are completely covered with surfactants because of the utilization of a high amount of surfactant. Initially, polymer chains are formed only in some droplets, as the initiation cannot be attained simultaneously in all micro-droplets.

Later, the osmotic and elastic influence of the chains destabilize the fragile micro-emulsions and typically lead to an increase in the particle size, the formation of empty micelles, and secondary nucleation [67]. Very small latexes, 5–50 nm in size, coexist with a majority of empty micelles in the final product (Figure 2.21).

2.19.14 Interfacial Polymerization

Interfacial polymerization (IP) is one of the well-established methods used for the preparation of PNPs. It involves step polymerization of two reactive monomers or agents, which are dissolved in two phases (i.e. continuous and dispersed phase), and the reaction takes place at the interface of the two liquids. The relative ease

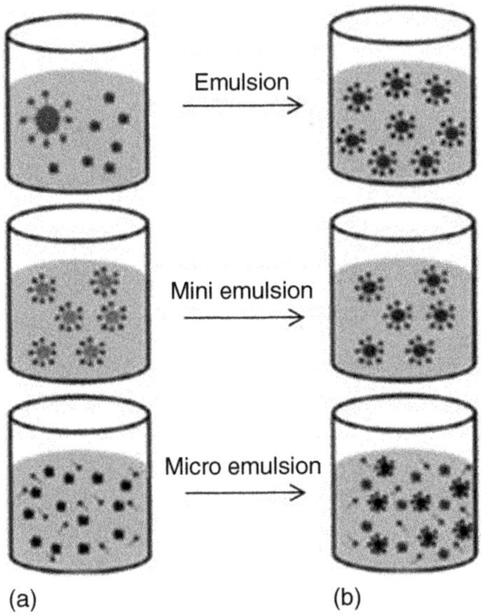

Figure 2.21 Differences among various heterophase polymerization methods. (a) Before and (b) after polymerization.

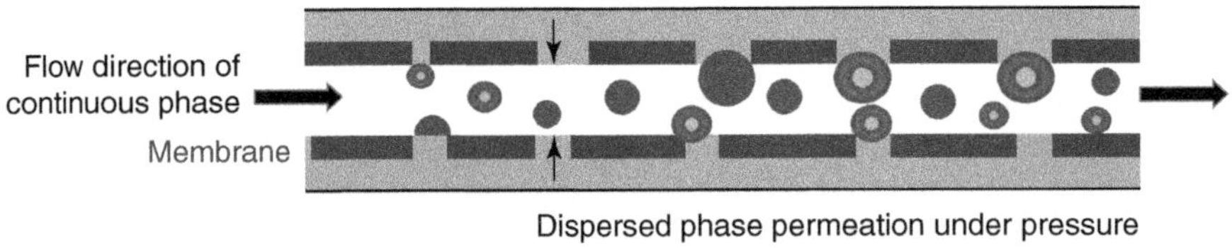

Figure 2.22 Schematic of membrane reactor for the preparation of PNPs.

of obtaining IP has made it a preferred technique in many fields, ranging from the encapsulation of pharmaceutical products to the preparation of conducting polymers [68] (Figure 2.22).

References

1 Jadhav, K.G. (2012). Synthesis of monomers for new conjugated polymers. Masters thesis, University of Massachusetts Amherst, p. 753.

2 Ciardelli, F., Passaglia, E., and Bronco, S. (2009). *Polymers and Their Synthesis, Encyclopedia of Life Support Systems*, Fundamentals of Chemistry, 379. Eolss Publishers.

3 Negrell, C., Voirin, C., Boutevin, B. et al. (2018). From monomer synthesis to polymers with pendant aldehyde groups. *Eur. Polym. J.* 106: 544–563.

4 Saldivar-Guerra, E. and Vivaldo-Lima, E. (2013). *Introduction to Polymers and Polymer Types*, Handbook of Polymer Synthesis, Characterization, and Processing. Wiley.

5 Stepto, R.F.T. (2009). Dispersity in polymer science. *Pure Appl. Chem.* 81: 351–353.

6 Odian, G. (2004). *Principles of Polymerization*. Hoboken, NJ: Wiley.

7 Nuyken, O. and Pask, S.D. (2013). Ring-opening polymerization - an introductory review. *Polymers* 5: 361–403.

8 Kricheldorf, H.R. (2002). Polycondensation. *Macromol. Symp.* 199: 1–13.

9 Marchildon, K. (2011). Polyamides – still strong after seventy years. *Macromol. React. Eng.* 5: 22–54.

10 Mizutani, M., Satoh, K., and Kamigaito, M. (2010). Metal-catalyzed simultaneous chain- and step-growth radical polymerization: marriage of vinyl polymers and polyesters. *J. Am. Chem. Soc.* 132: 7498–7507.

11 Jochi, Y., Seki, T., Soejima, T. et al. (2018). Spontaneous synthesis of a homogeneous thermoresponsive polymer network composed of polymers with a narrow molecular weight distribution. *NPG Asia Mater.* 10: 840–848.

12 Tang, W., Fukuda, T., and Matyjaszewski, K. (2006). Reevaluation of persistent radical effect in NMP. *Macromolecules* 39: 4332–4337.

13 Zhao, Y.L., Cai, Q., Jiang, J. et al. (2002). Synthesis and thermal properties of novel star-shaped poly(L-lactide) with starburst PAMAM-OH dendrimer macroinitiator. *Polymer* 43: 5819–5825.

14 Pascualt, J.P. and Williams, R.J.J. (2010). *Epoxy Polymers: New Materials and Innovations*. Weinheim: Wiley-VCH.

15 Goto, A. and Fukuda, T. (2004). Kinetics of living radical polymerization. *Prog. Polym. Sci.* 229: 329–385.

16 Penczek, S. and Moad, G. (2008). Glossary of terms related to kinetics, thermodynamics, and mechanisms of polymerization. *Pure Appl. Chem.* 80: 2163–2193.

17 Worldwide Rubber Statistics/IISRP (2009) Houston: International Institute of Synthetic Rubber Producers.

18 Kostjuk, S.V. and Ganachaud, F. (2010). Cationic polymerization of vinyl monomers in aqueous media: from monofunctional oligomers to long-lived polymer chains. *Acc. Chem. Res.* 43: 357–367.

19 Matyjaszewski, K. and Sigwalt, P. (1994). Unified approach to living and non-living cationic polymerization of alkenes. *Polym. Int.* 35: 1–26.

20 Matyjaszewski, K. and Spanswick, J. (2005). Controlled/living radical polymerization. *Mater. Today* 8: 26–33.

21 Moad, G., Rizzardo, E., and Thang, S.H. (2008). Toward living radical polymerization. *Acc. Chem. Res.* 41: 1133–1114.

22 Moad, G., Rizzardo, E., and Thang, S.H. (2009). Living radical polymerization by the RAFT process – a second update. *Aust. J. Chem.* 62: 1402–1472.

23 Moad, G., Chiefari, J., Chong, Y.K. et al. (2000). Living free radical polymerization with reversible addition-fragmentation chain transfer (the life of RAFT). *Polym. Int.* 49: 993–1001.

24 De Greef, T.F.A., Smulders, M.M.J., Wolffs, M. et al. (2009). Supramolecular polymerization. *Chem. Rev.* 109: 5687–5754.

25 Pol, B. (2017). Controlling supramolecular polymerization through multicomponent self-assembly. *J. Polym. Sci. A* 55: 34–78.

26 Cheremisinoff NP (2001) *Condensed Encyclopedia of Polymer Engineering Term.* Butterworth-Heinemann, Bostonhttps://doi.org/10.1016/b978-0-08-050282-3.50007-x.

27 Bhat, G. and Kandagor, V. (2014). *Synthetic polymer fibers and their processing requirements*, Advances in Filament Yarn Spinning of Text and Polymers, 3–30. https://doi.org/10.1533/9780857099174.1.3.

28 Deng, F., Luo, X.B., Ding, L., and Luo, S.L. (2019). *Application of Nanomaterials and Nanotechnology in the Reutilization of Metal Ion from Wastewater*, Nanomaterials for the Removal of Pollutants and Resource Reutilization, 149–178. https://doi.org/10.1016/b978-0-12-814837-2.00005-6.

29 Argenta, D.F., dos Santos, T.C., Campos, A.M., and Caon, T. (2019). *Chapter 3 - Hydrogel Nanocomposite Systems: Physico-Chemical Characterization and Application for Drug-Delivery Systems*, Nanocarriers for Drug Delivery, 81–131.

30 Christopher, C.I. (2011). *Thermoplastic Materials - Properties, Manufacturing Methods, and Applications.* Boca Raton: CRC Press https://doi.org/10.1201/b13623.

31 Sellergren, B. and Hall, A.J. (2001). Fundamental aspects on the synthesis and characterization of imprinted network polymers. *Tech. Instru. Anal. Chem.* 23: 21–57.

32 Abd Al-Rhman Magdy, A.Y. (2019). Solution & bulk polymerization. https://doi.org/10.13140/RG.2.2.16472.96001/2.

33 Ebnesajjad, S. and Morgan, R.A. (2012). *Manufacturing and Properties of Fluoroelastomer-Based Additives*, Fluoropolymer Additives, 53–65. https://doi.org/10.1016/b978-1-4377-3461-4.00005-2.

34 Mayes, A.G. (2001). Polymerisation techniques for the formation of imprinted beads. *Tech. Instru. Anal. Chem.* 23: 305–324.

35 Cengiz, U., Gengec, N.A., Kaya, N.U. et al. (2011). Mechanical and thermal properties of perfluoroalkyl ethyl methacrylate–methyl methacrylate statistical copolymers synthesized in supercritical carbon dioxide. *J. Fluorine Chem.* 132: 348.

36 Atiqullah, M., Tinkl, M., Pfaendner, R. et al. (2010). Synthesis of functional polyolefins using metallocenes: a comprehensive review. *Polym. Rev.* 50: 178–230.

37 Uhrig, D., Schlegel, R., Weidisch, R., and Mays, J. (2011). Multigraft copolymer superelastomers: synthesis morphology, and properties. *Eur. Polym. J.* 47: 560–568.

38 Bonilla-Cruz, J., Guerrero-Sanchez, C., Schubert, U.S., and Saldivar-Guerra, E. (2010). Controlled "Grafting-from" of poly[styrene-*co*-maleic anhydride] onto polydienes using nitroxide chemistry. *Eur. Polym. J.* 46: 298–312.

39 Oliveira, P.C., Guimaraes, A., Cavaille, J.Y. et al. (2005). Poly(dimethylaminoethyl methacrylate) grafted natural rubber from seeded emulsion polymerization. *Polymer* 46: 1105–1111.

40 Wang, Y., Shen, Y., Pei, X. et al. (2008). In situ synthesis of polystyrene/SiO_2 hybrid composites via a 'grafting through' strategy based on nitroxide-mediated radical polymerisation. *Polym. Polym. Compos.* 16: 621–626.

41 Bonilla-Cruz, J., Lara-Ceniceros, T.E., Saldivar-Guerra, E., and Jimenez-Regalado, E. (2007). Towards controlled graft polymerization of poly[styrene-co -(maleic anhydride)] on functionalized silica mediated by oxoaminium bromide salt. Facile synthetic pathway using nitroxide chemistry. *Macromol. Rapid Commun.* 28: 1397–1403.

42 Kade, M.J., Burke, D.J., and Hawker, C.J. (2010). The power of thiol-ene chemistry. *J. Polym. Sci. A* 48: 743–750.

43 Cao, X., Shi, Y., Wang, X. et al. (2016). Design a highly reactive trifunctional core molecule to obtain hyperbranched polymers with over a million molecular weight in one-pot click polymerization. *Macromolecules* 49: 760–766.

44 Li, H., Sun, J.Z., Qin, A., and Tang, B.Z. (2012). Azide-alkyne click polymerization: an update. *Chin. J. Polym. Sci.* 30: 1–15.

45 Bonilla-Cruz, J., Dehonor, M., Saldivar-Guerra, E., and Gonzalez-Montiel, A. (2013). *Chapter 10, Polymer Modification: Functionalization and Grafting,* Handbook of Polymer Synthesis, Characterization, and Processing. Wiley.

46 Khalil, I.R., Burns, A.T.H., Radecka, I. et al. (2017). Bacterial-derived polymer poly-γ-glutamic acid (γ-PGA)-based micro/nanoparticles as a delivery system for antimicrobials and other biomedical applications. *Int. J. Mol. Sci.* 18: 313–335.

47 Mallakpour, S. and Behranvand, V. (2016). Polymeric nanoparticles: recent development in synthesis and application. *eXPRESS Polym. Lett.* 10: 895–913.

48 Prasad Rao, J.P. and Geckeler, K.E. (2011). Polymer nanoparticles: preparation techniques and size-control parameters. *Prog. Polym. Sci.* 36: 887–913.

49 Geckeler, K.E. and Stirn, J. (1993). Polyreactions--mechanisms, taxonomy, relevance. *Naturwissenschaften* 80: 487–500.

50 Anton, N., Benoit, J.P., and Saulnier, P. (2008). Design and production of nanoparticles formulated from nano-emulsion templates-a review. *J. Controlled Release* 128: 185–199.

51 Ganachaud, F. and Katz, J.L. (2005). Nanoparticles and nanocapsules created using the ouzo effect: spontaneous emulsification as an alternative to ultrasonic and high-shear devices. *Chem. Phys. Chem.* 6: 209–216.

52 Nguyen, C.A., Allemann, E., Schwach, G. et al. (2003). Synthesis of a novel fluorescent poly(D,L-lactide) end-capped with 1-pyrenebutanol used for the preparation of nanoparticles. *Eur. J. Pharm. Sci.* 20: 217–222.

53 Mishra, B., Patel, B.B., and Tiwari, S. (2010). Colloidal nanocarriers: a review on formulation technology, types, and applications toward targeted drug delivery. *Nanomedicine: NBM* 6: 9–24.

54 Nassar, T., Rom, A., Nyska, A., and Benita, S. (2009). Novel double-coated nanocapsules for intestinal delivery and enhanced oral bioavailability of tacrolimus, a P-gp substrate drug. *J. Controlled Release* 133: 77–84.

55 Raj, R., Alwani, S., and Badae, I. (2019). Polymeric nanoparticles in gene therapy: new avenues of design and optimization for delivery applications. *Polymer* 11: 745.

56 Kostog, M., Kohler, S., Liebert, T., and Heinze, T. (2010). Pure cellulose nanoparticles from trimethylsilyl cellulose. *Macromol. Symp.* 294: 96–106.

57 Mishima, K. (2008). Biodegradable particle formation for drug and gene delivery using supercritical fluid and dense gas. *Adv. Drug Delivery Rev.* 60: 411–432.

58 Weber, M. and Thies, M.C. (2002). Understanding the RESS process. In: *Supercritical Fluid Technology in Materials Science and Engineering: Synthesis, Properties, and Applications* (ed. Y.P. Sun), 387–437. New York: Marcel Dekker.

59 Meziani, M.J., Pathak, P., Hurezeanu, R. et al. (2004). Supercritical-fluid processing technique for nanoscale polymer particles. *Angew. Chem. Int. Ed.* 43: 704–707.

60 Thickett, S.C. and Gilbert, R.G. (2007). Emulsion polymerization: state of the art in kinetics and mechanisms. *Polymer* 48: 6965–6991.

61 Asua, J.M. (2004). Emulsion polymerization: from fundamental mechanisms to process developments. *J. Polym. Sci. Part A* 42: 1025–1041.

62 Bertholon, I., Ponchel, G., Labarre, D. et al. (2006). Bioadhesive properties of poly(alkyl cyanoacrylate) nanoparticles coated with polysaccharide. *J. Nanosci. Nanotechnol.* 6: 3102–3109.

63 Akgol, S., Ozturk, N., and Denizli, A. (2010). New generation polymeric nanospheres for lysozyme adsorption. *J. Appl. Polym. Sci.* 115: 1608–1615.

64 Liu, G. and Liu, P. (2010). Synthesis of monodispersed crosslinked nanoparticles decorated with surface carboxyl groups via soapless emulsion polymerization. *Colloid Surf. A* 354: 377–381.

65 Wu, M., Dellacherie, E., Durand, A., and Marie, E. (2009). Poly(n-butyl cyanoacrylate) nanoparticles via miniemulsion polymerization dextran-based surfactants. *Colloid Surf. B* 69: 141–146.

66 Baruch-Sharon, S. and Margel, S. (2010). Synthesis and characterization of poly-chloromethyl styrene nanoparticles of narrow size distribution by emulsion and miniemulsion polymerization processes. *Colloid. Polym. Sci.* 288: 869–877.

67 Jiang, X., Dausend, J., Hafner, M. et al. (2010). Specific effects of surface amines on polystyrene nanoparticles in their interactions with mesenchymal stem cells. *Biomacromolecules* 11: 748–753.

68 Landfester, K., Musyanovych, A., and Mailander, V. (2010). From polymeric particles to multifunctional nanocapsules for biomedical applications using the miniemulsion process. *J. Polym. Sci. A* 48: 493–515.

3

Characterization of Polymer Materials

3.1 Introduction

The polymer characterization studies focus on establishing the molecular structure, mass, and morphology of synthetic and natural polymeric materials. These are comprised of concurrent structures of covalently bonded macromolecules. The thermodynamic and mechanical properties of these structures vary depending on their chemical or biological composition, but they typically emerge as high-versatile materials with a range of interdisciplinary applications. Natural polymers include materials such as wool and rubber, while synthetic polymers are primarily comprised of plastics such as polythene and polystyrene (PS).

Several characterization techniques have been used to analyze basic aspects such as the elucidation of a molecular structure, composition, configuration, and conformation of the chemical groups within the polymer. By analyzing the composition of a polymer molecule, the nature of the atoms in the polymer chain and the type of bonding can be inferred. The configuration gives an idea about the chemical state of the polymer, the spatial order of the chemical groups, and the optical characteristics and possible behavior of the whole molecule. The conformation characterizes the geometrical state of a polymer [1].

Particle size and size distribution are the key parameters used for evaluating the physical properties of polymer nanoparticles (PNPs). Photon correlation spectroscopy (PCS) also known as dynamic light scattering (DLS), laser diffraction (LD), and Coulter counter are commonly used to measure the size and size distribution of PNPs. PCS/DLS is widely used to determine the size and size distribution of nanosized particles suspended in a liquid medium. The mean particle size and size distribution indicated as the polydispersity index (PDI) is the typically measured parameter for this technique. A PDI value range from 0.1 to 0.25 indicates a narrow size distribution, while a PDI more than 0.5 refers to broad distribution. This technique is not useful for measuring the size of dry powders of PNPs [2]. Although PCS, LD, and Coulter counter techniques provide the rapid measurement of particle size and size distribution, they cannot be used

Polymer Nanocomposites for Energy Applications, First Edition.
T. Daniel Thangadurai, Manjubaashini Nandhakumar, Sabu Thomas, and Ange Nzihou.

to evaluate particle morphology. In this case, scanning electron microscopy (SEM), transmission electron microscopy (TEM), and atomic force microscopy (AFM) are widely used for the observation of particle morphology. AFM can provide a three-dimensional surface profile. Additionally, samples viewed by AFM do not require any special treatments such as metal/carbon coatings that would irreversibly change or damage the sample; most of the AFM modes can work perfectly well in ambient air or even a liquid environment. Importantly, AFM provides a higher resolution than SEM. The zeta potential of PNPs is commonly measured by laser Doppler electrophoresis, which evaluates the electrophoretic mobility of suspended particles in the medium. It is a rule of thumb that an absolute value of zeta potential above 60 mV yields excellent stability, while 30, 20, and less than 5 mV generally result in good stability, acceptable short-term stability, and fast particle aggregation, respectively, for pure electrostatic stabilization or in combination with low-molecular-weight surfactants, but not for high-molecular-weight stabilizers present in the medium [3].

3.2 UV–Visible Spectroscopy

The example of the application of UV–visible spectroscopy for determining the copolymer composition of poly(methyl methacrylate) (PMMA), with different chlorophenyl methacrylates, is analyzed. Namely, chlorophenyl methacrylates show intensive $\pi \rightarrow \pi^*$ transitions at wavelengths lower than 250 nm and $n \rightarrow \pi^*$ transitions located at wavelengths higher than 260 nm, which shows redshift with increasing of Cl substitution. In the spectra of copolymers, the absorption of PMMA moiety at wavelengths longer than 260 nm is negligible; thus, the absorption peaks at these wavelengths are the only contribution of poly(chlorophenyl methacrylates) moieties. This allows the calculation of copolymer composition directly from the intensity of the appropriate absorption bands at around 260 nm. The calibration curves were calculated from the absorption maxima in the UV spectra of different concentrations of individual homopolymers in polymer blends. The linear relationships between the absorbance and the polymer concentration confirm that Lambert–Beer's law was valid in the selected concentration range [4]. In the UV spectra of the copolymer with unknown composition, the absorbance at 260 nm was determined, for which value from the respected calibration curve the concentration of poly(chlorophenyl methacrylate) was evaluated. The copolymer compositions obtained in a presented way showed an excellent agreement with the composition obtained by chlorine analysis. If the polymers contain conjugated double bonds, such as polyenes, then the wavelength of the $\pi \rightarrow \pi^*$ absorption will depend on the length of the conjugated sequences. This makes the UV–visible spectroscopy suitable for the estimation of the polymer chain length from the known λ_{max} and according to the Schrodinger equation, solved in one dimension [5]. The UV spectra could be used for quantitative analysis of copolymers; however, it is limited to the case when the individual polymer constituents show strong absorption independent of each other.

3.3 Elemental Analysis

An elemental analyzer, based on the Dumas combustion method, proved to be a valuable solution for the elemental analysis (EA) in polymers and plastics in terms of accuracy, repeatability, and sensitivity of results. Its automation, high speed of analysis, and the elimination of the sample preparation process allowed efficient analysis and helped to reduce overall operational costs. The analysis was performed without matrix effect when changing the sample and element content indicating the complete combustion of the samples [6]. With an upgrade, the elemental analyzer could determine nitrogen only, simultaneous CHNS, NCS, CHN, and oxygen. Trace sulfur determination was achieved when the system was coupled with the FPD detector.

Combustion techniques, such as pyrolysis, are one of the most common analytical methods of identification of the constituents of any sample; the structure of the monomers or any other added molecules used during the polymer synthesis can then be subsequently confirmed by spectroscopic techniques. Analytical methods usually involve, for example, the burning of a sample in an oxygen-containing atmosphere, to determine amounts of carbon, hydrogen, nitrogen, sulfur, halogens, and oxygen. Also, dry-ashing, fusion, bomb, and acid digestion can be used to remove organic material and trace metal residues, which can be further analyzed by spectroscopic techniques. The elemental composition and content of some specific elements is an important analytical tool for polymer characterization, mainly for the characterization of copolymers and polymer blends and the determination of the molecular weight of homopolymers [7]. The knowledge of the elemental composition of a polymer can be a useful indicator of the identity of the polymer (Figure 3.1).

Determination of copolymer composition, EA is a convenient method for the determination of copolymer and blend composition if one homopolymer contains an element not present in the second one. For example, EA can be properly used to quantify nitrogen in copolymers containing acrylonitrile units and oxygen in polymeric surfactants such as poly(oxyalkylene). The commonly used techniques to determine the molecular weight of polymers are osmometry, viscosimetry, light scattering, gel permeation chromatography (GPC), NMR, etc. In all of them, the sample must be soluble in organic solvents or water [8]. However, several kinds of polymers, such as the new and intelligent materials, especially highly thermostable or conductive polymers such as poly(phenylene sulfides), poly(*p*-xylylidene), or polypyrrole, are barely soluble or even insoluble in typical solvents. In such cases, EA is a promising and useful method. The molecular weight of polymers could be obtained from the analysis of some elements present in the polymer chain, for example, heavy metal salts from the analysis of the metal content.

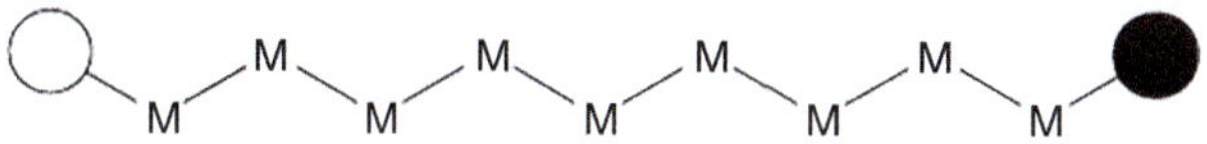

Figure 3.1 Representation of linear polymer chain. M represents the repeating unit of the monomer that will be repeated n several times. Clear and black circles can both be either the initiator fragment remaining on each side of the molecule or an end group deliberately added to functionalize the polymer.

3.4 Infrared Spectroscopy

Infrared spectroscopy is a technique based on the vibrations of atoms of a molecule. To get a spectrum, the sample is placed in a sample holder and then an infrared ray is passed through the sample. The signals or peaks that can be appreciated in an IR spectrum correspond to the energy absorbed by the sample at specific frequencies that depend on the molecule's structure. To detect a signal, molecules must change their electric dipole during irradiation, which implies the generation of specific movements between atoms and chemical bonds. Interactions between IR radiation and molecules involve changes in molecular dipoles associated with vibrations and rotations. The atoms in molecules can move and bond lengths can vary; also, one atom can move out of its present plane in stretching and bending movements. Vibrations can involve a change in either bond length (stretching) or bond angle (bending). Depending on the type of movement, there is also symmetric and asymmetric stretching; each of them is represented as an individual absorption signal. Even for simple molecules, there will be many vibrational signals. A simple molecule can generate a complex spectrum. In a polymer, the repeating unit represents the simple molecule that will be generating a pool of signals or bands. Bands of vibrations associated with the presence of characteristic functional groups are called skeletal vibrations, and these skeletal vibrations are likely to constitute a pattern or fingerprint of the molecule as a whole. In practice, the radiation emerging from a source is passed through an interferometer and then through the sample before reaching the detector. The most common interferometer used in FT-IR is the Michelson interferometer, which consists of two perpendicular plane mirrors, one of which can travel in a direction perpendicular to the plane. A semi-reflecting film, the beam splitter, bisects the planes of these two mirrors. The beam splitter material has to be chosen according to the infrared region to be examined [9].

3.5 Qualitative Analysis of Polymers

Most organic molecules (including polymers) show absorption bands from the interaction between the IR radiation and the atoms in a chemical bond in the mid-IR region. Most IR studies are related to the analysis of vibrations in the mid-IR region, but near- and far-IR regions also provide important information about certain materials. The mid-IR spectrum (4000–400 cm^{-1}) can be divided into four main regions, and the nature of a group frequency may generally be determined by the region in which it is located. Each band in an IR spectrum can be assigned to a particular movement in a molecule or group of atoms in a chemical bond. If this is applied to any bond in a molecule, there will be multiple bands even for similar molecules [10]. A spectrum may have a hundred or more absorption bands, and there is no need to assign the vast majority; this can be regarded as the "fingerprint" of the molecule [11] (Figure 3.2).

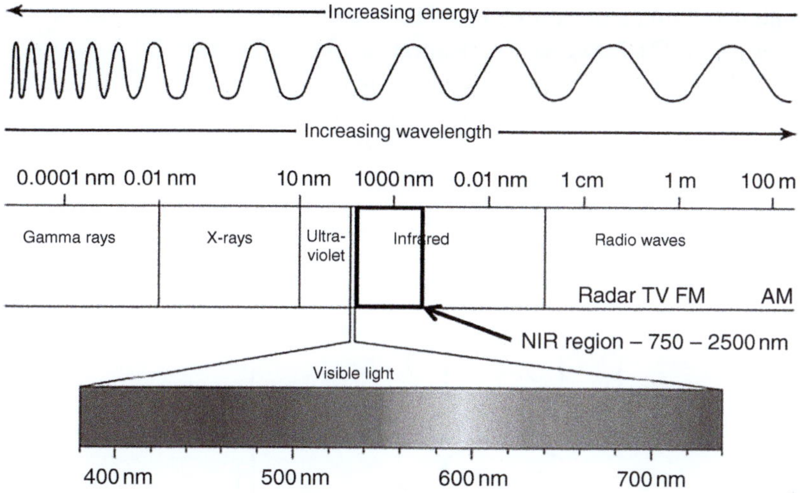

Figure 3.2 IR regions for near, middle, and far spectrum.

3.6 Spectral Analysis for Polyethylene and Polystyrene

The IR spectrum for low-density polyethylene (PE) has four high-intensity absorption bands. The first band (A) corresponds to the symmetric and asymmetric stretchings of methyl and methylene groups. Then, symmetrical bending of methyl (B) and methylene (C) groups of the polymer is observed, and the last band (D)

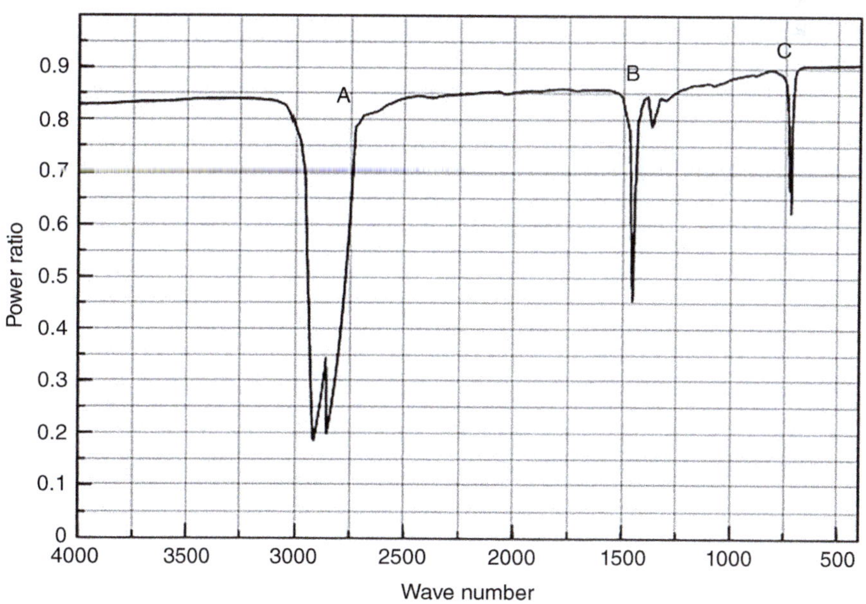

Figure 3.3 IR spectrum for high-density polyethylene.

corresponds to the stretching movement of the methylene groups in the backbone of the polymer chain [12]. However, the IR spectrum for high-density PE in Figure 3.3 shows some differences, since there are three main signals for a molecule having the same chemical composition. Band (A) represents the symmetrical stretching for methyl and methylene groups and band (B) corresponds to the symmetric methylene bending. The last one (C) is ascribed to the absorption of the methylene groups [13].

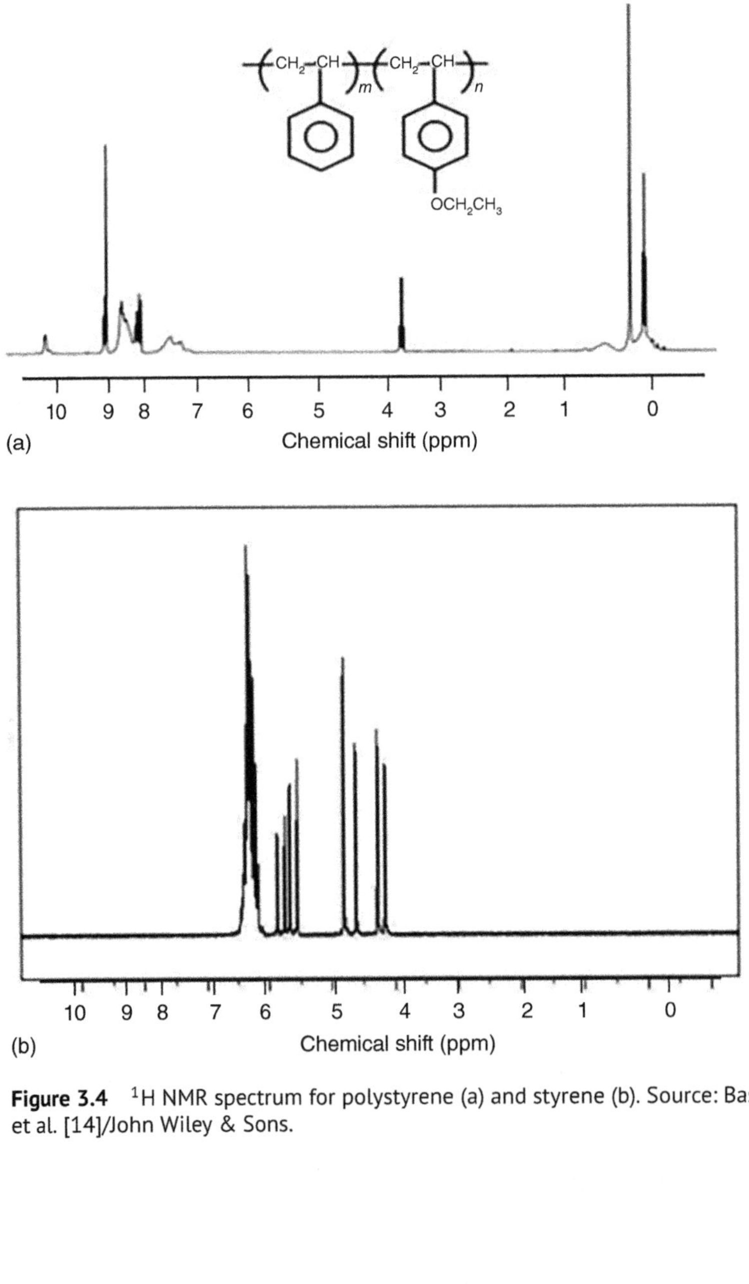

Figure 3.4 ^{1}H NMR spectrum for polystyrene (a) and styrene (b). Source: Based on Shethji et al. [14]/John Wiley & Sons.

Foe qualitative analysis of polymers, NMR spectroscopy is an analytical tool extremely useful to follow the synthesis of polymers or supramolecular systems, from starting materials, intermediates, to products, and their three-dimensional conformation. The rapid development of advanced techniques has substantially broadened the application of NMR to the study of polymers. Examples of ^{1}H NMR spectra of PS and its monomer styrene are shown in Figure 3.4 [15]. The faster decay of the signal from PS than that from styrene is due to faster relaxation of the spins or shorter relaxation times. The difference is reflected as a difference in peak width.

3.7 Determination of Molecular Weight and Thermodynamic Properties

As shown in Figure 3.5, in the limit of small angles, the effective volume that is probing is very large in comparison with the volume occupied by a single macromolecule. Therefore, the number of polymer molecules within this volume is considerably large. As we approach $\theta \approx 0$, this scale becomes quasi-macroscopic and can reduce the calculation of the fluctuations in the dielectric constant to that of the fluctuation of the thermodynamic degrees of freedom of the system [17].

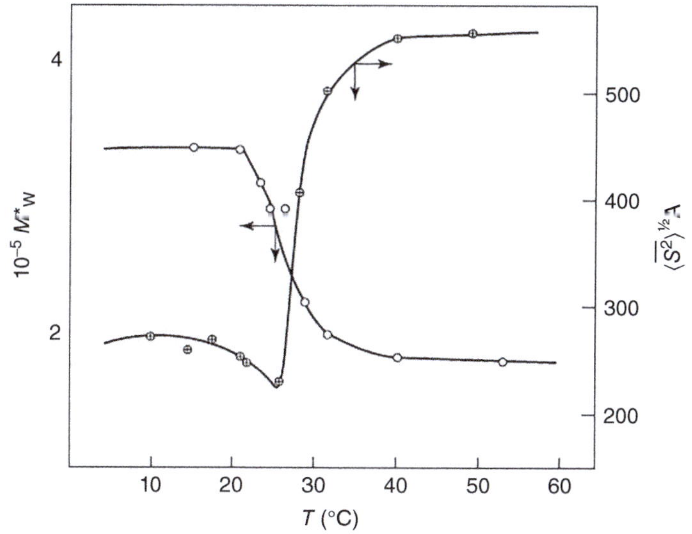

Figure 3.5 Helix–coil transitions of poly (L-benzyl glutamate) in a mixture of dichloroacetic acid and heptane showing the variation of the apparent molecular weight and the radius of gyration. Source: Cowie [16]/Walter de Gruyter.

3.8 Differential Scanning Colorimetry (DSC) Analysis

A heat flux differential scanning colorimetry (DSC) (i.e. containing a single heating block) is comprised of a thermal cell with a sensor that registers the temperature difference between the sample-filled pan and a similar reference pan that contains only air. When the sample evolves heat through some thermal process, such as a crosslinking reaction, the DSC plot shows an increase in heat flow (Figure 3.6). This is indicative of an exothermic event because the temperature registered by the sample sensor is higher than that sensed for the reference. If the sample is undergoing a thermal event that causes it to absorb more heat than the reference does (such as melting), the DSC plot shows a decrease in heat flow. This is called an endotherm and, in these cases, the temperature sensor measures a lower temperature for the sample compared to the reference. For example, a material can be heated at a controlled steady rate, such as 10 °C per minute, and the heat flow can be monitored to characterize the thermal events of the sample as a function of increasing temperature. Figure 3.6 shows a DSC plot for a polyethylene terephthalate (PET) sample which had been cooled from the melt at an extremely high rate. The plot illustrates both exothermic and endothermic thermal events that occurred during a temperature scan from 50 to 300 °C. The endothermic step change (glass transition) occurs first in the scan, followed by an exothermic peak due to cold crystallization, which is then followed by the endothermic peak due to melting. Many modern DSC instruments can measure absolute heat flow. This is done by dividing the signal by the measured heating rate, which converts it into a heat capacity signal [18]. Monitoring the heat-capacity-related signal as a function of the applied experimental conditions (such as a heating ramp) can determine how the heat capacity of the sample changes as it undergoes a phase change or a chemical reaction.

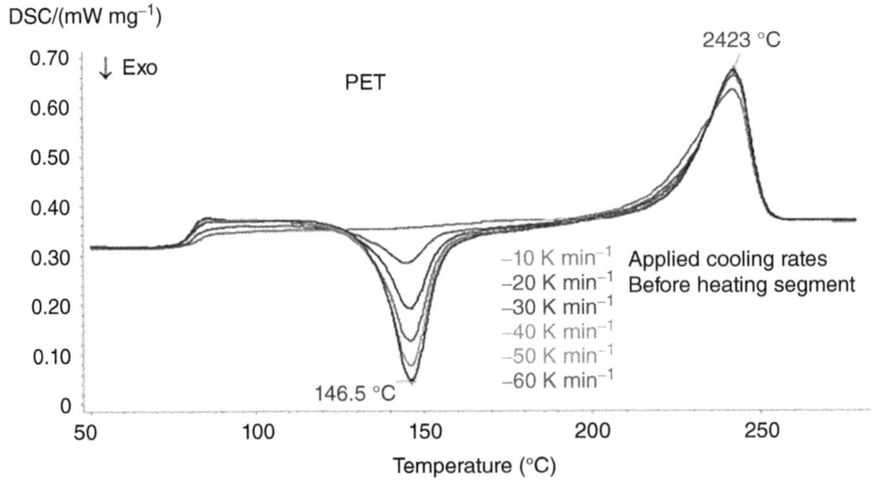

Figure 3.6 DSC scan of polyethylene terephthalate.

3.9 Thermogravimetric Assays (TGAs)

Thermogravimetric assays (TGAs) are another type of calorimetric technique that measures the weight loss of the samples. It is a useful technique to determine the amount of nano-conjugation, since the change in the nanomaterial composition produces changes in temperature weight loss. For example, researchers used TGA analysis to confirm the covalent attachment of a dendritic carbosilane wedge to PNPs. Since both calorimetric techniques have similar performance, their advantages and disadvantages have been summarized together. Calorimetric techniques are advantageous in terms of the small amount of sample required, high precision, and sensitivity [19].

3.10 Gel Permeation Chromatography (GPC)

Gel permeation chromatography (GPC) is a widely used technique to determine the molecular weight of materials dissolved in organic solvents as well as the physical stability of assembled nanomaterials. The nanomaterials are eluted as a function of their molecular weight: the bigger the nanomaterials, the faster the elution. The quantification of the eluted samples is performed utilizing UV–vis absorption or changes in the refractive index. This technique could be useful, for example, to study the stability of a polymer dissolved in an organic solvent. After various periods, the dissolved polymer should be analyzed through the GPC and its molecular weight assessed by using a calibration curve [20]. Other examples of the use of GPC in the nanoscale could be the assessment of the polymer molecular weight, the degree of polymerization of a synthesized polymer, or even the number of monomer subunits that a polymer contains. In some cases, GPC has also been used for the purification of quantum dots or carbon nanotubes. GPC is advantageous in terms of short-time experiments. However, an important drawback of this technique is the possible interaction between the nanomaterials and the column filling, which could interfere with the size assessment [21].

3.11 High-Performance Liquid Chromatography (HPLC)

High-performance liquid chromatography (HPLC) is the most used type of chromatography not only for colloidal nanosystem studies but also for other types of materials (e.g. proteins). In the vast majority of studies, it is used for the fine quantification and separation (purification) of actives, such as drugs. Briefly, it consists of the injection of the liquid sample using a pump that introduces it to a flow (mobile phase) that passes through a separate column (stationary phase), which entraps the molecules depending on their nature. The more interactions the molecules have with the column filling, the later they will be eluted [22]. Further, molecules are eluted in a characteristic pattern for each compound. It results in a

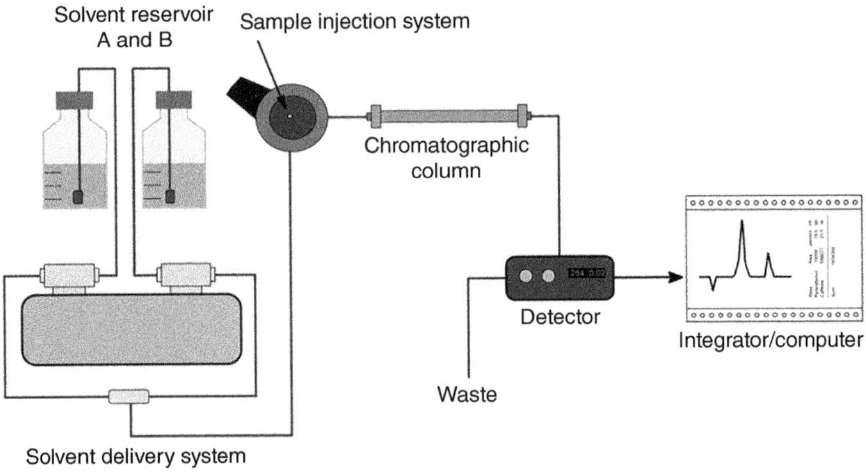

Figure 3.7 Schematic representation of an HPLC system.

chromatogram with the peaks of each compound (Figure 3.7). The quantification of the actives is required in any study of the encapsulation efficiency of drugs in the nanosystems or their release kinetics, as well as the percentage of conjugation to some nanosystems. Examples of studies using HPLC for drug quantification exists are numerous, for example, the encapsulation and release kinetics of dexamethasone (an anti-inflammatory drug) from PNPs. They were able to determine very low concentrations of the drug in a release study receptor solution due to the high sensibility that offers the HPLC technique. The resolution of the HPLC depends on the filling of the column (on the stationary phase properties), which is commonly composed of silica with attached alkyl chains, being the reversed-phase C_{18}-type columns the most widely used since it enables a differential retentionship depending on the polarity of the compounds. The advantages of HPLC are the high resolution, the low volumes required, and easy, rapid, and economic manipulation [23].

3.12 Size Exclusion Chromatography (SEC)

Size exclusion chromatography (SEC), together with ion exchange and affinity chromatography are classified as low-pressure liquid chromatography. It is useful to characterize and separate nanomaterials with different properties, dispersed in an aqueous solution. For example, it has been useful to separate antibody-conjugated nanoparticles from the free antibody and nanoparticles. Another example of its use is the characterization of the molecular weight of nano-objects, such as proteins or polymers [24]. The separation of the compounds depends not only on their molecular weight but also on their 3D dimensions, due to differences in pore permeation. It has the same advantages as other chromatographic techniques [25].

3.13 Raman Spectroscopy

Raman spectroscopy is a useful technique for direct monitoring of polymerization reactions because the C=C double bond has stretching vibration in different positions for different monomers. The intensity of a band in Raman spectra is linearly proportional to the concentration of the scattering units, which allows quantitative measurements. It is necessary to perform quantification using known amounts or concentrations of reference substances. Hence, the technique is helpful for copolymer composition determination, which is well illustrated in the example of composition determination of terpolymer methyl methacrylate (MMA)/butyl acrylate (BA)/styrene (S) (the system contains di-unsaturated crosslinker) (Figure 3.8).

The procedure is composed of calculating the intensity (I) ratios of the characteristic peaks of the individual polymers and presenting them in a plot as a function of their known mass in the copolymer [27]. Firstly, the intensity ratio of styrene ring mode (1603 cm^{-1}) to carbonyl stretch of MMA and BA (1731 cm^{-1}) is presented vs. the mass ratio of $M_S/(M_{MMA} + M_{BA})$, resulting in a linear calibration curve, which allows the calculation of the mass ratio of styrene to total acrylics. Similarly, the intensity ratios of 842 cm^{-1} characteristics of BA and 812 cm^{-1} characteristics of MMA are plotted vs. BA/MMA mass ratios, again yielding a linear calibration curve. The calculated mass ratio of the acrylics (BA/MMA) from the last calibration curve and the previously calculated S/(MMA + BA) mass ratio allow the evaluation of the total composition of an unknown copolymer. One of the useful features of Raman spectroscopy in the characterization of polymers is the investigation of their morphology. The position of the characteristic vibrations is sensitive to the orientation of the bonds and the neighboring environment as a result of the particular property of symmetry that some modes of bond vibration [26]. This gives practical information on the polymer orientation and subsequently on the mechanical and other

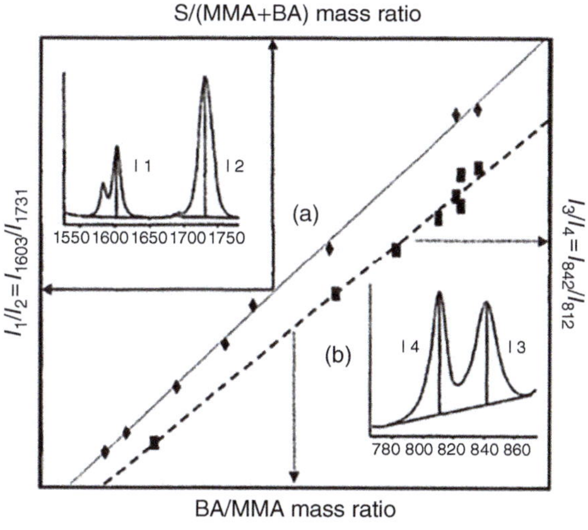

Figure 3.8 Calibration curves of the terpolymer system MMA/BA/S. (a) Ratio S/(MMA + BA) and (b) ratio BA/MMA. Source: Everall and King [26]/John Wiley & Sons.

physical properties of the polymer. Under the increased stress, the characteristic Raman shifts move toward lower wave numbers, proportionally to the applied stress.

Raman spectroscopy method's high sensitivity to polymer primary and secondary structure, as well as to solvation state, has allowed for detailed information to be obtained on a wide variety of soft materials. Furthermore, the portability of the technique allows for its use in online process monitoring. Raman spectroscopy has great utility in the characterization of polymeric fibers because of its low fluorescence background and the ability to obtain both second- and fourth-order orientation functions from polarized data [28]. Thus, Raman spectroscopy was used to monitor the effects of firing temperature and polymer drawing during manufacturing. Cis/trans ratios in polymeric composites have been studied as a function of temperature. PE production has been monitored for amorphous phase orientation and deformation in uniaxially drawn material and when peroxide-oxidized precursors are used. Similarly, the stress in a composite multifiber has been observed and related to manufacturing flaws, ultimately leading to design improvements [29].

Online Raman spectroscopy of crystallization has allowed for corrections to be made during polymer fabrication. The use of Raman spectra to determine polymer conformation has been the subject of several publications [30]. Raman has been used to observe conformational disorder in waxes, solvent reorientation in polyacrylamide gels, and strain in-homogeneities in highly oriented gel-spun PE and for monitoring curing in divinyl ethers and epoxies. A wide range of side products in the formation of PMDA-ODA polyimide has been observed by FT-Raman with a near-IR detector. Iodine doping of natural rubber and pseudo-interpenetrating polymer lead to fingerprint resonance Raman spectra of $>1000\,cm^{-1}$ [31].

3.13.1 Polyethylene Density

PE is a key polymer available in two general forms, namely, low-density polyethylene (LDPE) and high-density polyethylene (HDPE). PE with a density range of 0.920–0.935 $g\,cm^{-3}$ is defined as LDPE, while with a density range of 0.955–0.970 $g\,cm^{-3}$ is defined as HDPE. Less creep, less gas permeability, and high strength are the characteristics of HDPE, whereas LDPE is more translucent and flexible. Raman spectra of PE samples with densities of 0.882, 0.910, and 0.960 $g\,cm^{-3}$ are presented in Figure 3.9. A calibration model can be created using curve fitting techniques to correlate a PE sample's Raman spectrum to its density.

3.13.2 Polybutadiene Microstructure

Polybutadiene characterization can be effectively performed using Raman spectroscopy due to its ability to differentiate all fundamental structural units through the ν (C=C) bond stretching of 1,4-cis, 1,4-trans, and 1,2-vinyl units, which provide Raman bands at 1639, 1650, and 1664 cm^{-1}, respectively [33]. Additionally, polybutadiene samples can be quantitatively analyzed using the Raman band intensities

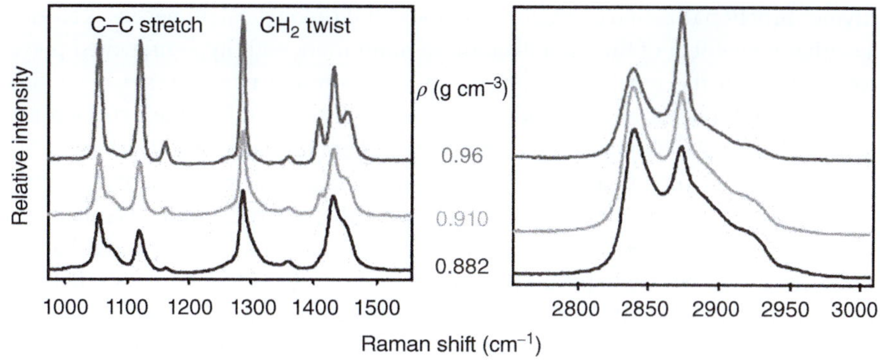

Figure 3.9 Raman spectra comparing polyethylene samples of various densities. Source: Based on Koenig [32].

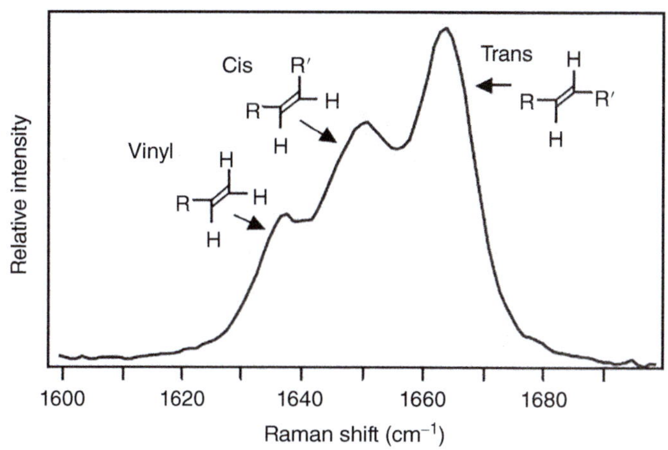

Figure 3.10 Raman spectral region used to identify microstructure of polybutadienes.

as they vary in proportion to the concentration of the samples. It is necessary to use curve fitting techniques due to the partial overlapping of the Raman bands. The resulting Raman data is reproducible to within ±1% (Figure 3.10).

3.14 Mechanical Testing and Rheometry

The biocompatibility of polymeric biomaterials could be influenced by mechanical properties. The most important mechanical parameters in biomedical applications can be determined via tensile testing. Depending on the application of biomaterials, the influence of these parameters will be different. The viscoelastic behavior and rheological properties of polymeric biomaterial commonly could be evaluated by rheometry, which measures viscosity and storage modulus. Mechanical properties of polymeric materials include the degree of polymerization, crosslinking, polymer structure, and functionality. Examples are presented to demonstrate the effect of polymer functionalization on their mechanical and rheological properties. Recently, the enhancing effect of functionalization of nanofillers with hyperbranched polyglycerol (PG) on the mechanical properties of composite nanofibers was studied. Nanodiamonds (NDs), as nanofillers, were decorated with PG on their surface, and poly(vinyl alcohol) (PVA) was selected as a model polymer matrix [34]. Unmodified ND/PVA and PG-modified ND/PVA composite nanofibers were prepared by the electrospinning technique, and their mechanical properties were characterized by tensile tests. The grafting of PG on the ND surface improved the mechanical properties of ND/PVA composite nanofiber membranes, which was revealed by a 60% increase of tensile strength, 51% enhancement of Young's modulus, and 59% boost of toughness at 3 wt% filler content. Therefore, the enhanced mechanical performance of ND-PG/PVA nanocomposites could be attributed to the introduced dense hydroxyl groups of the grafted PG, which improved the dispersion status of nanofillers within a composite matrix and strengthened filler matrix interactions [35].

Rheological measurements exhibited higher viscosity and storage modulus compared to the unmodified samples. When the chain-extender (GMA) concentrations are increased, the increase of viscosity and the storage modulus become more obvious. These viscoelastic properties are related to the molecular structure of the modified polymers. The effect of reactive compatibilization of GMA on the PLA/PBAT blends has been confirmed using tensile tests by the improvement of mechanical properties like a strain at break and Young's modulus. According to the tensile mechanical results, the elongation at break of PLA/PBAT (80/20) blends with 0.25% and 0.5% wt GMA achieved 116% and 135%, respectively, higher than for unmodified PLA/PBAT blends (50%). Compared to unmodified samples, the presence of reactive epoxy functions showed an increase in the tensile modulus (from 820 MPa (PLA/PBAT blends) to 1095 MPa (PLA/PBAT/GMA)), which indicates a reactivity control at the interface due to the formation of ester linkages between PLA, PBAT, and GMA [36]. Generally, the incorporation of multifunctional epoxy compounds such as GMA highlighted an improvement of mechanical and rheological properties of PLA/PBAT blends. The functionalization of hybrid chitosan (CTS) with GMA shows the rheological analysis for the characterization of these dental restoratives. Rheological properties of CTS/GMA suspensions increased by several orders of magnitude in the elastic modulus (G′) and there was a reduction in the viscous modulus (G″) when compared to pure CTS. The latter indicates that CTS/GMA suspensions have enhanced elastic behavior. In contrast, pure CTS suspensions behave like a viscous fluid. Dynamical rheological tests concluded that CTS-GMA solutions act as physical hydrogels [37].

3.15 Nuclear Magnetic Resonance Spectroscopy

Polymer NMR does not refer to a single methodology, owing to the wide range of structural and dynamic features of synthetic as well as natural polymers. Depending on whether just the chemical make-up or actual properties of a specific material and its unique morphology are to be characterized, one must employ widely different NMR methods [38].

In addition to the usual examination of polymers in solution and the molten state, a new surge of interest in the application of carbon (^{13}C) NMR to the examination of polymers in the solid state by magic angle spinning (MAS) has been observed. This technique is providing new and sometimes different structural information about polymers because it is examining them in their natural states [39]. Thus, with the use of MAS, along with the more common methods of examination, NMR has taken on an even more important role in its application for the determination of polymer structure, proton (^{1}H) NMR to be a fertile field to characterize the microstructure of hydrocarbon polymers, and fluorine (^{19}F) NMR for the characterization of fluoropolymers.

In many applications, in particular, those of rubbers made of simple homopolymers or statistical copolymers, the chemical resolution provided by HR-MAS is not of primary relevance. Rather, D_{res} as a measure of crosslink density is of interest. On a qualitative level, D_{res} is governing the transverse (T_2) relaxation measured in a Hahn-echo or related experiment, with non-refocused dipolar dephasing being the dominant source of signal loss, rendering ^{1}H T_2 experiments a good choice to obtain the desired insights. However, the most reliable measurement of D_{res} itself is certainly achieved by static multiple-quantum NMR, which will be addressed [40].

3.16 X-ray Diffraction

X-ray Diffraction (XRD) analysis can assist in the assessment and quantification of the crystalline phases (polymorphism), polytypes, and all types of solid-state molecular arrangements [41]. If the polymer is crystalline, then the XRD diffraction pattern is a result of a crystal structure (as related by Bragg's law); the pattern can be indexed and represented by a stick pattern of positions and intensities. To aid the identification of semicrystalline polymers in a polymers sample using the position and relative intensities of the X-ray fingerprint of the crystalline phases of the polymer [42] is used. Polymers can be processed into fibers and films and can be molded and extruded. Each of these processes can orient the molecules and the diffraction can be used to measure the orientation in both crystalline and noncrystalline materials.

The degree of the crystallinity of crystalline polymer is a structural parameter that has much to do with their physical properties in general [43]. The degree of crystallinity is calculated as a ratio of crystal-derived scattering intensity to total scattering intensity [44] (Figure 3.11).

Degree of crystallinity (%) = Crystal derived scattering intensity/(Crystal-derived scattering intensity + Noncrystal-derived scattering intensity) × 100

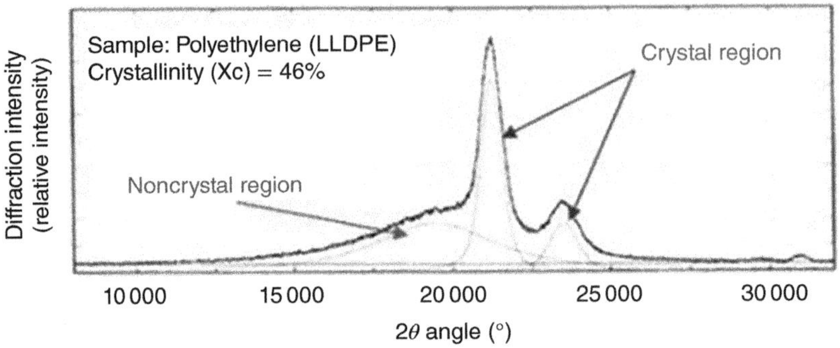

Figure 3.11 Indication of crystal and noncrystal region of polyethylene XRD analysis.

3.17 Molar Mass and Molar Mass Distribution

Polymers are mixtures of molecules of different sizes. Contrary to proteins, where the chain length is constant, synthetic polymers (rubbers and plastics) and polysaccharides always possess a certain polydispersity, which is intrinsic to the polymerization process [45].

An important step for polymer characterization is to measure the number and weight of macromolecular chains at the size (length) s, at the size (s + 1), at the size (s + 2), at the size (s + 3), and so on. This is referred to as molar mass distribution (MMD) determination. The MMD of a polymer is of prime importance in its application. In most instances, there is a molar mass (MM) range for which a given polymer property will be optimal for a particular application. The control of MM (expressed in g/mol) and of its distribution is essential for the practical application of a polymerization process, since its utility is greatly reduced unless the reaction can be carried out to yield a polymer of a sufficiently high and specified MM [46].

3.18 Osmometry

In osmometry, one measures the osmotic pressure, $\prod$, of a polymeric solution at various dilutions. The quantity $\prod/RTC$ is then computed and plotted vs. C (C is the concentration of the polymer solution, T is the temperature, and R is the gas constant). The resulting plot is a straight line and its intercept is $1/\acute{M}n$, since $\prod$ is related to as follows [47]:

$$\prod/RT = C/\acute{M}n + A_2C^2 + A_3C^3$$

where A_2 and A_3 are the second and the third coefficients of the virial expansion.

The field of application of osmometry is limited to polymers with masses below 100 000 since the precision depends on the mass, and therefore it becomes inaccurate at high masses [48]. Furthermore, being a colligative method, osmometry yields $\acute{M}n$ but not $\acute{M}w$, and therefore it cannot be used to discriminate a narrow MMD from a broad MMD nor a unimodal MMD from a bimodal MMD.

3.19 Mass Spectrometry

Mass spectroscopy methods have experienced a steadily increasing use in polymer analyses due to their high sensitivity. Mass spectrometry is useful also in those cases in which the scope of the analysis is not simply to measure $\acute{M}_n$ and $\acute{M}_w$ of the sample, but the determination of the entire MMD [49]. For instance, in the case of polymers obtained by pulsed laser polymerization, the determination of $\acute{M}_n$ and $\acute{M}_w$ of the sample is not the primary goal. One has to determine the point of inflection M_{inf} and MS can be used for this purpose. One plot the oligomer's abundance against mass and then takes the second derivative. The point at which the second derivative changes from negative to positive is the point of inflection. Also in the case of polymers made principally of cyclic molecules, the determination of the sample is not the primary goal. One has to determine the "trend" followed by the MMD and whether the abundance of cyclic molecules changes as the ring size grows. MS can be used for this purpose, by fitting the logarithm of the abundance against log (n) and fitting the points with a straight line. The slope of the line is the exponent [50].

The analysis of polydisperse polymers by MS methods poses some problems that have been only recently solved. Polydisperse polymers (including many industrial polymers) are made of a mixture of macromolecular chains that have quite different sizes, ranging from dimers and trimers, up to chains with thousands of units [51]. Using MS methods, all the oligomer chains can be ionized, but the ratio between the number of ions of a given size and the number of molecules of that size (i.e. the ion yield as a function of chain size) is not constant. The structures and morphologies of polymers have been under investigation employing electron microscopy (EM) [52].

3.20 Scanning Electron Microscopy (SEM)

EM can be divided into the techniques of TEM and SEM [53]. A step in studying structures of materials came with the development of scanning probe microscopy, in particular with a modified, for polymers interesting technique of AFM. In general, all of the different types of microscopes can be classified according to whether imaging is achieved by irradiation of the object with a "lamp" or by feeling the surface with a "finger" or "needle" [54].

The relatively easy preparation of polymer surfaces makes SEM a very valuable tool to study larger and even smaller polymer structures. Often, fracture surfaces of polymers give information on large structural details and structural defects, which are the source of damage and fracture of polymeric materials [55]. After selective etching of a smooth surface, many structures are visible in SEM. A well-developed spherulite with a central part of parallel-packed lamellae and bent lamellae around the so-called eyes of a spherulite. Additional details of the structure of semicrystalline polymers can be determined on selectively etched surfaces [56]. The two main types of isotactic polypropylene (PP) are α-modification with the cross-hatched arrangement of the crystalline lamellae; the main lamellae are growing radially from an initial site (center of the spherulite), whereas the "secondary lamellae" are formed by an epitaxial growth onto them. In β-modified PP, the lamellae form a more sheaf-like superstructure with a parallel arrangement of bundles of lamellae [57].

3.21 Transmission Electron Microscopy (TEM)

In practice, most TEM investigations of polymers make use of mass thickness contrast, that is, specimen parts containing elements with a higher atomic number and/or which are thicker scatter electrons stronger, but not contributing to image formation. To increase the density of particular structures in polymers, usually local chemical staining is applied [58, 59]. The amorphous parts and particularly the boundaries of the lamellae are chemically stained, appearing dark, whereas the crystalline lamellae are bright. Crystalline samples can be studied using diffraction contrast by diffraction patterns and dark field and bright field contrast [60].

3.22 Atomic Force Microscopy (AFM)

AFM is uniquely suited to characterize polymer materials on the nanoscale revealing structures and morphology without the need for extensive sample preparation or vacuum environment. Unlike its EM counterparts, the interaction between probe and sample in AFM is mechanical-based, making it especially suited to provide contrast on polymeric type samples [61].

Within the field of polymer sciences, AFM has been used to quantify the entropic elasticity of single-polymer chains, the elastic moduli of nanowires, single-polymer chain elongation, molecular stiffness of hyperbranched macromolecules, friction of single polymers on surfaces, the influence of temperature on the stability of single-chain conformation, and surface glass transition temperature [62]. It has also been used to perform stretching experiments on single carboxy-methylated amylase and to differentiate between sugar isomers [63].

3.23 Optical Microscopy (OM)

Optical microscopy (OM) is a highly flexible imaging technique commonly used to study the crystal growth behavior and kinetics of polymeric materials. In the case of polymer nanocomposites, OM is used to study the clay-induced crystal growth of polymeric matrices. OM can also be used to study the degree of dispersion of silicate layers in polymer nanocomposites in the molten state of the polymer matrix [64]. If the silicate layers are dispersed at the nanoscale, the nanocomposite sample should exhibit no turbidity, since the fundamental particle size is less than $\lambda/4$. Therefore, aggregates significantly less than $\sim$1 μm in size are not resolvable by

OM. The resolution of OM has traditionally been limited by the wavelength of visible light; however, recent near-field scanning techniques (NSOMs) pushed the resolution of OM significantly below this limit [65].

References

1 de los Gladys, S.-V. and Elizalde, L.E. (2013). Polymer spectroscopy and compositional analysis. In: *Handbook of Polymer Synthesis, Characterization, and Processing* (ed. E. Saldívar-Guerra and E. Vivaldo-Lima). Wiley.

2 Xiao, Y.L., Dao, C.W., Zheng, J.L., and Chen, G.Q. (2011). Polymer nanoparticles. *Prog. Mol. Biol. Transl. Sci.* 104: 299–323.

3 Cun, D., Jensen, D.K., Maltesen, M.J. et al. (2011). High loading efficiency and sustained release of siRNA encapsulated in PLGA nanoparticles: quality by design optimization and characterization. *Eur. J. Pharm. Biopharm.* 77: 26–35.

4 Ahmed, R.M. (2009). Optical study on poly(methyl methacrylate)/poly(vinyl acetate) blends. *Int. J. Photoenergy* Article ID 150389, 7 pages.

5 Young, R.J. and Lovell, P.A. (2011). *Introduction to Polymers*. Boca Raton: Taylor and Francis Group.

6 Mihaljev, Z.A., Jaksic, S.M., Prica, N.B. et al. (2015). Comparison of the Kjeldahl method, Dumas method, and NIR method for total nitrogen determination in meat and meat products. *J. Agroaliment. Proc. Technol.* 21: 365–370.

7 Xiao-Dong, H., Jenkins, S.E., Min, B.G. et al. (2003). Rigid-rod polymers: synthesis, processing, simulation, structure, and properties. *Macromol. Mater. Eng.* 288: 823–843.

8 Leue, M., Gerke, H.H., and Ellerbrock, R.H. (2011). Correcting microtopography effects on DRIFT mapping signals of organic matter at intact soil aggregate surfaces. *Soil Sci. Soc. Am. J.* 75: 1626.

9 Silverstein, R.M., Bassler, G.C., and Morrill, T.C. (1991). *Spectrometric Identification of Organic Compounds*. New York: Wiley.

10 Chalmers, J.M. and Everall, N.J. (2007). Qualitative and quantitative analysis of plastics, polymers, and rubbers by vibrational spectroscopy. In: *Handbook of Vibrational Spectroscopy* (ed. J.M. Chalmers and N.J. Everall). Wiley.

11 Pereira, H.L., Machado, F., Lima, L.E., and Pinto, J.C. (2011). In-line monitoring of vinyl acetate/acrylic acid batch copolymerization through near-infrared spectroscopy. *Macromol. Symp.* 299 (3): 1–9.

12 Matyjaszewski, K., Gnanou, Y., and Leibler, L. (ed.) (2007). *Macromolecular Engineering: Precise Synthesis, Materials Properties, Applications*, 1937. Weinheim Germany: Wiley-VCH.

13 Ibett, R.N. (ed.) (1993). *NMR Spectroscopy of Polymers*. UK: Blackie Academic & Professional.

14 Shethji JK Stephen MC Ritchie (2015) Microfiltration membrane functionalized with multiple styrenic homopolymer and block copolymer grafts. *J. Appl. Polym. Sci.* https://doi.org/10.1002/APP.42501.

15 Hatada, K. and Kitayama, T. (2004). *NMR Spectroscopy of Polymers*. Berlin: Springer.

16 Cowie, J.M.G. (1970). The application of light scattering to the investigation of preferential absorption effects in quasi-ternary systems. *Pure Appl. Chem.* 23: 355–374.

17 Hammouda, B. (2010). SANS from polymers – review of the recent literature. *J. Macromol. Sci., Polym. Rev.* 50: 14–39.

18 Haq, I., Chowdhry, B.Z., and Jenkins, T.C. (2001). Calorimetric techniques in the study of high-order DNA-drug interactions. *Methods Enzymol.* 340: 109–149.

19 Sapsford, K.E., Tyner, K.M., Dair, B.J. et al. (2011). Analyzing nanomaterial bioconjugates: a review of current and emerging purification and characterization techniques. *Anal. Chem.* 83: 4453–4488.

20 Fornaguera, C. and Solans, C. (2018). Characterization of polymeric nanoparticle dispersions for biomedical applications: size, surface charge, and stability. *Pharma Nanotechnol.* 6: 147–164.

21 Rahimi, A., Ulbrich, A., Coon, J.J., and Stah, S.S. (2014). Formic-acid induced depolymerization of oxidized lignin to aromatics. *Nature* 515: 249–252.

22 Flavel, B.S., Moore, K.E., Pfoh, M. et al. (2014). Separation of single-walled carbon nanotubes with a gel permeation chromatography system. *ACS Nano* 8: 1817–1826.

23 Cielecka-Piontek, J., Zalewski, P., Jelińska, A., and Garbacki, P. (2013). UHPLC: the greenings face of liquid chromatography. *Chromatographia* 76: 1429–1437.

24 Wagner, S., Rothweiler, F., Anhorn, M.G. et al. (2010). Enhanced drug targeting by attachment of an anti αv integrin antibody to doxorubicin-loaded human serum albumin nanoparticles. *Biomaterials* 31: 2388–2398.

25 Sadao, M. and Barth, H.G. (2013). *Size Exclusion Chromatography*. Springer Science and Business Media.

26 Everall, N. and King, B. (1999). Raman spectroscopy for polymer characterization in an industrial environment. *Macromol. Symp.* 141: 103–116.

27 Elizalde, O., Asua, J.M., and Leiza, J.R. (2005). Monitoring of high solids content starved-semi-batch emulsion copolymerization reactions by Fourier transform Raman spectroscopy. *Appl. Spectrosc.* 59: 1270–1279.

28 Andrew Lyon, L., Keating, C.D., Fox, A.P. et al. (1998). Raman spectroscopy. *Anal. Chem.* 70: 341R–361R.

29 Dieing, T., Hollricher, O., and Toporski, J. (2011). *Confocal Raman Microscopy*. Berlin Heidelberg: Springer-Verlag https://doi.org/10.1007/978-3-642-12522-5.

30 Bruckmoser, K., Resch, K., Kisslinger, T., and Lucyshyn, T. (2015). Measurement of interdiffusion in polymeric materials by applying Raman spectroscopy. *Polym. Test.* 46: 122–133.

31 Colomban, P. (2002). Analysis of strain and stress in ceramic, polymer, and metal matrix composites by Raman spectroscopy. *Adv. Eng. Mater.* 4: 535–542.

32 Koenig, J.L. (2001). *Infrared and Raman Spectroscopy of Polymers*, vol. 12. Rapra Technology Limited.

33 Yang, X., Su, Z., Wu, D. et al. (1997). Raman analysis of a conformational distribution of poly(ethylene oxide) and its model compound in the liquid state. *Macromolecules* 30: 3796–3802.

34 Hollister, S.J. (2005). Porous scaffold design for tissue engineering. *Nat. Mater.* 4: 518–524.

35 Cai, N., Li, C., Luo, X. et al. (2016). A strategy for improving mechanical properties of composite nanofibers through surface functionalization of fillers with hyperbranched polyglycerol. *J. Mater. Sci.* 51: 797–808.

36 Al-Itry, R., Lamnawar, K., and Maazouz, A. (2012). Improvement of thermal stability, rheological and mechanical properties of PLA, PBAT and their blends by reactive extrusion with functionalized epoxy. *Polym. Degrad. Stab.* 97: 1898–1914.

37 Flores-Ramirez, N., Elizalde-Pen, E.A., Vasquez-Garcia, S.R. et al. (2005). Characterization and degradation of functionalized chitosan with glycidyl methacrylate. *J. Biomater. Sci. Polym. Ed.* 16: 473–488.

38 Saalwachter, K. (2019). Applications of NMR in polymer characterization – an introduction. In: *NMR Methods for Characterization of Synthetic and Natural Polymers*, 1–22. https://doi.org/10.1039/9781788016483-00001.

39 Brame, E.G. (1983). *NMR Spectroscopy in the Characterization of Polymers*, Polymer Characterization, 419–420. https://doi.org/10.1021/ba-1983-0203.ch023.

40 Saalwachter, K. (2007). Proton multiple-quantum NMR for the study of chain dynamics and structural constraints in polymeric soft materials. *Prog. Nucl. Magn. Reson. Spectrosc.* 51: 1.

41 Chauhan, A. and Chauhan, P. (2017). XRD: a pioneer technique for characterizing the polymer and fiber. *J. Text. Sci. Eng.* 7: 3.

42 Inan, T.Y. (2017). Thermoplastic-based nanoblends. In: *Recent Developments in Polymer Macro, Micro and Nano Blends* (ed. P.M. Visakh, G. Markovic and D. Pasquini), 17–56. Woodhead Publishing.

43 Laura, P., Debora, P., Luigi, T. et al. (2014). Processing of nanostructured polymers and advanced polymeric based nanocomposites. *Mater. Sci. Eng. R.* 85: 1–46.

44 Robert, W.G. (1990) Advances in X-ray analysis. *Thirty-Ninth Annual Conference on Applications of X-ray Analysis*, Colorado, pp. 459–463.

45 Painter, P.C. and Coleman, M.M. (1997). *Fundamentals of Polymer Science*. Technomic Pub: Lancaster.

46 Hakkarainen, M. (ed.) (2012). *Mass Spectrometry of Polymers – New Techniques*. Berlin Heidelberg: Springer-Verlag https://doi.org/10.1007/978-3-642-28041-2.

47 Cleverdon, D. and Laker, D. (2007). The osmometry of high polymers. I. Technique of osmometry. *J. Appl. Chem.* 1: 2–6.

48 Podzimek, S. (2005). *Polymers Synthetic, Encyclopedia of Analytical Science*, 257–266.

49 Giorgio, M. and Robert, P.L. (2002). *Mass Spectrometry of Polymers*. CRC Press LLC.

50 Sato, H., Nakamura, S., Teramoto, K., and Sato, T. (2014). Structural characterization of polymers by MALDI spiral-TOF mass spectrometry combined with Kendrick mass defect analysis. *J. Am. Soc. Mass. Spectrom.* 25: 1346–1355.

51 Gruendling, T., Steffen, W., Falkenhagen, F., and Barner-Kowollik, C. (2010). Mass spectrometry in polymer chemistry: a state-of-the-art up-date. *Polym. Chem.* 1: 599–617.

52 Michler, G.H. (2008). *Electron Microscopy of Polymers*. Berlin, Heidelberg: Springer-Verlag.

53 Kern, M. and Trempler, J. (2008). *Observation and Measurement Microscopy in Material Science*. Berlin Heidelberg: Brünne-Verlag.

54 Goodhew, P.J., Humphreys, F.J., and Beanland, R. (2000). *Electron Microscopy and Analysis*, 3rde. London: Taylor & Francis.

55 Poelt, P., Ingolic, E., Gahleitner, M. et al. (2000). Characterization of modified polypropylene by scanning electron microscopy. *J. Appl. Polym. Sci.* 78: 1152.

56 Karger-Kocsis, J. (1995). Polypropylene: structure, blends and composites. In: *Structure and Morphology*, vol. 1 (ed. G. Hinrichsen). London: Chapman & Hall.

57 Schonherr, H. and Vancso, G.J. (2010). *Scanning Force Microscopy of Polymers*. Berlin, Heidelberg: Springer-Verlag.

58 Zhang, X.F. and Zhang, Z. (ed.) (2001). *Progress in Transmission Electron Microscopy, Ch. 1: Concepts and Techniques*. Berlin: Springer.

59 Li, Z.R. (2003). *Industrial Application of Electron Microscopy*. New York: Marcel Dekker Inc.

60 Michler, G.H. and Lebek, W. (2016). Electron microscopy of polymers. In: *Polymer Morphology*, 37–53.

61 Maver, U., Maver, T., Peršin, Z. et al. (2013). Polymer characterization with the atomic force microscope. In: *Polymer Science* (ed. F. Yılmaz). Intech Open https://doi.org/10.5772/51060.

62 Magonov, S.N. (2000). Atomic force microscopy in analysis of polymers. In: *Encyclopedia of Analytical Chemistry*. https://doi.org/10.1002/9780470027318.a2003.

63 Shanmugham, S., Jeong, J., Alkhateeb, A., and Aston, D.E. (2005). Polymer nanowire elastic moduli measured with digital pulsed force mode AFM. *Langmuir* 21: 10214–10218.

64 Sinha Ray, S. (2013). Structure and morphology characterization techniques. In: *Clay-Containing Polymer Nanocomposites*, 39–66.

65 Coceancigh, H., Higgins, D.A., and Ito, T. (2019). Optical microscopic techniques for synthetic polymer characterization. *Anal. Chem.* 91: 405–424.

4

Diverse Applications of Polymer Materials

4.1 Board Area of Polymer Applications

Polymers play a very important role in human life. Our body is made of a lot of polymers, for example, proteins and enzymes. Other naturally occurring polymers like wood, rubber, leather, and silk are serving humankind for many centuries now. Modern scientific tools revolutionized the processing of polymers, thus available synthetic polymers like useful plastics, rubbers, and fiber materials [1]. Polymers have a wide range of applications in agriculture, sports, adhesives, coatings, foams, packaging materials, textile, and industrial fibers, *composites*, electronic devices, energy and storage, biomedical devices, optical devices, and precursors for many newly developed high-tech ceramics [2]. Some broad areas of polymer applications are detailed in this chapter as follows:

1. Polymeric material-based flexible and stretchable electronics.
2. Functional polymers and their composites for sensors and actuators.
3. Polymer-based Light-emitting diode (LED).
4. Polymeric membranes.
5. Eco-friendly polymeric materials.
6. Functional polymeric textiles.
7. Polymeric self-healing materials.
8. Functional polymeric surfaces.
9. Applications of polymeric materials and their composites in energy generation and storage.
10. 3D/4D printing of polymeric materials and their composites [3].

Polymers are a highly versatile class of material that is found in all areas of engineering from avionics to biomedical devices, and the development and implementation of these rely on polymer applications and data provided through rigorous testing [4]. The applications of polymeric materials and their composites are still growing rapidly due to their low cost and ease of manufacture. This in turn fuels further advances in research and development. A better understanding of the properties of the materials in differing environments and temperature ranges is central to sourcing the correct polymer materials to suit the application [5].

Polymer Nanocomposites for Energy Applications, First Edition.
T. Daniel Thangadurai, Manjubaashini Nandhakumar, Sabu Thomas, and Ange Nzihou.

The importance of polymers in advanced technology is a key factor in the future of materials development, as indicated in the following applications:

I. Polymer dielectrics in electronics offer the basis for the smallest circuits and the highest speed of operation.
II. Conducting polymers have been commercialized in rechargeable batteries and offer the greatest promise for high energy storage with low weight [6].
III. Polymer sensors exist for chemical species, thermal and acoustic radiation, temperature, pressure, humidity, ionizing radiation, electric charge, and more.
IV. Buildings can be equipped with a network of optical fibers linking remote locations with a management console. The polymer sensors can be built into the optical fibers to report the presence of toxic gases or to turn off superfluous lights to conserve energy.
V. Implanted sensors can detect the glucose level in blood and call for insulin injections using an implanted pump, as needed.
VI. Electromagnetic shielding will become increasingly necessary, and conducting polymers offer solutions that are conveniently fabricated in complex shapes.
VII. Polymer resists are the basis for the microlithography that makes integrated circuit electronics possible. They are also the basis for the emerging field of micromechanics, which could produce machines smaller than a human cell.
VIII. High-density information storage is available through compact disk technology, and improved polymers will enhance the performance of this medium. In the future, polymer-based holographic devices could revolutionize the storage and manipulation of information [7].
IX. Polymers offer solutions to critical economic problems facing the introduction of photonics, the light analog of electronics. The couplers, splitters, and other elements of photonic circuit boards all admit to polymeric solutions that may provide the economic breakthrough needed for the photonics revolution. Broadband communications can be brought directly to the home and office by polymer or glass fibers, using polymeric photonic circuits.
X. Smart windows based on polymeric materials could reflect light when the sun is too bright and transmit light when it is not.
XI. The fabrication of liquid crystal display devices for computers and television can be facilitated and the robustness of the product enhanced by the incorporation of conductive, transparent polymer films.
XII. Light-emitting diodes based on flexible polymeric films have been fabricated and are likely to find diverse applications in the future.
XIII. Electrophotography is now based on polymeric photoactive materials, and these have made possible many improvements, such as compact and convenient machine architecture, the durability of machines, and long-term print quality.
XIV. Polymers are now the recording medium of choice for holography in many applications. This technology offers the promise of ultrahigh density information storage [8].

4.2 Polymers in Biotechnology

Polymers play a major role in all aspects of biological processes. It is legitimate to proclaim that polymers are the molecular basis of life. The genetically inherited information required for the growth and health of living systems is encoded in the macromolecule deoxyribonucleic acid (DNA), the backbone of which forms the famous double helix. The molecular genetic code uses only four purine and pyrimidine bases to dictate the structure of the proteins that make up so much of living systems. DNA directs the assembly of about 20 amino acids in complex sequences that become proteins. These proteins are polypeptide polymers that differ from one another only in the sequence of their constituent amino acids. All enzymes, which control the reaction rates in biological systems, are proteins. Collagen proteins from fibers and connective tissue are found in tendons, cartilage, blood vessels, skin, and bone. Elastin, an elastic substance found in ligaments and the walls of blood vessels, is a protein. Other polymers such as polysaccharides are also important. They make up chains of sugar units present as a major constituent in all connective tissue. Ribonucleic acid (RNA) molecules also carry information and can serve protein-like functions [9]. Thus informational, chemical, mechanical, and other properties of living systems find their origin in the molecular structure of their component polymers.

The development of biodegradable polymeric materials for biomedical applications has advanced significantly. Biodegradable polymeric materials are favored in the development of therapeutic devices, including temporary implants and three-dimensional scaffolds for tissue engineering. Further advancements have occurred in the utilization of biodegradable polymeric materials for pharmacological applications such as delivery vehicles for controlled/sustained drug release. These applications require particular physicochemical, biological, and degradation properties of the materials to deliver effective therapy. As a result, a wide range of natural or synthetic polymers able to undergo hydrolytic or enzymatic degradation is being studied for biomedical applications [10].

The application area for nanochannels and nanoslits is DNA stretching/linearization. Analysis of DNA linearization can provide insights into the physical properties of DNA, which is interesting for basic polymer physics as well as understanding the regulation of gene expression. Linearization of the randomly coiled DNA structure in microscale environments is also useful for mapping locations of certain sequences within a strand of DNA as well as for the direct reading approaches for DNA sequencing. Examples of types of analysis involving DNA linearization include molecular combing, DNA direct linear analysis (DLA), optical mapping, and nano-confinement. DNA and optical mapping both utilize the shear stretching of molecules in small channels for linearization. Molecular combing performs shear stretching of DNA without the use of channels but by using a moving air–liquid meniscus. Shear stretching has the advantage that submicron or micron-scale channels or even no channel setups can be used to obtain a relatively large degree of DNA linearization [11].

Ultrahigh molecular weight polyethylene (UHMWPE) is an engineering polymer that varies from high-density polyethylene (HDPE) in terms of average molecular weight and average chain length. UHMWPE is a semicrystalline polymer that contains fully crystalline and fully amorphous phases as an interfacial all-trans phase. In the crystalline phase, the particular lamellar shape of crystallite is due to the chain folding with the chain axis, which enlarges the chain fold area [12]. In the amorphous phase, the chains are interconnected through occasional crosslinks and random entanglements instead of proper chain folding. The relations between amorphous and crystalline phases are provided by tie molecules. The crystallinity of UHMWPE depends on its volumetric percentage of crystallites. The properties of UHMWPE are determined by the connections between amorphous and crystalline phases, i.e. tie molecules, crystallinity, the degree of crosslinks and entanglements, and the positions of the crystallites [13].

UHMWPE has high wear resistance, toughness, durability, and biocompatibility. Therefore, it is commonly used as a bearing material with ceramic or metallic counter surfaces in joint arthroplasty. UHMWPE's significance for achieving outstanding performance in total joint arthroplasties is unquestionable. For long-term clinical applications, its tribological performance and lifetime are key aspects. However, UHMWPE implants have limited life due to their wear complications [14]. When the UHMWPE is used in the periprosthetic environment, it induces osteolysis followed by loosening of the implant. This implant loosening is joined with fatigue causes the aseptic loosening which ultimately causes the implant's failure. Many methods such as improving crosslinking or crystallinity percentage through irradiation, surface modification through plasma treatment, or introducing effective textures, and reinforcements with particles or fibers have been used for enhancing the properties of UHMWPE [15].

4.3 Polymer Dielectrics for Electronics

Synthetic polymers have inspired a broad range of applications because of their design flexibility, strong mechanical strength, and solution processability. However, fossil fuel sources are used to prepare synthetic polymers, leading to serious environmental pollution. To address this issue, renewable polymers have received extensive interest over the past decade, which has been driven by their environmentally friendly value as renewable biomaterials and as a sustainable source of materials [16] (Figure 4.1).

The development of a smaller, lighter, and flexible display is a continuing trend for handheld soft devices. To meet this demand, polymeric substrates are one of the potential candidates to replace glass, but most plastic materials have a large coefficient of thermal expansion (CTE), in the order of 50–200 ppm K^{-1}. As a result, the functional materials deposited onto the plastic substrates would be destroyed from thermal processes because of the mismatch between the CTEs from different materials. However, reports show that cellulose nanofibers with CTE values as low as 0.1 ppm K^{-1} along the axial direction. Besides, bacterial cellulose (BC) has been used as a reinforcement to enhance the strength and decrease the CTE for transparent

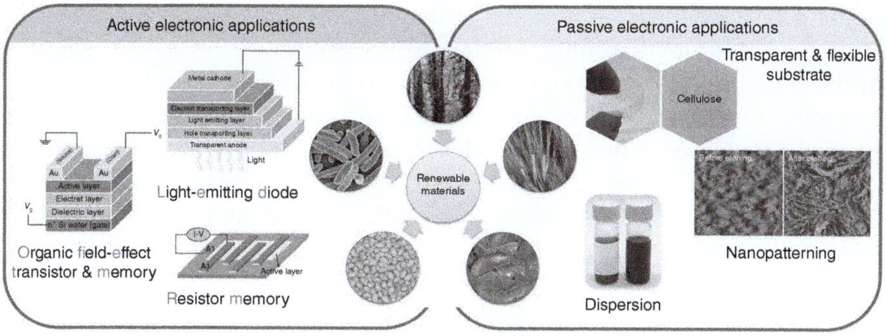

Figure 4.1 Schematic of renewable polymers for electronic applications. Source: Sun et al. [5]/Springer Nature.

plastics. Thus, researchers discovered a foldable transparent BC composite film with a low CTE (of 4 ppm K^{-1}) and utilized it as a substrate in an organic LEDs device [17].

4.3.1 Luminescent Layers in Light-Emitting Diodes

Organic LEDs are emissive electroluminescent devices that are widely used in the digital display industry to fabricate television screens, computer monitors, and mobile phones. Usually, the substrates, electrodes, electron-transporting layers, hole-transporting layers, and light-emitting layers in common organic LEDs configurations are made of petroleum-based products rather than materials derived from renewable sources. However, researchers reported the first synthesis of fluorescent polymers based on cellulose modified with alkyl carbazole derivatives

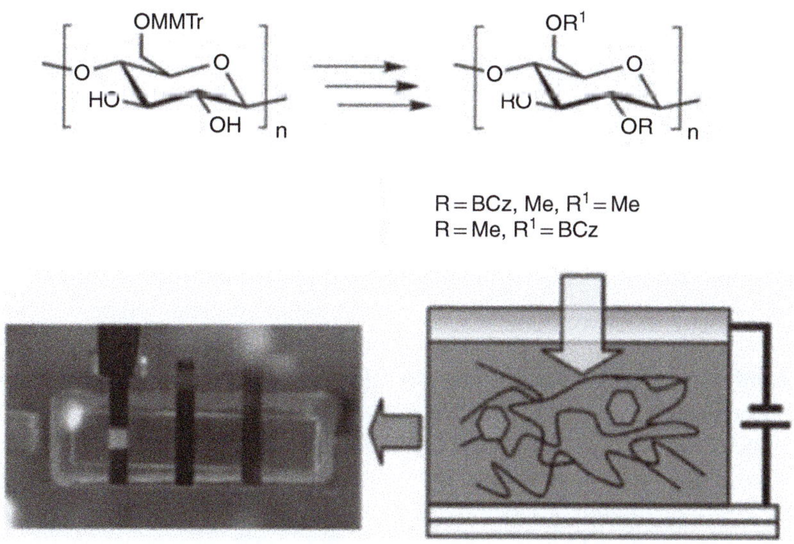

Figure 4.2 (a) Chemical structures and (b) fabricated organic light-emitting diode devices of fluorescent cellulose derivatives. Source: Sun et al. [5]/Springer Nature.

(Figure 4.2). The ionization potential of the cellulose derivative is approximately 5.8 eV. The derivatives exhibit UV-blue emissions at 350–450 nm, which is attributed to the carbazole groups and similar to emissions from neat poly(*N*-vinyl carbazole). Furthermore, the cellulose materials that were integrated into an organic LED showed a high current density and a low turn-on voltage of 6.5 V for 0.1 cd m^{-2} light emission. These results indicate that carbazole-functionalized cellulose has the potential to serve as the hole-transporting material in organic LEDs [18].

4.4 Smart and Self-healing Coatings

Reconstructable polymer surfaces form a toolbox for the rapidly developing field of smart coatings. The structure of the coatings can be programmed in the formulation. After deposition, external stimuli affect the phase separation of the ingredients to self-assemble into a coating with programmed properties. For example, colloidal particles prepared by the emulsion copolymerization of acrylate and fluorinated acrylate monomers can form stratified film morphologies, where the fluorinated phase can be driven to the film/air or film/substrate interfaces. As a consequence, static and kinetic coefficients of friction can be controlled at the film–air interface, resulting in superhydrophobic surfaces. In another example of programmed behavior, colloidal particles are stabilized in an acidic aqueous solution by grafted PS-b-P2VPb-PEO triblock copolymers. The casting of the particle suspension at higher pH results in a film consisting of particle aggregates and textured coating. The coating becomes superhydrophobic when heated to above the glass transition temperature of polystyrene blocks, because these blocks migrate to the topmost layer of the coating [19].

4.5 Polymeric Biocides and Herbicides

A new technique has recently emerged for the controlled release formulations designed to avoid or reduce the possible side effects accompanying the use of biologically active agents. The purpose of this technique includes protecting the supply of the agent, allowing the automatic release of the agent to the target at controlled rates, and maintaining its concentration within the optimum limits over a specified period, thereby producing a great specificity and persistence [20]. There are two different approaches to combining biological agents with polymeric materials: either by physical combination to act as a rate-controlling device or by chemical combination to act as a carrier for the agent.

The polymeric biocide has many advantages and its potential benefits include are as follows: (i) it allows lower amounts than conventional biocides to be used as it releases the required amount of active agent over a long period; (ii) number of applications is reduced because of long period of activity by a single application, (iii) it eliminates the time and cost of repeated over applications because less active materials is needed, reduction of toxicity; (iv) it eliminates the need for widespread distribution of large amount of biocide levels in the surrounding environment; (v) reduction of evaporation and degradation loses by environmental forces or leaching by rain into the soil or waterways due to the macromolecular nature [21], (vi) it extends the duration of activity of less or nonpersistent biocides which are unstable under an aquatic environment by protecting them from environmental degradation and hence enhances the practical applicability of these materials, and (vii) reduction of phytotoxicity by lowering the high mobility of the biocides in the soil and hence reduces its residue in the food web, extension of herbicide selectivity to additional crops by providing a continuous amount of herbicide at a level sufficient to control weeds but without injury to the crop [22].

4.6 Polymers for Soil Remediation

Contamination of soils with toxic metal elements is of great concern to scientists and the general public. Long-term intake of contaminant metals by humans may lead to chronic effects, although maximum acceptable limits in food were already established for several toxic elements by the national agency for food, drugs, and administration and control, European Food Service Authority, and the US Food and Drug Administration [23].

The effects of metals on ecosystems and biological resources are also increasingly recognized. Metals do not degrade as organic compounds do and have long residence times in soils. They can however exist in different forms, which include water-soluble (ionic and chelated with soluble compounds), adsorbed on soil surfaces, chelated by insoluble organic matter, precipitated, occluded by soil oxides and hydroxides, present in living organisms or residues, and as part of primary and secondary minerals. Ideally, contaminated soil should be restored to regain its original potential, but this can be a very expensive process, and thus depends not only on the expected benefit of the cleanup and future value of the soil but also on political and public awareness of the problem [24]. Conventional remedial approaches to severely metal-contaminated soils involve removal and replacement of soil with clean materials or capping the soil with an impermeable layer to reduce exposure to contaminants, and chelating agent ethylenediaminetetraacetic acid (EDTA) has been used extensively for heavy metals extraction from soil. Much work has been done on the recovery of metal-loaded EDTA by electrochemical or chemical processes. Another chelating agent, pyridine-2,6-dicarboxylic acid, has also been shown to be effective in heavy metals, although these are not considered the most economically or environmentally sound solutions available [25]. Only through the establishment of a vegetation cover to stabilize metal-contaminated soils will a successful long-term rehabilitation be achieved.

4.7 Benefits of Polymers in Fabric and Home Care Formulations

A detergent, in general, is a cleaning agent composed predominantly of a surfactant or a mixture of surfactants whose main task is to remove water-insoluble substances like dirt and grease from permeable surfaces (i.e. fabrics and clothes) and/or hard surfaces (i.e. metals, plastics, and ceramics) [26]. Other than surfactant, a modern detergent formulation normally consists of builders and chelants, co-builders, and other polymer additives used for specific washing effects such as soil release and anti-redeposition, dye transfer inhibition, and rheology modification. The incorporation of polymers in fabric and home care formulations has provided numerous benefits in enhancing the action and efficiency of detergents, particularly the phosphate-free types [27]. Consequently, the use of these materials, especially the recently developed specialty polymers, has gained increasing attention over the years, because they provide specific benefits at a very low percent weight (1% or less). Examples of these types of polymers are polyesters based on terephthalic acid which serves as soil release agents, ethylene/propylene glycol-based polymers as anti-redeposition agents, and poly(vinylpyrrolidone) as DTIs.

4.8 Polymeric Materials for Photonics

Photonics is a technology analogous to electronics in which the photon replaces the electron as the working particle. Many of the applications now accomplished electronically, including transmission, switching, amplification, and modulation can also be realized using photonics, and there are advantages to be gained by converting to a photon-based technology in some areas. Transmission of light in fiber optic systems is the direct analogy of electrical transmission in coaxial cable systems [28].

Fiber optic systems are now in place all over the world, and they handle much of the world's long-distance telephone traffic. The transmission medium of the fibers employed is based on inorganic glasses, but polymers are used for protective coatings and in cabling structures. Polymers can also be made into optical fibers, but the loss is considerably larger than with the inorganic fibers and only short-distance applications are realistic. The main advantage of polymer fibers is their flexibility when made in larger diameters, which are easier to splice. Today, fiber optic cables are generally terminated at the area substation level, where the optical signal is converted back to an electrical signal for transmission to the customer. This conversion process is necessary because the optical components needed to reach the individual telephone or terminals are not available at a sufficiently low cost at this time. Polymeric organic materials will play a major role in the realization of optical technology as fiber to the home becomes a reality [29].

4.9 Polymers for Electrophotography

One of the major applications of polymers with tailored electronic and optical properties has been in electrophotography for copier, duplicator, and printer applications. In this application, an electroactive polymer is used as one component of the light-sensitive element used for creating the latent electrostatic image of an original subject. The image source can be light reflected from a document and focused onto the surface of the photoreceptor or a digital file of an original image, which is used to control a laser beam that is scanned over the surface of the photoreceptor. The electrostatic image is rendered visible by dusting the surface of the photoreceptor with an electrostatic powder composed of a pigment-loaded thermoplastic polymer. The latent image can then be transferred to paper by a combination of pressure and electrical bias and then fused to the paper by heating [30].

The photoreceptor itself was the key invention that enabled the development of electrophotography as a commercial success. The original photoreceptor materials were based on selenium and its alloys as well as group II–VI and other semiconductor materials. Because of the poor mechanical properties of selenium and its alloys, photoreceptors had to be fabricated on rigid metallic drums. This, in turn, dictated relatively cumbersome and expensive copier machine architectures [31]. These materials had several shortcomings, including degradation of photoconductive properties, instabilities in surface properties leading to incomplete toner transfer, and catastrophic abrasion.

4.10 Polymers in Energy Applications

Polymers have been studied widely due to their versatile and adjustable chemical and physical properties. The three-dimensional network structures of polymers decide that they can be employed as the template to fabricate mesoporous materials or be used as a polymeric matrix in the solid electrolyte; the high catalytic activity for I_3^- reduction makes it potential as counter electrode materials. The diversified functional groups make it possible for the polymers to regulate the perovskite morphologies from bulk and interface aspects; the high carrier mobilities enable polymers to act as electron and hole transfer materials; the functional groups in polymers decide that they can be used as the interface layers to passivate defects, adjust the work function of the metal electrode [32], and improve the device performances; the diversified structures and functional modification also equip polymers with various optical adsorption properties and variable electron mobility, being used as the photoactive layer or buffer layer in organic photovoltaic (OPV); the processability of polymer also makes it possible to fabricate polymer-based micro/nanostructure devices. The partial polymers with good conductivity, named conductive polymers, were widely used in many fields.

The diversity of nanostructures obtained from organic polymerization is limited when compared to the huge amount of documented inorganic nanostructures. Researchers develop a synergistic mechanism between *in situ* inorganic salt hydrolysis and vapor-phase polymerization for the metal oxide-poly (3,4-ethylene dioxythiophene) (PEDOT) hybrid nanostructure growth. The steady state polymer growth and kinetically controlled hydrolysis enable homogeneous deposition of high aspect ratios crystal phases such as β-FeOOH, TeO_2, and SnO_2 coated by a conducting polymer. By controlling the hydrolysis kinetics, the hybrid material is synthesized in one step with morphologies controlled from 1D nanofiber to 2D nanoflowers and nanostructures from monolithic to core-shell. This fundamental understanding of the connection between hydrolysis and polymerization allows the future development of nanostructured inorganic, polymeric, and inorganic–organic hybrid materials. Electrodes for energy storage are fabricated with different PEDOT morphologies, and their structure–property relationships are discussed. The 1D fibrillar structure shows a higher capacitance of 185 $F\,g^{-1}$ at 25 $mV\,s^{-1}$ compared to 2D nanoflowers because this morphology enhances electrolyte diffusion kinetics that facilitates PEDOT doping and de-doping, leading to a lower internal resistance [33].

The chromic polymer platforms can also be adopted as colorimetric sensors against specific stimulus (i.e. with UV sensor using photochromic, chemo-sensor using chemochromic, or even stretching sensor using mechanochromic) for an early warning system. Important factors in these applications will be chrominance before and after exposure, and especially the surface areas. Smart patches with a large surface area will be beneficial to the fast detection of the external stimulus [34]. Electrospinning is one of the best methods to prepare a nonwoven nanofiber patch, and it has also been reported previously for the fabrication of a highly functionalizable nonwoven mat using our azido-TPU. SP-TPU can be also adopted in electrospinning without any tedious optimization process as indicated in Figure 4.3a. Because SP-TPUs with different concentrations of organic dyes can be introduced, nonwoven patches with different chromic after UV exposure are obtained as illustrated in Figure 4.3b. These chromic patches with the different signals via the same external stimuli; in other words, one can also generate a certain level of signals *via* different external stimuli by controlling the concentration of dyes or adopting different dyes can be expanded to detect the quantitative level of the signal [35] (Figure 4.4).

Fluorescent probes possess a low limit, high sensitivity, and wide response range, etc., but they have some disadvantages including contamination to the detection

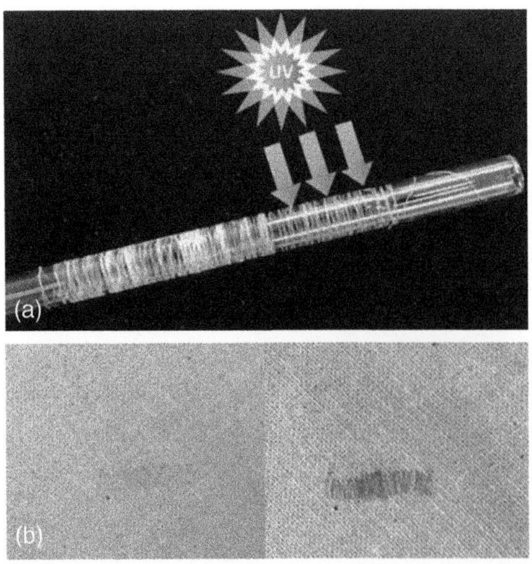

Figure 4.3 Wet-jetting process: (a) photograph of wound fiber on a glass rod and (b) stitched SP-TPU02 fiber on fabric before and after UV irradiation. Source: Seo et al. [8]/Springer Nature.

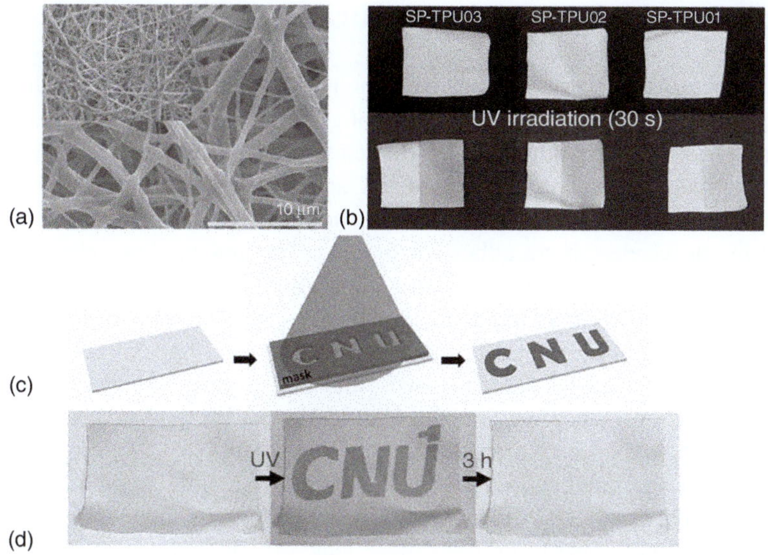

Figure 4.4 Electrospinning process: (a) SEM images of fiber, (b) photograph of three types of concentrations fiber mat after UV irradiation, (c) scheme of masking process on fiber mat, and (d) reversible chromic patterning on SP-TPU02. Source: Seo et al. [8]/ Springer Nature.

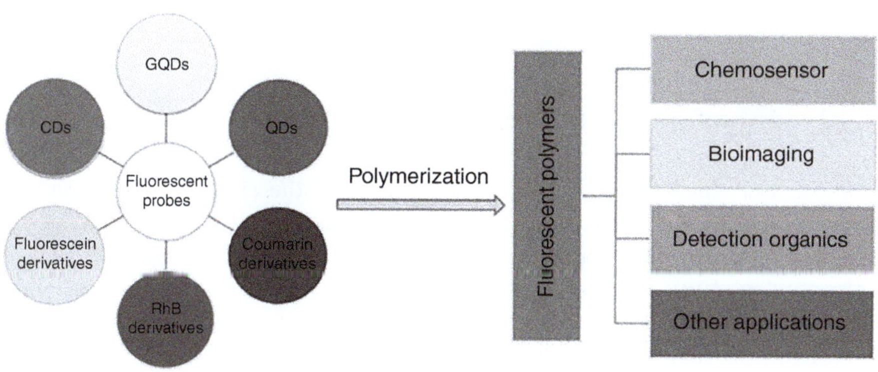

Figure 4.5 Applications of fluorescent polymers as fluorescent probes.

system, limitation of repeated use. Polymers combined with fluorescent probes could overcome these drawbacks and they have been widely used in chemosensor, biological imaging, and pH/temperature sensor for their unique favorable mechanical properties and good workability. Researchers mainly reviewed organic and inorganic fluorescent probe polymers, focusing on synthetic methods, properties, and applications. With further advances in the design and synthesis of multifunctional fluorescent polymers with high quality, the widespread application of them could be expected in advanced bioimaging, ultrasensitive molecular diagnosis, and novel light-emitting nanodevices [36] (Figure 4.5).

4.11 Polymers in Construction Applications

Polymers find extensive usage and account for the highest growth area as building materials in the construction industry, for both structural and nonstructural applications [37]. The major advantage of this class of material is the variety of characteristics that can be tailored for various applications. There are various types of polymers classified as thermoplastics, thermosets, elastomers, plastics, rubbers, thermoplastic elastomers, adhesives, foams, paints and windows, cladding, rainwater, pipes, membranes, seals, glazing, and insulation. With abounding commercially available polymers constantly being researched and improvised, new applications are continuously emerging and being practically applied. The field of construction is such that all materials used must have proven mechanical, and tensile strength and resilience, so that due confidence in the performance and properties of materials used is assured [38].

Properties of various engineering polymers are high strength or modulus to weight ratios (lightweight but comparatively stiff and strong), toughness, mechanical strength, resilience, corrosion resistance, lack of conductivity (heat and electrical), color, transparency, processing, and cost-effectiveness. Polymers are used in the construction industry extensively, for both structural and nonstructural purposes. Structurally, they can be used in ropes, grids, rebars-fire reinforced plastic (FRP) reinforcement bars for concrete, prestressing tendons for concrete members, and FRP sheets can be used to increase flexural strength in weakened or under-designed members [39].

4.12 Polymers in Automobile Applications

Polymer and allied materials are rapidly capturing the share of metal usage in automotive. This is in turn getting rid of the traditional approach of product design and development [40]. Major polymers used in light vehicles include 90 lb of polypropylene (PP), 58 lb of polyurethanes, 46 lb of Nylon, 30 lb of polyvinyl chloride (PVC), 24 lb of acrylonitrile-butadiene-styrene (ABS), 20 lb of polycarbonate resins, and 18 lb of polyethylene resins.

ABS is a durable thermoplastic. It is a copolymer built by polymerizing styrene and acrylonitrile in the presence of polybutadiene. The styrene gives the plastic a shiny and impervious surface. The butadiene, a rubbery substance, provides resilience even at low temperatures [41]. A variety of modifications can be made to improve impact resistance, toughness, heat resistance, weather, and some chemical resistance. Used in automotive body parts, dashboards, wheel covers manufacture of housings, covers, and linings.

Polyamide (PA) is known as nylon 6.6 or nylon 6. It is a general-purpose nylon that can be both molded and extruded. Nylon 6/6 has good mechanical properties and wear resistance. They are frequently used when a low-cost, high-mechanical strength, rigid and stable material is required [42]. They also absorb water easily, and components in wet or humid conditions will expand, precluding their use in applications where dimensional stability is required. The main application of PA is the manufacture of parts that are under the engine hood, mainly using the types of PA reinforced by fiberglass [43]. It is used in gears, bushes, cams, bearings, and weather-proof coatings.

References

1 Namazi, H. (2017). Polymers in our daily life. *Bioimpacts* 7: 73–74.

2 Anamica and Pande, P.P. (2017). Polymer hydrogels and their applications. *Int. J. Mater. Sci.* 12: 11–14.

3 Candlin, J.P. (2008). Polymeric materials: composition, uses, and applications. *Compr. Anal. Chem.* 53: 65–119.

4 Brinson, H.F. and Catherine Brinson, L. (2008). *Polymer Engineering Science and Viscoelasticity*. Boston, MA: Springer.

5 Sun, H.S., Chiu, Y.C., and Wen, C.C. (2017). Renewable polymeric materials for electronic applications. *Polym. J.* 49: 61–73.

6 Stuart, M.A.C., Huck, W.T.S., Genzer, J. et al. (2010). Emerging applications of stimuli-responsive polymer materials. *Nat. Mater.* 9: 101–113.

7 Kurahatti, R.V., Surendranathan, A.O., Kori, S.A. et al. (2010). Defence applications of polymer nanocomposites. *Defence Sci. J.* 60: 551–563.

8 Seo, E., Choi, J., Lee, B. et al. (2019). Dye clicked thermoplastic polyurethane as a generic platform toward chromic-polymer applications. *Sci. Rep.* 9: 18648.

9 Gopanna, A., Rajan, K.P., Thomas, S.P., and Chavali, M. (2019). Polyethylene and polypropylene matrix composites for biomedical applications. *Mater. Biomed. Eng.* 175–216. Elsevier.

10 Song, R., Murphy, M., Li, C. et al. (2018). Current development of biodegradable polymeric materials for biomedical applications. *Drug Des. Dev. Ther.* 12: 3117–3145.

11 Chantiwas, R., Park, S., Soper, S.A. et al. (2011). Flexible fabrication and applications of polymer nanochannels and nanoslits. *Chem. Soc. Rev.* 40: 3677.

12 Atwood, S.A., Van Citters, D.W., Patten, E.W. et al. (2011). Tradeoffs amongst fatigue, wear, and oxidation resistance of cross-linked ultra-high molecular weight polyethylene. *J. Mech. Behav. Biomed. Mater.* 4: 1033–1045.

13 Zeman, J., Ranusa, M., Vrbka, M. et al. (2018). UHMWPE acetabular cup creep deformation during the run-in phase of THA's life cycle. *J. Mech. Behav. Biomed. Mater.* 87: 30–39.

14 Ruggiero, A., Damato, R., Gomez, E., and Merola, M. (2016). Experimental comparison on tribological pairs UHMWPE/TIAL6V4 alloy, UHMWPE/AISI316L austenitic stainless and UHMWPE/Al_2O_3 ceramic, under dry and lubricated conditions. *Tribol. Int.* 96: 349–360.

15 Hussain, M., Ali Naqvi, R., Abbas, N. et al. (2020). Ultra-high-molecular-weight-polyethylene (UHMWPE) as a promising polymer material for biomedical applications: a concise review. *Polymers* 12: 323–350.

16 Nogi, M. and Yano, H. (2008). Transparent nanocomposites based on cellulose produced by bacteria offer potential innovation in the electronics device industry. *Adv. Mater.* 20: 1849–1852.

17 Hinestroza, J.P. and Netravali, A.N. (ed.) (2014). *Cellulose Based Composites: New Green Nanomaterials*. Weinheim, Germany: Wiley-VCH.

18 Karakawa, M., Chikamatsu, M., Vakamoto, C. et al. (2007). Organic light-emitting diode application of fluorescent cellulose as a natural polymer. *Macromol. Chem. Phys.* 208: 2000–2006.

19 Motornov, M., Sheparovych, R., Lupitskyy, R. et al. (2008). Superhydrophobic surfaces generated from water-borne dispersions of hierarchically assembled nanoparticles coated with a reversibly switchable shell. *Adv. Mater.* 20: 200–205.

20 Milani, P., Franca, D., Balieiro, A.G., and Faez, R. (2017). Polymers and its applications in agriculture. *Polímeros* 27: 256–266.

21 Ekebafe, L.O., Ogbeifun, D.E., and Okieimen, F.E. (2011). Polymer applications in agriculture. *Biokemistri* 23: 81–89.

22 Russo, R., Giuliani, A., Immirzi, B. et al. (2004). Alginate/polyvinyl alcohol blends for agricultural applications: structure-properties correlation, mechanical properties, and greenhouse effect evaluation. *Macromol. Symp.* 218: 241–250.

23 Kzlkaya, R., Askn, T., Bayrakl, B., and Saglam, M. (2004). Microbiological characteristics of soils contaminated with heavy metals. *Eur. J. Soil Biol.* 40: 95–102.

24 Vig, E.K. and Hu, H. (2000). Lead toxicity in older adults. *J. Am. Geriatr. Soc.* 48: 1501–1506.

25 Perez-de-Mora, A.P., Burgos, E., Madejon, F. et al. (2006). Microbial community structure and function in a soil contaminated by heavy metals: effects of plant growth and different amendments. *Soil Biol. Biochem.* 38: 327–341.

26 Watson, R.A. (2006). Liquid laundry detergents. In: *Handbook of Detergents, Part D: Formulation*, 51–104. Boca Raton, FL: Taylor & Francis Group.

27 Paderes, M., Ahirwal, D., and Prieto, S.F. (2017). Natural and synthetic polymers in fabric and home care applications. *Phys. Sci. Rev.* 2: 1–20.

28 Ana, M.R. and Pinto, L.-A.M. (2012). Photonic crystal fibers for sensing applications. *J. Sensors* Article ID: 598178, 21 pages.

29 Xiong, R., Luan, J., Kang, S. et al. (2020). Biopolymeric photonic structures: design, fabrication, and emerging applications. *Chem. Soc. Rev.* 49: 983–1031.

30 Izdebska, J. and Thomas, S. (2016). *Fundamentals and Applications. Printing on Polymers*, 1–20. Elsevier.

31 Wu, J.T., Lien-Chung Hsu, S., Tsai, M.H. et al. (2012). Direct ink-jet printing of silver nitrate-silver nanowire hybrid inks to fabricate silver conductive lines. *J. Mater. Chem.* 22: 15599–15605.

32 Hou, W., Xiao, Y., Han, G., and Lin, J.Y. (2019). The applications of polymers in solar cells: a review. *Polymers* 11: 143.

33 Wang, H., Diao, Y., Rubin, M. et al. (2018). Metal oxide-assisted PEDOT nanostructures via hydrolysis-assisted vapor-phase polymerization for energy storage. *ACS Appl. Nano Mater.* 1: 1219–1227.

34 Wang, K., Lu, H., Liu, B., and Yang, J. (2018). A high-efficiency and low-cost AEE polyurethane chemo-sensor for Fe^{3+} and explosives detection. *Tetrahedron Lett.* 59: 4191–4195.

35 Hu, Y. and Zheng, Z. (2018). Progress in textile-based triboelectric nanogenerators for smart fabrics. *Nano Energy* 56: 16–24.

36 Ye, Q., Yan, F., Kong, D. et al. (2016). Synthesis and applications of fluorescent polymers as fluorescent probes. *Curr. Org. Chem.* 20: 266–288.

37 Mousumi, B.J. (2019). Polymers in civil engineering: review of alternative materials for superior performance. *J. Appl. Sci. Comput.* 6: 1076–5131.

38 Brinson, H.F. and Catherine Brinson, L. (2008). *Polymer Engineering Science and Viscoelasticity, An Introduction*, 55–97. LLC: Springer Science+ Business Media.

39 Humphreys, F. and Matthew (2019) ResearchGate. *The use of polymer composites in construction.*

40 Vivek, S. and Rajeev, S. (2013). Advances in automotive polymer applications and recycling. *Int. J. Innovative Res. Sci. Eng. Technol.* 2: 744–746.

41 Akshat, P., Arun, P., and Rajesh, P. (2017). An overview of polymeric materials for automotive applications. *Mater. Today: Proc.* 4: 3807–3815.

42 Hotta, S. and Paul, D.R. (2003). Modeling properties of nylon 6/clay nanocomposites using composite theories. *Polymer* 44: 4993–5013.

43 Mativetsky, J.M. and Datars, W.R. (2002). Morphology and electrical properties of template-synthesized polypyrrole nanocylinders. *Physica B* 324: 191–204.

5

Introduction to Nanomaterials

5.1 Nanotechnology

Nanotechnology encompasses the creation and utilization of materials, devices, and systems at the level of atoms and molecules, cutting across such disciplines as chemistry, physics, and biology, engineering, and material science. Nanoscience is acknowledged to be a new frontier of research in science and technology that influences the level of materials in the range of atomic, molecular, and macromolecular, and their properties differ considerably from larger scale to nanoscale. Nanotechnology is defined as "the understanding and control of matter at dimensions of roughly 1–100 nm, where unique phenomena enable novel applications [1]. The physical, chemical, and biological properties of structures and systems at the nanoscale are substantially different from macroscale counterparts due to the interactions of individual atoms and molecules thereby offering unique and novel functional applications [2]. As the size of the particles gets reduced to the nanoscale range, there is an immense increase in the surface-to-volume ratio which increases reactivity and changes the mechanical, electrical, and optical properties of the particles.

However, the term "nano" allows researchers to highlight the fact that processes or material structures are designed and optimized to use specific properties and behaviors at lengths of 10^{-7} to 10^{-9} m. The potential benefits of nanotechnology have been reported by many industries, and many commercial products are already being manufactured, such as microelectronics, aerospace, cosmetics, and pharmaceutical industries [3]. Developments in these industries are driven by fundamental and applied research in physics, chemistry, biology, engineering, and materials science (Figure 5.1).

5.2 Nanomaterials

Nanomaterials have morphological features smaller than one-tenth of a micrometer in dimension with zero-, one-, two-, and three-dimensional confinements. Nanomaterials reveal unique physical and electrochemical properties for manufacturing materials that would be stronger, harder, and more wear-resistant.

Polymer Nanocomposites for Energy Applications, First Edition.
T. Daniel Thangadurai, Manjubaashini Nandhakumar, Sabu Thomas, and Ange Nzihou.

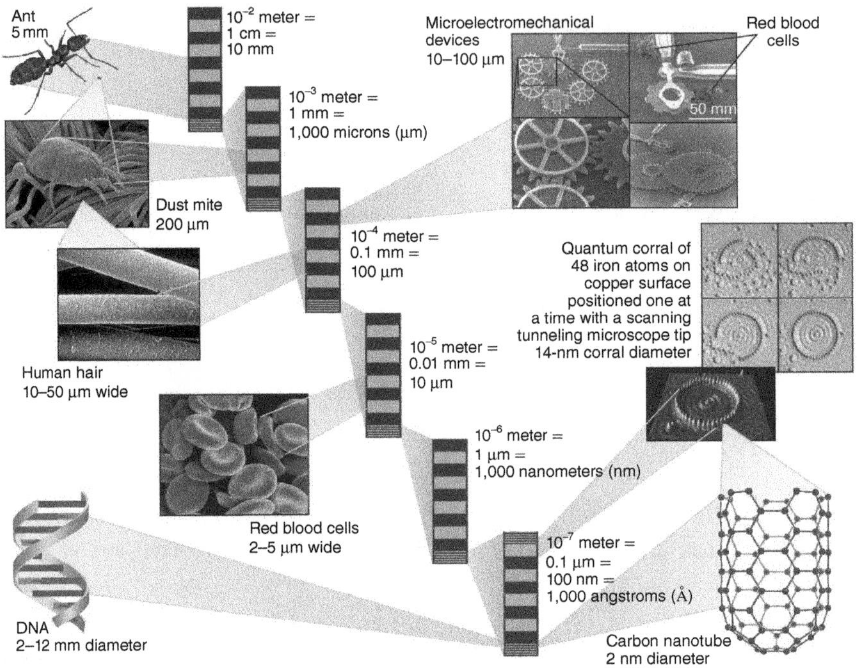

Figure 5.1 Nanotechnology examples from biological and mechanical realms illustrate various "orders of magnitude" (powers of 10), from 10^{-2} m down to 10^{-7} m. Source: Dobson et al. [4]/Encyclopedia Britannica Inc.

Interface and colloid science has given rise to many materials that may be useful in nanotechnologies, such as carbon nanotubes, other fullerenes, various nanoparticles (NPs), and nanorods [5]. The increase in surface area-to-volume ratio at nanoscale alters the mechanical, thermal, and catalytic properties of materials. For example, due to an increase in surface area, melting points can change, opaque substances become transparent (copper), inert materials become catalysts (platinum), stable materials turn combustible (aluminum), solids turn into liquids at room temperature (gold), and insulators become conductors (silicon). A material such as gold, which is chemically inert at normal scales, can serve as a potent chemical catalyst at the nanoscale [6].

Generally, nanomaterials have size- and surface-dependent properties; thus, employing nanomaterials for a sensing platform can further improve the merits of sensing properties. Low sample consumption and miniaturization of a device are also possible because of their much higher surface-area-to-volume ratios than their bulk counterparts. Metal nanoparticles (MNPs) are an important class of nanomaterials with fascinating optical, electronic, and magnetic properties [7]. MNPs are attracting much attention due to their applications in catalysis, medicine, and industrial fields. Enhanced permeability and retention are the unique properties of nanoparticles to accumulate and interact with the tumor cells. NPs such as dendrimers, quantum dots, polymer gels, and gold nanoparticles (AuNPs) have

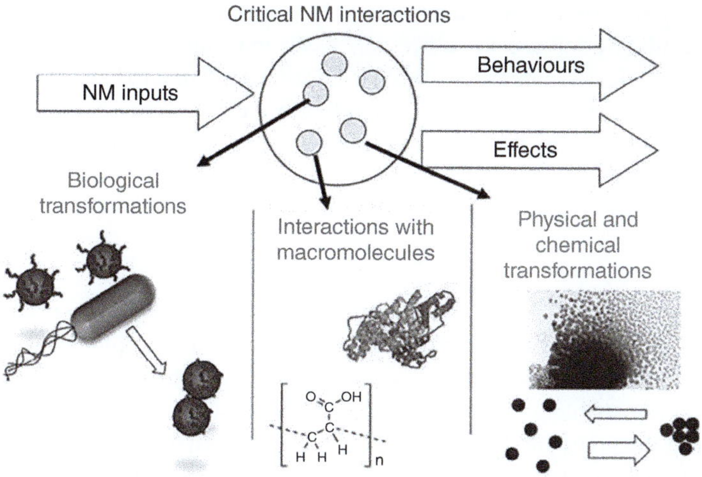

Figure 5.2 Nanomaterial transformations are critical processes affecting nanomaterial interactions.

more unique properties and are widely used in some applications such as drug delivery systems and imaging [8] (Figure 5.2).

5.3 Types of Nanomaterials

5.3.1 Quantum Dots

Quantum dots discovered by Louis E Brus constitute an important class of nanomaterials. They are semiconductor nanocrystals whose excitons are confined in all three spatial dimensions. The crystals are composed of periodic groups of II-VI, III-V, or IV-VI materials, and thus they have properties that are intermediate between those of bulk semiconductors and discrete molecules [9]. Different-sized quantum emits light of different colors due to quantum confinement. They are used in computing, photovoltaic devices, biological research, and light-emitting diodes (LEDs). Colloidal quantum dots are ideal in the application as emitting layers in LEDs due to their tunable colors, bright emission, solution processability, and stability.

5.3.2 Organic Materials

Sustainable and eco-friendly organic preparations have attracted the attention of synthetic organic chemists across the globe because of environmental issues with most of the traditional synthetic methodologies. Organic electronic devices such as organic field-effect transistors (OFETs), organic light-emitting diodes (OLEDs), organic photovoltaic cells (OPCs), and chemical and photosensors have been developed as a result of intense research in both industrial and academic sectors [10]. These electronic devices have numerous advantages such as the capability to realize flexible and large-area applications with low cost and simple fabrication processes using graphic arts printing techniques. Now the organic devices are a strong potential candidate for use in future soft electronics and wearable smart devices. Organic non-volatile memory (ONVM) is another fundamental element in the construction of electronic systems [11]. Therefore, much work has been done toward developing high-density, low-cost and non-volatile solid-state data storage devices.

5.3.3 Metal Oxides

MNPs are purely made of metal precursors. Due to well-known localized surface plasmon resonance (LSPR) characteristics, these NPs possess unique optoelectrical properties. NPs are alkali and noble metals, i.e. Cu, Ag, and Au, which have a broad absorption band in the visible zone of the electromagnetic solar spectrum. Metal oxides like TiO_2 and ZnO NPs are the most studied materials as a photocatalyst, but due to their higher bandgap (3.2 eV), the effective utilization of solar light is not possible. The sunlight encompasses only 4% of UV light and almost 40% to 45% of the visible portion. AuNPs are widely used in biomedical applications due to their unique properties such as rapid and simple synthesis, large surface area, strong adsorption ability, and facile conjugation to various biomolecules. The remarkable photophysical properties of AuNPs as biosensors have excellent biocompatibility, conductivity, and catalytic properties. AuNPs play a key role in the targeted sensing of selective biomolecules using functionalized AuNPs [12]. AuNPs-based sensorsare expected to change the foundations of sensing and detecting biomolecules and the use of surface-functionalized AuNPs for smart sensor fabrication leading to the detection of specific biomolecules. In addition to sensing, AuNPs are an attractive candidate for photothermal therapeutic, diagnostic, and drug delivery applications. Bioimaging and therapeutic applications of these unique nanomaterials will be described of their tunable optical properties, which strongly depend on the particle size, shape, composition, and surface coating. AgNPs-based nanofluids have been studied mainly as thermal exchange media, showing in all cases an improvement of the thermal properties concerning the base fluid [13].

5.3.4 Carbon Nanotubes

Carbon-based materials have attracted considerable interest in many energy-related applications due to their abundance, chemical and thermal stability, processability, and the possibility of tuning their textural and structural characteristics to fulfill the requirements of specific applications. In particular, activated carbons stand out for their large surface area (>1000 m^2 g^{-1}), pore volume (>0.5 cm^3 g^{-1}), and relatively low cost. Thus, they have a long history as adsorbents for the removal of impurities from gases and liquids. The properties of the CNTs, be it high conduction or semiconducting, depend on the precise offset of the graphite sheet as it is wrapped into a tube (chirality). Then, the artificial "transport molecules" or nano-cars have been made by attaching C_{60} molecules to rod-like molecules [14]. The resulting devices show directional motion across surfaces and may form a basis for constructing inorganic systems to transport molecules in a manner analogous to biological systems. The more advanced generations of nanotechnology are anticipated to lead to ways of mimicking biological systems and functions. Recent advances in designing nanostructured carbon materials and methods for fine-tuning their pore structures have broadened the range of carbon materials of potential interest for biomedical applications [15]. It was shown that finely tuned carbide-derived carbons (CDCs) can selectively remove inflammatory cytokines from the blood.

5.3.5 Polymeric Nanomaterials

Polymeric nanomaterials have revolutionized how medicine is approached and executed. According to Duncan and Vicent, "polymer therapeutics" encompass polymeric drugs, polymer conjugates of proteins, drugs, aptamers, block copolymer micelles, and multicomponent nonviral vectors with covalent linkages [16]. Polymers are of interest in therapeutic applications owing to the design flexibility based on functionalization, macromolecular synthesis methods, and polymer diversity. Initial uses of polymer nanoparticles (PNPs) for nano-based applications were based on non-biodegradable polymers such as polymethylmethacrylate, polyacrylamide, polystyrene, and polyacrylates. PNPs are particularly promising delivery vehicles for encapsulated proteins due to their high protein-loading capacity and porous nanostructure.

PNPs can promote stability through physical confinement of the protein and enable increased activity by serving as a pH buffer that provides proteins with an ideal chemical environment. Techniques utilized to fabricate nanometer features on polymers include e-beam lithography, polymer demixing, chemical etching, cast-mold techniques, and the use of spin casting. For those that have been applied to bone regeneration, chemical etching followed by mold casting and polymer demixing techniques have received the most attention [17]. The imprinted polymers function by mimicking the function of biological receptors.

5.4 Synthesis of Nanoparticles

Various forms of nanostructures can be synthesized using the synthesis process shown in Figure 5.3. The nanostructure can be defined as a structure with at least one or two dimensions in the 1–100 nm range. NPs can be synthesized using a variety of techniques. Dry particles and NPs in a liquid dispersion are synthesized by the techniques shown in Figure [18]. Nanostructures can be generated by building from atoms or by diminishing the size from microparticle to nanoparticle.

5.4.1 Coprecipitation

The coprecipitation technique involves the simultaneous occurrence of nucleation, growth, coarsening, and/or agglomeration processes. Coprecipitation reactions exhibit the following characteristics: (i) the products are generally insoluble species

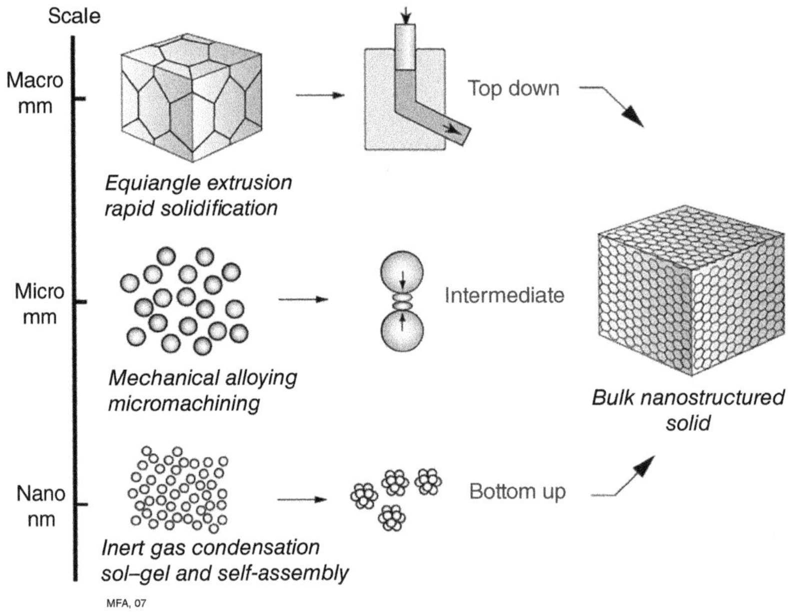

Figure 5.3 Plausible methods for the synthesis of nanoparticles.

formed under conditions of high supersaturation, (ii) nucleation is a key step, and a large number of small particles will be formed, (iii) secondary processes, such as Ostwald ripening and aggregation, dramatically affect the size, morphology, and properties of the products [19].

5.4.2 Hydrothermal Technique

The hydrothermal technique has been most admired, garnering attention from scientists and technologists from different disciplines. The hydrothermal synthesis has been successful for the preparation of important solids, such as microporous crystals, superionic conductors, chemical sensing oxides, electronically conducting solids, complex oxide ceramic and fluorides, magnetic materials, and luminescence phosphors. It is also a route to unique condensed material, including nanometer particles, gels, thin films, distinguished helical and chiral structures, and particularly stacking sequence materials [20].

5.4.3 Inert Gas Condensation

Inert gas condensation (IGC) is the most widely used method for the synthesis of metal nanoparticles. In IGC, metals are evaporated in an ultrahigh vacuum chamber filled with helium or argon gas at very high pressure. The evaporated metal atoms lose their kinetic energy by collisions with the gas and condense into small particles [21]. These particles then grow by Brownian coagulation and coalescence and finally form nanocrystals (Figure 5.4).

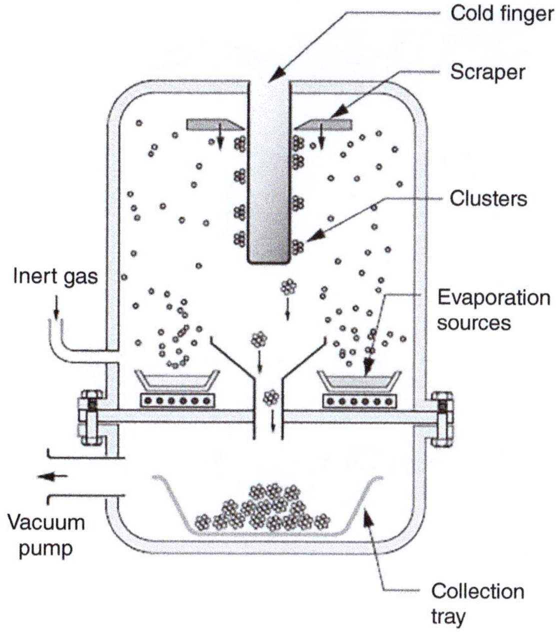

Figure 5.4 Schematic representation of inert gas condensation.

5.4.4 Sonochemical

In sonochemical processing, ultrasound is used to nucleate a chemical reaction. Ultrasound spans the frequencies in the range of 15 kHz to 1 GHz. A magnetostrictive or piezoelectric transducer is used to generate ultrasonic waves with a wavelength of 1–10 000 μm in a liquid-filled reaction vessel. These are not molecular dimensions, so there is no direct coupling of the acoustic field with the chemical species. The reaction comes about because of cavitation. The tensile part of the wave is intense enough to pull the liquid apart and form a tiny cavity. The compression part of the wave then compresses it, but before it does, some reactants vaporize inside it. The next tensile wave re-expands the bubble, which oscillates in volume at the frequency of the soundwaves, pumping it up as more vapors enter during the expansionpart of the cycle. When the bubble reaches a critical size, it collapses. The collapse is adiabatic because the very fast collapse rate leaves no time for heat flow, generating a tiny, localized hot spot. The temperatures are very high (as high as 5000 °C, near that of the surface temperature of the sun) and so too are the pressures (around 2000 atm, roughly those at the bottom of deep oceans), triggering reactions that create a nanoparticle within the spot. The size of the spot determines the size of the resulting particles. The technique can be used to produce a large volume of material for industrial applications (Figure 5.5).

5.4.5 Microemulsion

The microemulsion method is one of the ideal techniques for the preparation of inorganic nanoparticles, yet the mechanism of nanoparticle formation in the

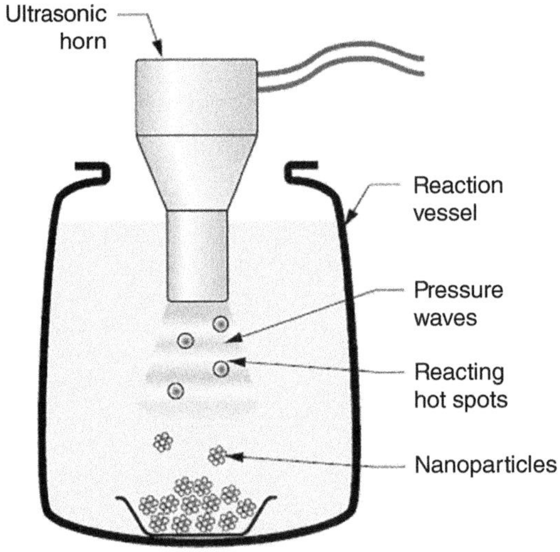

Figure 5.5 Sonochemical processing for nanoparticle synthesis.

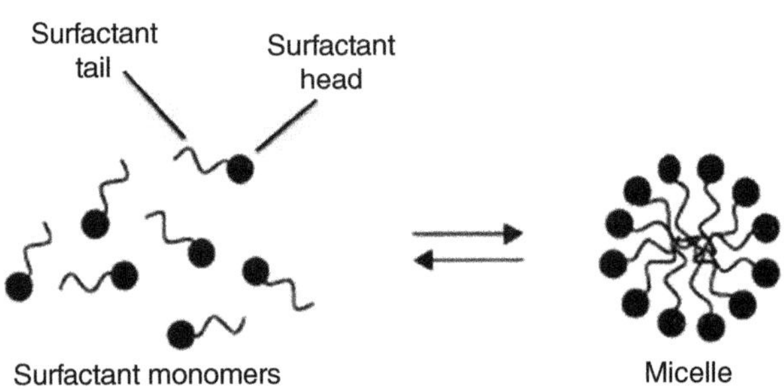

Figure 5.6 Spontaneous self-assembly of surfactants into micelles in aqueous solution.

microemulsion has not yet been understood well. However, some researchers have suggested a mechanism for nanoparticle synthesis within microemulsions. When the microemulsion materials, including reactants, are mixed, reactant exchange takes place during the collision of water droplets in the microemulsion. The reactant exchange is too fast and a precipitation reaction occurs in the nanodroplets, which is followed by nucleation growth and coagulation of primary particles, resulting in the final nanoparticles surrounded by water and/or stabilized by surfactants [22] (Figure 5.6).

5.4.6 Microwave-Assisted

Microwave-assisted synthesis is popular in areas ranging from biochemical processes to nanotechnology. Chemical reactions are often faster than traditional convection heating methods and have high yields and fewer side products. Microwave reactors provide excellent control over reaction mixing, withstanding high temperatures and pressures, and demonstrate reproducibility from reaction to reaction. Microwave-assisted techniques provide improved engineering control over the separation of the nucleation and growth stages of nanomaterial synthesis when the reaction is initiated from room temperature. Microwave-assisted heating could provide some selectivity in activating the nanomaterial precursor materials,

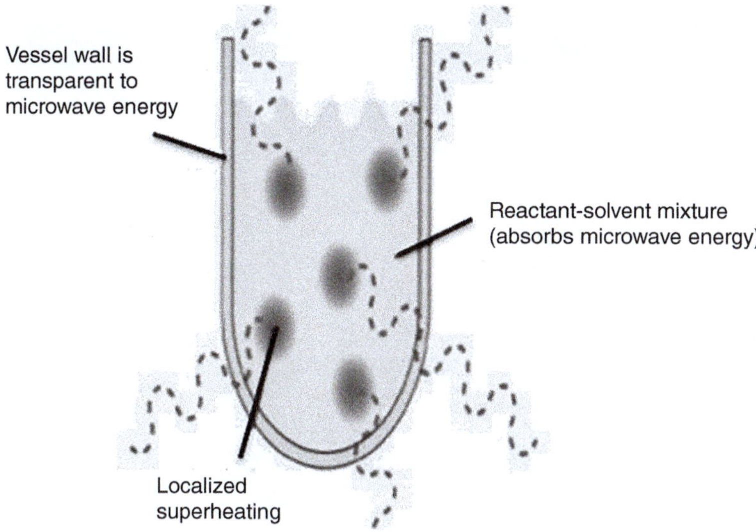

Figure 5.7 Microwave heating of the reaction mixture.

which is important for scalability [23]. Microwave synthesis has the potential to selectively heat either the solvent or the precursor molecules for nanomaterials preparation (Figure 5.7).

5.4.7 Laser Ablation

Laser ablation is the process of removing material from a solid or occasionally liquid surface by irradiating it with a low- and/or high-energy laser beam. When the material is heated by the absorbed laser energy, it evaporates or sublimates at low laser flux and is typically converted to plasma at high laser flux [24]. Besides, it is possible to ablate material in laser ablation with a continuous wave laser beam if the laser intensity is high enough (Figure 5.8).

5.4.8 Sol–Gel

Sol–gel is a method for producing solid materials from small molecules. In this chemical procedure, the sol (or solution) gradually evolves toward the formation of a gel-like diphasic system containing a liquid phase and a solid phase, the morphologies of which range from discrete particles to continuous polymer networks. Ultrafine and uniform ceramic powders can be formed by precipitation. These powders of single- and multicomponent composition can be produced on a nanoscale particle size for dental and biomedical applications [25] (Figure 5.9).

5.4.9 Spark Discharge

The synthesis of nanoparticles using spark discharge is a promising method for low-cost, industrial-scale nanofabrication of advanced materials. An electric spark

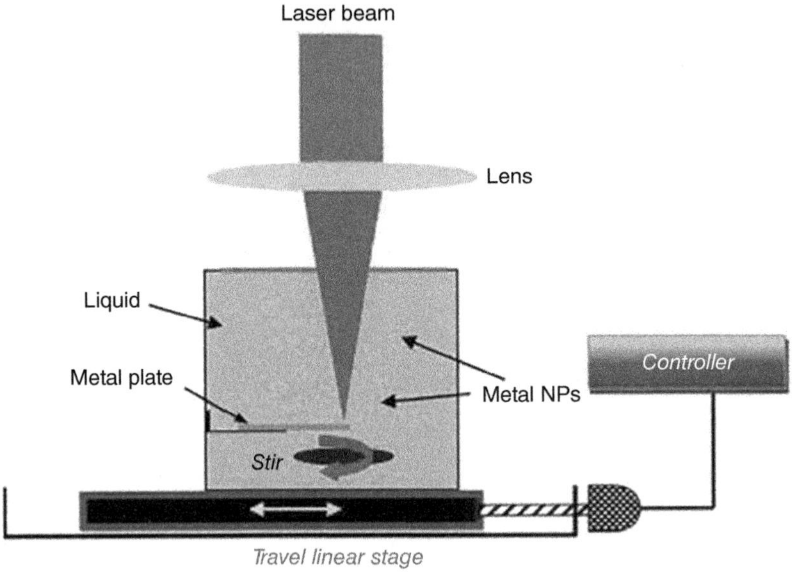

Figure 5.8 Laser ablation setup for the preparation of metal nanoparticles contains a travel linear stage, a high-power laser, a metal plate, and a liquid tank.

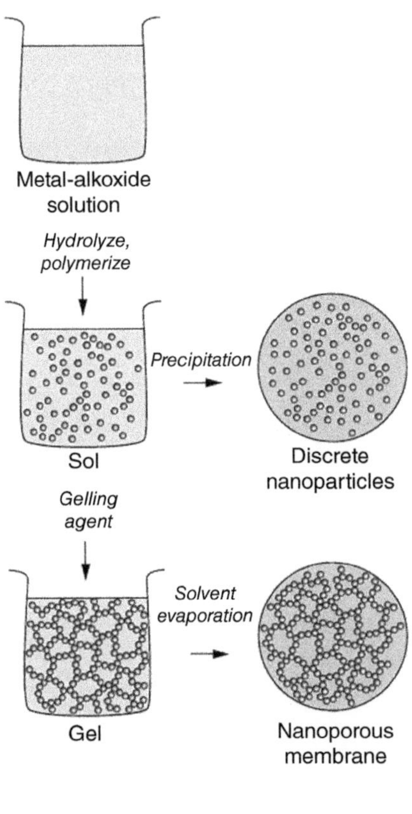

Figure 5.9 Sol–gel method for nanoparticle preparation.

is an abrupt electrical discharge that occurs when a sufficiently high electric field creates an ionized, electrically conductive channel through a normally insulating medium, often air or other gases or gas mixtures [26].

5.4.10 Template Synthesis

With the increasing emphasis on green chemistry, it is becoming more important to develop an environmentally friendly, facile method for the synthesis of nanoparticles. It has been reported that template synthesis is one of the most promising methods for the preparation of monodispersed inorganic nanoparticles, in which the uniform void spaces of porous materials are used as hosts to confine the synthesized nanoparticles as guests. The role of the template is twofold [27]. First, it allows the reproduction of the structure with the best possible reproducibility and plays the role of a skeleton to organize the different functions of a device, the active components, and the different interfaces (Figure 5.10).

5.4.11 Biological Synthesis

The biological synthesis of nanoparticles is a green chemistry approach that interconnects nanotechnology and biotechnology. However, despite the stability, biological nanoparticles are not monodispersed and the rate of synthesis is slow. The concentration of synthesized macromolecules or components involved in the nucleation of particles varies with time and prolongs the nucleation period, which causes the polydispersity of nanoparticles and subsequent decreased rate

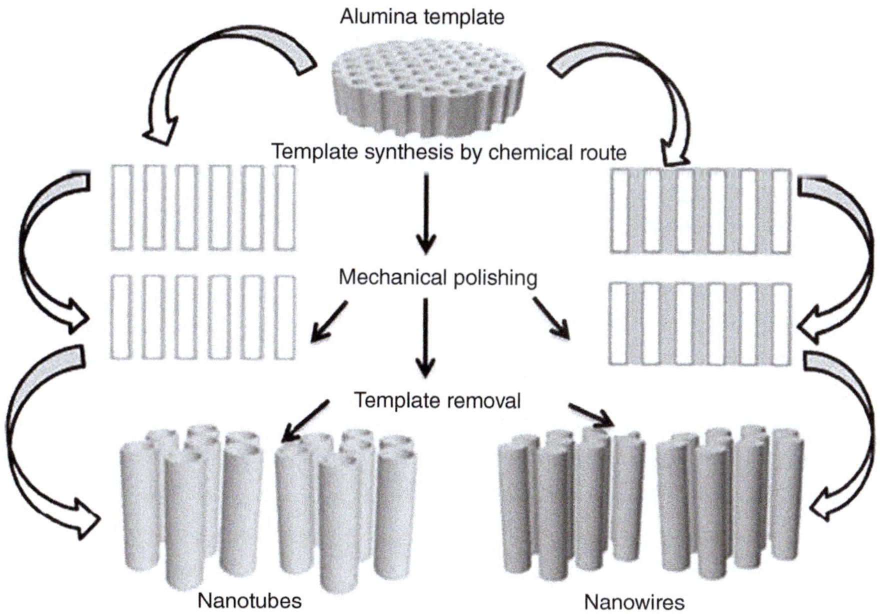

Figure 5.10 Template synthesis coupled with other methods.

of synthesis. To overcome these problems, several methods, such as microbial cultivation methods and extraction techniques, have to be optimized, and a combinatorial approach, such as photobiological methods, may be used [28]. Cellular, biochemical, and molecular mechanisms that mediate the synthesis of biological nanoparticles should be studied in detail to increase the rate of synthesis and improve the properties of the nanoparticles. Owing to the rich biodiversity of plants and microbes, the potential as biological materials for nanoparticle synthesis is yet to be fully explored.

5.5 Applications of Nanotechnology

Nanotechnology offers a wide range of opportunities for the development and application of structures, materials, or systems with new properties in various areas like agriculture, food, environment, sensors, pharmaceutical, cosmetic, electronics, energy, and biomedical, etc. [29].

5.5.1 Nanotechnology in Energy Sector

Nanotechnology is now more effectively used to develop renewable and environmentally friendly energy sources, e.g. wind, geothermal, tidal, and solar energy [30]. Energy applications are now focusing on tailoring the nanoscale effects for efficient energy technologies such as photovoltaics, photochemical solar cells, thermoelectric, fuel cells, batteries, and so forth. Solar energy is well recognized as one of the most promising choices to meet the enormous energy demand in the future. Thus, developing efficient solar cells with high power conversion efficiency (PCE) needs efficient charge separation, transport, and collection processes beneficial for the improvement of PCE, which requires optimization at both material and configuration levels. On this account, one-dimensional nanomaterials share various advantages based on their unique configuration, large surface-to-volume ratio, which enables rapid charge transport along the axial direction, making them suitable for efficient solar cells [31].

Organic–inorganic halide perovskite solar cells (OIHPSCs) show promise as renewable energy sources. A distinct feature of OIHPSCs is their combined merits of low-cost solution processability and extraordinary optoelectronic properties, including long carrier lifetime, large absorbing coefficient, high tolerance for trap states, and high dielectric constant. Within a quite short period, the device performance of perovskite solar cells was rapidly improved from 3.8% to 22.7% as a result of intensive research on compositions, device architecture designs, morphologies, and interfacial engineering [32].

5.5.2 Nanotechnology in Textile

The use of nanotechnology in textile materials including nanofibers, nanocomposite fibers, and nano-finished textiles imparts multi-properties providing their application in the sports clothing market. Several fabric companies have utilized nanotechnology to develop a wide range of sports apparel. Scholler, a Swiss company, has developed a nano-based technology to produce clothing with an optimal balance of comfort, air permeability, wind and water resistance, and self-cleaning property for extreme cold weather sports such as mountaineering and skiing [33]. Efforts are being made worldwide to create smart and intelligent textiles by incorporating various nanoparticles or creating nanostructured surfaces and nanofibers which lead to an unprecedented level of textile performance, such as stain-resistant, self-cleaning, antistatic, UV protective properties, etc. [34].

The nanotechnology research in the textile arena mainly centers on creating unique properties in everyday fabrics such as self-cleaning, water and oil repellency, stain-proof nature, antibacterial property, UV protective nature, antistatic property, improved moisture regaining, and comfort in synthetic-based textiles, but without compromising the original hand breathability and durability of the fabric. It also shows applications in advanced textile materials such as nanocomposite fibers, nanofibers, and other nanomaterial incorporated in textiles for medical, defense, aerospace, and other technical textile applications such as filtration, protective clothing besides a range of smart and intelligent textiles [35].

5.5.3 Nanotechnology in Agriculture

Nanotechnology is the most remarkable technology in modernistic agriculture, and this technology is expected to be soon an important driving force in food production. Nanotechnology in the agricultural sector concentrates on sustainable food production and the protection of food crops for both human nutrition and animal feeding against pests and diseases [36]. Nanotechnology provides new agrochemical opportunities to improve food crop production and reduce adverse environmental effects by reducing the use of pesticides and chemical fertilizers.

Nanotechnology monitors a leading agricultural controlling process, especially by its miniature dimension. Additionally, many potential benefits such as enhancement of food quality and safety, reduction of agricultural inputs, enrichment of absorbing nanoscale nutrients from the soil, etc., allow the application of nanotechnology to be a resonant encumbrance. Hence, agriculture, food, and natural resources are a part of those challenges like sustainability, susceptibility, and human health. The ambition of nanomaterials in agriculture is to reduce the number of chemicals used, minimize nutrient losses in fertilization, and increase yield through pest and nutrient management [37]. Nanotechnology has the prospects to improve the agriculture and food industry with novel nanotools for controlling rapid diseases, diagnostic purposes, enhancing the capacity of plants to absorb nutrients among others. The significant interests of using nanotechnology in agriculture include specific applications like nanofertilizers and nanopesticides to increase productivity without decontamination of soils, waters, and protection against several insects, pests, and microbial diseases. Nanotechnology may act as sensors for monitoring the soil quality of the agricultural field and thus maintaining the health of crops [38].

5.5.4 Nanotechnology in Electronics

Quantum-dot cellular automata (QCA) is identified as the new upcoming nanotechnology for the design of digital circuits. The need for such a technology arises from the fact that the scaling of conventional complementary metal–oxide–semiconductor (CMOS) is taken down to the nano-regime. This new technology called QCA uses the quantum effects to its advantage and thus is more suitable for nanotechnology-based applications. QCA calls for a departure from the conventional current and voltage paradigms and is governed by the laws of physics such as columbic repulsions and quantum mechanical tunneling [39]. Organic nanomaterials such as organic nanowires/nanofibers also have recently been started to utilize as electric materials for fabricating CMOS circuits and wearable power generators. Intrinsic deformability of organic nanomaterials, solution processability, and low cost make them promising for wearable devices [40].

5.5.5 Nanotechnology in Cosmetics

Nanomaterials have been extensively researched and used as a vehicle for the delivery of bioactive agents to the skin within the cosmetic arena. Nanotechnology confers small sizes ranging from 1 to 100 nm and a very large surface-area-to-volume ratio, facilitating their interaction with the target organ [41]. Nanomaterials can also be designed to deliver established or developed formulations of drugs in a sustained, controlled, and targeted manner to avoid adverse systemic side effects. This liberation system not only allows for the enhanced delivery of active substances to the epidermis of the skin but also permits prolonged contact at the site of action properties that bring a variety of benefits to cosmetic products. Furthermore, due to their small size, optical transparency is also enhanced. TiO_2 microparticles, the particles frequently used in sunscreens, are capable of penetrating the hair follicle [42]. Research demonstrates that nanomaterials have entered just about every personal hair care product on the market.

Shampoos have incorporated nanomaterials to optimize resident contact time with the scalp and hair follicle, allowing active agents time to form a protective film, sealing moisture within the cuticles. Typically, during washing, the cuticle layer is opened by hot water, exposing the hydro-lipid emulsion layer, whose function is to enable external water absorption and prevent internal water loss [43]. Nanomaterials hold the potential to improve hair cosmesis by promoting increased contact with the hair shaft and follicle, increasing the quantity of incorporated active ingredients reaching the target site.

5.5.6 Nanotechnology in Medical Field

Nanotechnology has the potential to make significant contributions to disease detection, diagnosis, therapy, and prevention. Nanotechnology could have a profound influence on disease prevention efforts, because it offers innovative tools for understanding the cell as well as the differences between normal and abnormal cells [44]. Medical devices have already benefited from recent developments in nanotechnology like surgical tools with enhanced material properties enabling better handling, microcantilevers for label-free assays used in molecular *in vitro* diagnostics, bone replacement materials obtained by nanostructured materials allowing better implant integration and biocompatibility, pacemakers, and hearing aids based on spintronic technology enabling size reduction and power enhancement of these medical devices, DNA/protein microarrays and lab-on-a-chip devices for molecular *in vitro* diagnostics, wound dressings and textiles incorporating nanocrystalline particles with antibacterial and fungicidal activity, micro-needle-based systems for minimally invasive drug administration or blood substance monitoring limiting tissue damage and pain sensation [45].

References

1 Rani, R. (2017). Nanotechnology: big things from a miniature globe. *Int. J. Educ. Sci. Res.* 4: 79–85.

2 Akula, K.K. (2013). Revisiting neuroscience with nanomedicine, a renaissance in remedy. *Inno Pharm. Pharmacol.* 1: 1–60.

3 Chadha, S. (2013). Nanotechnology and its application. *Int. J. Agri. Food Sci. Technol.* 4: 1011–1018.

4 Dobson, P., King, S. and Jarvie, H. (2009). Nanoparticle. *Encyclopedia Britannica*, 14 May. https://www.britannica.com/science/nanoparticle. Accessed 29 November 2021.

5 Behari, J. (2010). Principle of nanoscience: an overview. *Ind. J. Exper. Biol.* 48: 1008–1019.

6 Sau, T.K., Rogach, A.L., Jackel, F. et al. (2010). Properties and applications of colloidal nonspherical noble metal nanoparticle. *Adv. Mater.* 22: 1805–1825.

7 Rajamanikandan, R. and Ilanchelian, M. (2017). Simple and visual approach for highly selective biosensing of vitamin B1 based on glutathione coated silver nanoparticles as a colorimetric probe. *Sens. Actuators, B* 244: 380–386.

8 Matea, C.T., Mocan, T., Tabaran, F. et al. (2017). Quantum dots in imaging, drug delivery, and sensor applications. *Int. J. Nanomed.* 12: 5421–5431.

9 Ghosh, T., Panicker, J.S., and Nair, V.C. (2017). Self-assembled organic materials for photovoltaic application. *Polymers* 9: 112–152.

10 Khan, I., Saeed, K., and Khan, I. (2017). Nanoparticles: properties, applications, and toxicities. *Arabian J. Chem.* 12: 908–931.

11 Cai, Y., Tan, J., YeFan, L. et al. (2016). A flexible organic resistance memory device for wearable biomedical applications. *Nanotechnology* 27: 275206–275212.

12 Yu, X., Jiao, Y., and Chai, Q. (2016). Applications of gold nanoparticles in biosensors. *Nano LIFE* 6: 1642001–1642012.

13 Varshney, K. (2014). Carbon nanotubes: a review on synthesis, properties, and applications. *Int. J. Eng. Res. Gen. Sci.* 2: 660–677.

14 Seredych, M., Mikhalovska, L., Mikhalovsky, S., and Gogotsi, Y. (2018). Adsorption of bovine serum albumin on carbon-based materials. *J. Carbon Res.* 4: 3–14.

15 Han, J., Zhao, D., Li, D. et al. (2018). Polymer-based nanomaterials and applications for vaccines and drugs. *Polymers* 10: 31–44.

16 Smith, J., Sprenger, K.G., Liao, R. et al. (2017). Determining dominant driving forces affecting controlled protein release from polymeric nanoparticles. *Biointerphases* 12: Article No. 02D412.

17 Lai, R.W.S., Yeung, K.W.Y., Yung, M.M.N. et al. (2018). Regulation of engineered nanomaterials: current challenges, insights and future directions. *Environ. Sci. Pollut. Res.* 25: 3060–3077.

18 Rane, A.V., Kanny, K., Abitha, V.K., and Thomas, S. (2018). Methods for synthesis of nanoparticles and fabrication of nanocomposites. *Synth. Inorg. Nanomater* 121–139.

19 D'Costa, G., Pisal, D.S. and Rane, A.V. (2012). Report on synthesis of nanoparticles and functionalization: co-precipitation method https://doi.org/10.1016/B978-0-08-101975-7.00005-1.

20 Meskin, P.E., Ivanov, V.K., Baranchikov, A.E. et al. (2006). Ultrasonically assisted hydrothermal synthesis of nanocrystalline ZrO_2, TiO_2, $NiFe_2O_4$, and $Ni_{0.5}Zn_{0.5}Fe_2O_4$ powders. *Ultrason. Sonochem.* 13: 47–53.

21 Ward, M.B., Brydson, R., and Cochrane, R.F. (2006). Mn nanoparticles produced by inert gas condensation. *J. Phys. Conf. Ser.* 26: 296–299.

22 Nafisi, S. and Maibach, H.I. (2017). Nanotechnology in cosmetics. In: *Cosmetic Science and Technology*, 337–369. https://doi.org/10.1016/B978-0-12-802005-0.00022-7.

23 Chikan, V. and McLaurin, E.J. (2016). Rapid nanoparticle synthesis by magnetic and microwave heating. *Nanomater* 6: 85.

24 Wang, S. and Gao, L. (2019). Laser-driven nanomaterials and laser-enabled nanofabrication for industrial applications. *Ind. Appl. Nanomater.* 181–203.

25 Jain, R., Chaurasia, S.K. and Kalga, A. (2012). Report on sol–gel method for nanoparticle synthesis. https://doi.org/10.1016/B978-0-08-101975-7.00005-1.

26 Efimov, A.A., Lizunova, A.A., Volkov, I.A. et al. (2016). A new approach to the high-yield synthesis of nanoparticles by spark discharge. *J. Phys. Conf. Ser.* 741: 012035. http://dx.doi.org/10.1088/1742-6596/741/1/012035.

27 Liu, Y., Goebl, J., and Yin, Y. (2013). Templated synthesis of nanostructured materials. *Chem. Soc. Rev.* 42: 2610–2653.

28 Pantidos, N. and Horsfall, L.E. (2014). Biological synthesis of metallic nanoparticles by bacteria, fungi, and plants. *J. Nanomed. Nanotechnol.* 5: 233. http://dx.doi.org/10.4172/2157-7439.1000233.

29 Tan, K.Y., Li, C.Y., Li, Y.F. et al. (2017). Real-time monitoring ATP in mitochondrion of living cells: a specific fluorescent probe for ATP by dual recognition sites. *Anal. Chem.* 89: 1749–1756.

30 Qiao, R. and Zuo, L. (2018). Self-assembly monolayers boosting organic-inorganic halide perovskite solar cell performance. *J. Mater. Res.* 33: 387–400.

31 Srinivasan, K., Rajanikumar, K., Sheetal Bhardwaj, K. et al. (2018). Nanotechnology trends in fashion and textile engineering. *Curr. Trends Fashion Technol. Text. Eng.* 2: 555590–555594.

32 Banotra, M., Kumar, A., Sharma, B.C. et al. (2017). Prospectus of use of nanotechnology in agriculture – a review. *Inter. J. Curr. Microbiol. Appl. Sci.* 6: 1541–1551.

33 Prasad, R., Bhattacharyya, A., and Nguyen, Q.D. (2017). Nanotechnology in sustainable agriculture: recent developments, challenges, and perspectives. *Front. Microbiol.* 8: 1–13.

34 AKM, A.H.A. and Zayedul Hasan, M.D. (2018). Application of nanotechnology in modern textiles: a review. *Inter. J. Curr. Eng. Technol.* 8: 22–231.

35 Ahmad Wani, K. and Kothari, R. (2018). Agricultural nanotechnology: applications and challenges. *Ann. Plant Sci.* 7: 2146–2148.

36 Grillo, R., Chirakkuzhyil Abhilash, P., and Fernandes Fraceto, L. (2016). Nanotechnology applied to bio-encapsulation of pesticides. *J. Nanosci. Nanotechol.* 16: 1231–1234.

37 Qiao, L., Qian, S., Wang, Y., and Lin, H. (2018). A colorimetric sensor array based on sulfuric acid assisted $KMnO_4$ fading for the detection and identification of pesticides. *Talanta* 181: 305–310.

38 Banerjee, R. (2017). Nanocosmetics: the good, the bad, and the beautiful. *Trichol. Cosmetol. Open J.* 1: 9–11.

39 Chen, X. and Selloni, A. (2014). Introduction: titanium dioxide (TiO_2) nanomaterials. *Chem. Rev.* 114: 9281–9282.

40 Tripura Sundari, P. and Anushree, H. (2017). Novel delivery systems: current trend in cosmetic industry. *Eur. J. Pharm. Med. Res.* 4: 617–627.

41 Garimella, R. and Eltorai, A.E.M. (2017). Nanotechnology in orthopedics. *J Orthop.* 14: 30–33.

42 Gupta Swaroopa Rani, N. (2016). Exploration of various classes of nanoparticles used in cosmetic applications. *Inter. Res. J. Mater. Sci. Eng.* 4: 91–97.

43 Huang, X., Liu, Y., Yung, B. et al. (2017). Nanotechnology-enhanced no-wash biosensors for in vitro diagnostics of cancer. *ACS Nano* 11: 5238–5292.

44 Ramasamy, M. and Lee, J. (2016). Recent nanotechnology approaches for prevention and treatment of biofilm-associated infections on medical devices. *Bio. Med. Res. Inter.* Article no. 1851242.

45 Maas, M.B., Perold, W.J., and Dicks, L.M.T. (2017). Biosensors for the detection of *Escherichia coli*. *Water SA* 43: 701–721.

6

Introduction to Polymer Nanocomposites

6.1 Classes of Nanocomposites

Nanocomposites are materials in which at least one material should be in nanosize in at least one dimension. These nanocomposites have improved properties when compared to the individual component materials by the synergetic effect of both of the individual components [1].

There are two parts to nanocomposites, i.e. continuous phase and discontinuous reinforcing phase. Nanocomposite can be prepared from any combination of materials that can be categorized into three basic building blocks, i.e. metals, ceramics, and polymers. The nanocomposite, hence, can have a combination or have markedly different mechanical, electrochemical, electrical, catalytic, thermal, and optical properties from the component materials [2]. The nanocomposites have different phases as zero-dimensional (core-shell), one-dimensional (nanowires and nanotubes), two-dimensional (lamellar), and three-dimensional (metal matrix composites) [3]. Based on their structural characteristics, these nanocomposites are classified as nanolayered composites, nanofilamentary composites, and nanoparticle composites (Figure 6.1).

6.2 Different Types of Nanocomposites

Class	Examples
Metal	Co/Cr, Ni/Al_2O_3, Fe-Cr/Al_2O_3, Fe/MgO, Al/CNT, Mg/CNT
Ceramic	SiO_2/Ni, Al_2O_3/SiO_2, Al_2O_3/TiO_2, Al_2O_3/SiC, Al_2O_3/CNT
Polymer	Polyester/TiO_2, polymer/CNT, thermoplastic/thermoset polymer/layered silicates, polymer/layered double hydroxides

Polymer Nanocomposites for Energy Applications, First Edition.
T. Daniel Thangadurai, Manjubaashini Nandhakumar, Sabu Thomas, and Ange Nzihou.

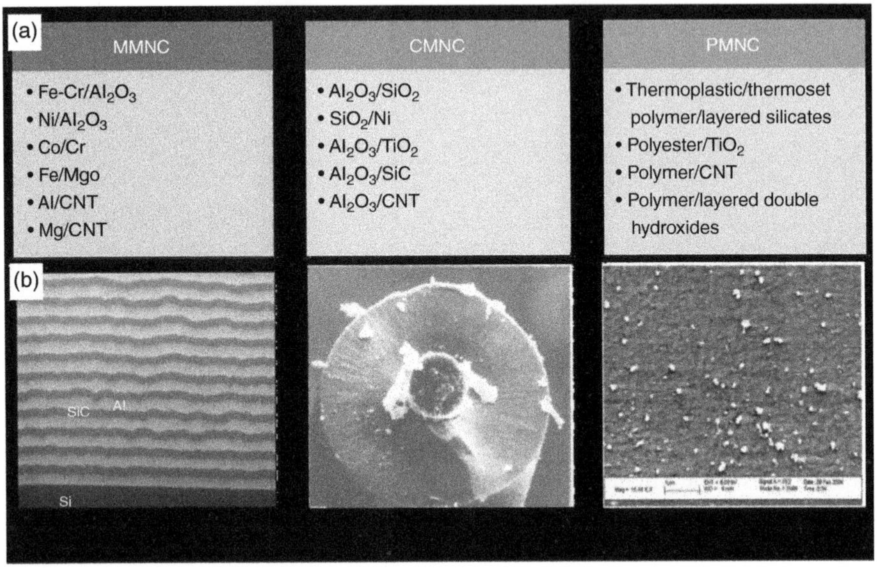

Figure 6.1 Examples of three types of nanocomposites (a) and their three classes (b).

6.2.1 Polymer-Based and Non-Polymer-Based Nanocomposites

Polymer-based nanocomposites

1. Polymer/Ceramic Nanocomposite
2. Inorganic/Organic Polymer Nanocomposite
3. Inorganic/Organic Hybrid Nanocomposite
4. Polymer/Layered Silicate Nanocomposite
5. Polymer/Polymer Nanocomposite
6. Biocomposites

Non-polymer-based nanocomposites

1. Metal/Metal Nanocomposite
2. Metal/Ceramic Nanocomposite
3. Ceramic/Ceramic Nanocomposite

6.2.1.1 Polymer/Ceramic Nanocomposite

Nanocomposites consist of single ceramic layers (1 nm thickness) homogeneously dispersed in a continuous matrix. The host ceramic layer tends to orient itself parallel to each other due to dipole–dipole interaction [4]. Natural bone is a nanocomposite consisting of approximately 30% matrix (collagen) material and 70% nanosized minerals (hydroxyapatite).

6.2.1.2 Inorganic/Organic Polymer Nanocomposites

Metal polymer nanocomposites attract attention because of the unique properties of metal clusters that are dispersed in the polymer matrix. The typical size of such a metal cluster is approximately 1–10 nm. The properties of clusters and nanoparticles (bandgap, spectral properties, and transport of electrons) are very different from those of bulk materials and individual atoms or molecules. The size and grains depend on the mobility of the metal atoms on the polymer surface. For example, in the case of polymethylmethacrylate (PMMA) polymer, the cluster size depends on the amount of crosslinking of the polymer, which changes the mobility of the metal atoms [5].

6.2.1.3 Inorganic/Organic Hybrid Nanocomposite

Hybrid inorganic/organic materials are not simply physical mixtures; they can be broadly defined as nanocomposites with organic and inorganic components intimately mixed. Indeed, hybrids are either homogenous systems derived from monomers and miscible organic/inorganic components or heterogeneous systems (nanocomposites) where at least one of the components has the scale of nanometer [6].

6.2.1.4 Polymer/Layered Silicate (PLS) Nanocomposites

Polymer/layered silicate (PLS) nanocompositesare gained significant attention in polymer science research. In recent years, the PLS nanocomposites have attracted great interest both in industry and academia because they often exhibit remarkable improvements in materials when compare with virgin polymer and conventional macro- and micro-composites. The most commonly used smectite-type-layered silicates are hactorite and montmorillonite for the preparation of the nanocomposites.

6.2.1.5 Polymer/Polymer Nanocomposites

Polymers are more than ever under pressure to be chip and offered property profiles. The gap between block copolymer self-assembly and offer nanostructured plastic endowed with still unexplored combinations of properties are getting narrower. Mixtures of different polymers often phase separate, even when their monomer is mixed homogeneously.

6.2.1.6 Biocomposites

Metals and metal alloys are used in orthopedics, dentistry, and other load-bearing applications. Ceramics are used with an emphasis on either their chemically inert nature or high bioactivity; all polymers are used for soft tissue replacements and used for many other nonstructural applications [7].

6.2.1.7 Ceramic Matrix Nanocomposites

Ceramic matrix nanocomposites mainly have Al_2O_3 or SiC systems. The strength of the Al_2O_3 matrix in the low volume fraction of SiC particles depends on the suitable size and hot pressing.

6.2.1.8 Metal Matrix Nanocomposites

Metal matrix nanocomposites (MMNCs) are referred to as materials consisting of a ductile metal or alloy matrix in which some nanosized reinforcement material is implanted. These materials combine metal and ceramic features [8].

6.2.1.9 Polymer Matrix Nanocomposites

Polymer matrix nanocomposites are broadly applied in industry for their ease of production, lightweight, and ductile nature. They have some drawbacks such as low modulus and strength compared to metals and ceramics.

6.3 Synthesis Methods of Nanocomposite

Polymer nanocomposites are the materials in which nanoscopic inorganic particles, usually 10–100 nm in at least one dimension, are dispersed in an organic polymer matrix to substantially improve the performance properties of the polymer. Such structures combine the best properties of each component through the synergistic effect of both the component materials. These include improved electronic, ionic, mechanical, chemical properties, etc., that may be exploited for different applications. Polymer nanocomposites are generally prepared by some basic methods [9], such as (i) solution casting, (ii) melt blending, (iii) *in situ* polymerization method, (iv) smelt intercalation, (v) template synthesis, and (vi) exfoliation adsorption.

6.3.1 Solution Casting Method

In the solution casting technique, the polymer matrix is dissolved in a suitable solvent. The solvent is selected in such a way that it should dissolve the polymer and it should be able to disperse the nanoparticle. To the homogeneous polymer solution, the required quantity of nanoparticles is added and thoroughly mixed by ultrasonication and stirred for a certain period. The obtained homogeneous mixture is then cast in a Teflon dish and allowed to dry for complete evaporation of the solvent and the formation of nanocomposite as a thin film [10].

6.3.2 Melt Blending Method

The melt blending method is superior over intercalation and *in situ* methods of synthesis of polymer nanocomposites in terms of environmental safety, since this method involves no solvent for the synthesis. This method involves annealing statically or under shear a mixture of polymer and nanoparticle beyond the softening point of the polymer. In this method, the polymer and nanoparticles are added into the extruder and subjected to intensive mixing for a certain period; finally, nanocomposite comes out from the die [11]. This method is generally used for the synthesis of polymer nanocomposites from polymers that are not suitable for the synthesis of nanocomposites by solution casting or *in situ* polymerization methods.

6.3.3 In situ Polymerization Method

In situ polymerization involves the swelling of the filler in the liquid monomer or monomer solution as the low-molecular-weight monomer seeps in between the interlayers causing the swelling. Polymerization starts using heat, radiation, initiator diffusion, or by organic initiator or catalyst fixed through a cationic exchange. The monomers then polymerize in between the interlayers forming intercalated or exfoliated nanocomposites. The advantage of this approach lies in the better exfoliation achieved compared to melt and exfoliation adsorption methods. The synthesis of nylon-6/clay nanocomposite via *in situ* polymerization in which clay is dispersed in caprolactam monomer and under polymerization conditions, an exfoliated nanocomposite is formed [12].

6.3.4 Exfoliation Adsorption Method

Exfoliation adsorption, also called polymer or prepolymer intercalation from solution, is merely based on a solvent in which the polymer or prepolymer is soluble. The layered silicate, for instance, is first swollen and dispersed in a solvent before mixing it with the polymer solution. The polymer chains then intercalate and displace the solvent within the silicate interlayers. Eventually, on the removal of the solvent, a multilayer structure is formed as the sheets reassemble trapping the polymer chains, as shown in Figure 6.2. This approach is widely used for water-soluble polymers to produce intercalated nanocomposites based on polymers with low or no polarity such as poly (vinyl alcohol), poly (ethylene oxide), poly (vinylpyrrolidone), or poly (acrylic acid). However, unlike melt intercalation, this method is environmentally non-friendly because of the usage of large amounts of solvents [13]. Emulsion polymerization is considered to be under this method as monomers, usually methyl methacrylate and styrene, are dispersed in water along with an emulsifier and different silicate concentrations. The monomer is polymerized with a part of silicate embedded inside the polymer particle and a part adsorbed on the particle surface, forming a nanocomposite.

6.3.5 Template Synthesis Method

Template synthesis, also known as sol–gel technology, is based on the opposite principle to the previous methods. This approach involves the formation of the inorganic

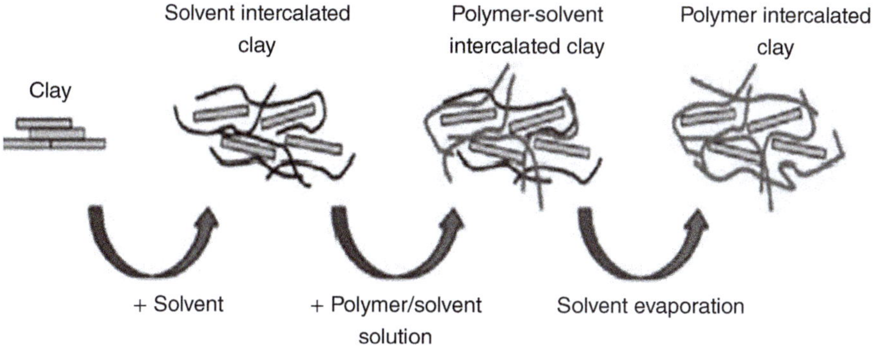

Figure 6.2 Illustration of the exfoliation adsorption process.

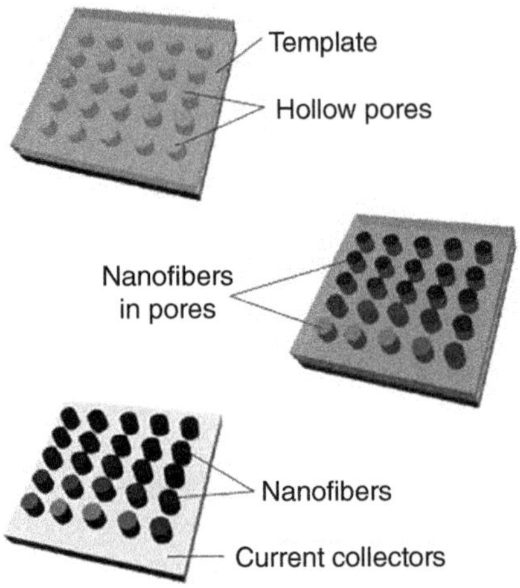

Figure 6.3 Synthesis of nanocomposites through the template method.

filler in an aqueous solution or gel containing the polymer and the filler building blocks. The polymer functions as a nucleating agent and upholds the growth of the inorganic filler crystals. As those filler crystals nurture, the polymer is ensnared within the layers and thus forms the nanocomposite. It is mainly used for the synthesis of double-layer hydroxide-based nanocomposite and is much less developed for the synthesis of layered silicates [14]. This is because of the high temperature used during synthesis, that degrades the polymer and the resulting aggregation tendency of the growing inorganic crystals (Figure 6.3).

6.4 Characterization Techniques for Nanocomposite

The structure of nanocomposites generally comprises the matrix material containing the nanosized strengthening components in the form of particles, whiskers, fibers, nanotubes, etc. Different methods have employed various equipment and techniques for the characterization of nanocomposites, including AFM, STM, FTIR, XPS, NMR, DSC, SEM/TEM, etc. For example, AFM is a powerful tool to study the surface property even down to the nanometer scale. Simultaneous small-angle X-ray scattering (SAXS) and XRD studies have been lately employed for the quantitative characterization of nanostructures and crystallite structures in some nanocomposites [15]. Moreover, theoretical calculations and simulations have been worked out to predict strength properties, including stress–strain curves of the nanocomposites [16]. Nanocomposites can dramatically improve properties like mechanical properties including strength, modulus, and dimensional stability, electrical conductivity: decreased gas, water, and hydrocarbon permeability, flame retardancy, thermal stability, and chemical resistance [17].

6.5 Applications of Nanocomposite Materials

Nanocomposite materials have potential applications in a wide range of fields which include information and communication technologies, healthcare, water decontamination, polymer electrolyte, agriculture, packaging, etc.

6.5.1 Automotive Industry

The use of polymer nanocomposites in automotive industries has tremendously reduced the weight and processing cost. Clay incorporated tires have excellent mechanical strength compared to ordinary tires, as well as improved gas barrier performance and flame retardance for tubeless tire applications [18]. Nanocomposites are also used in the manufacturing of car parts such as handles, rearview mirrors, timing belts, components of the gas tank, engine cover, bumper, etc.

6.5.2 Packaging Industry

By making nanocomposites, barrier properties of the polymers used in the packaging industry have been improved. Further, the polymer nanocomposite formation increases the thermal stability and flexibility and hence is extensively used in the food and beverage packaging industries [19]. Due to excellent solvent barrier properties, polymer nanocomposites have been used in chemical protective and surgical gloves to protect against chemical warfare agents and for avoiding contamination from medicine.

6.5.3 Catalysis

The high surface area of the nanocomposites is finding applications in catalysis fields to enhance the efficiency of the catalyst. Due to their high surface area and synergetic properties, nanocomposites enhance the rate of reaction and also show significant advantages such as activity, selectivity, recyclability, and lifetime in chemical transformations and electrocatalysis [20].

6.5.4 Solid Polymer Electrolyte

Polymer nanocomposites are having immense applications in polymer electrolytes. The incorporation of nanomaterials in polymers improves the conductivity of the polymer electrolyte by reducing the crystallinity of polymer and by improving the mechanical and interfacial properties, which are essential for electrolyte applications [21].

6.5.5 Water Treatment Applications

Nanocomposites have been used in the degradation of pollutants, dyes, and other organic contaminants discharged into the water by industries such as rubber vulcanization and textile industries. Further, nanocomposites are also used in the killing of microorganisms present in the contaminated water [22].

6.5.6 Aircrafts

Researchers have made relatively awesome discoveries on nanocomposites over the last decade, ever since the pioneering work on nanoclays by the company Toyota. The dispersion of the silicate nanolayers with its high aspect ratio, large surface area, and high stiffness within a polymer matrix results in significant improvement of the properties of polymeric materials, including mechanical properties, barrier properties, resistance to solvent swelling, ablation performance, thermal stability, fire retardancy, controlled release of drugs, anisotropic electrical conductivity, and photoactivity [23].

6.5.7 Electronics

Conductive nanocomposites are capable of conducting electric current well owing to the electric charges in their structure. Polycarbonates which is an insulator can be made conductive polycarbonates, the inexpensive plastics known for their excellent optical and mechanical properties, could in the future find applications into newer and more important horizons. Polycarbonates are tagged as poor electrical conductors, but a research team from the University of Houston (UH) has altered this property by adding carbon nanotubes to them thereby resulting in highly conductive nanocomposites. The team has come up with a strategy to achieve higher conductivities using carbon nanotubes in plastic hosts than what has been currently achieved. By combining nanotubes with polycarbonates, the team was able to reach a milestone of creating nanocomposites with ultra-high conductive properties. Dr. Shay Curran, an Associate Professor of Physics at UH, the United States, demonstrated ultra-high electrical conductive properties in these plastics by mixing them with just the right amount and type of carbon nanotubes [24]. As a result, the inexpensive plastic used to make optical disks will feature in high-end military aircraft to shield them against the build-up of electrical charges and pulses which can lead to significant failures.

6.5.8 Environmental Protection

Water-laden atmospheres have long been regarded as one of the most damaging environments, which polymeric materials can encounter. Thus, the capability to reduce the extent to which water is absorbed can be a major benefit. Available data indicate that a significant reduction of water absorption in a polymer could be achieved by nanoclay incorporation. Similar effects could also be achieved with polyamide-based nanocomposites. Specifically, increasing the aspect ratio diminishes substantially the amount of water absorbed, thus indicating the beneficial effects likely from nanoparticle incorporation compared to microparticle loading. Hydrophobicity enhancement would promote both improved nanocomposite properties and diminish the extent to which water would be transmitted through to an underlying substrate [25]. Thus, applications in which contact with water or moist environments is likely could benefit from materials incorporating nanoclay particles.

Nanocomposite denture teeth wear resistance is the most important physical property of denture teeth [26]. Porcelain denture teeth are most wear-resistant, but they are brittle, lack bonding to the denture base, and are difficult to polish. Acrylic resin denture teeth are easier to recontour but undergo excessive wear. Nanocomposite denture teeth comprise PMMA and uniformly dispersed nano-sized filler particles. Their advantages are highly polishable, stain and impact resistant material, lively surface structure, superior surface hardness, and wear resistance.

References

1 Thomas, S.P., Stephen, R., Bandyopadhyay, S., and Thomas, S. (2007). Polymer nanocomposites: Preparation properties and Applications. In: *Nanocomposites and Polymers with Analytical Methods*, vol. 2 (ed. J. Cuppoletti), 49–56. ISBN: 978-953-307-352-1.

2 Lateef, A. and Nazir, R. (2017). Metal nanocomposites: synthesis, sharacterization and their applications. In: *Science and applications of Tailored Nanostructures*, 239–256. One Centre Press.

3 Rao, C.N.R., Ramakrishna Matte, H.S.S., Voggua, R., and Govindaraj, A. (2012). Recent progress in the synthesis of inorganic nanoparticles. *Dalton Trans.* 41: 5089–5120.

4 Pedro Henrique Cury, C., Kestur Gundappa, S., and Fernando, W. (2009). Nanocomposites: synthesis, structure, properties, and new application apportunities. *Mater. Res.* 12: 1–39.

5 Tai, H., Jiang, Y., Xie, G. et al. (2008). Influence of polymerization temperature on NH_3 response of PANI/TiO_2 thin film gas sensor. *Sens. Actuators, B* 129: 319–326.

6 Liu, F.J., Huang, L.M., Wen, T.C. et al. (2007). Interfacial synthesis of platinum loaded polyaniline wires in poly(styrene sulfonic acid). *Mater. Lett.* 61: 4400–4405.

7 Pandya S (2015) *Nanocomposites and its Application-Review* https://doi.org/10.13140/RG.2.1.2798.9840.

8 Dutta, A., Das, D., Grilli, M.L. et al. (2003). Preparation of sol-gel nanocomposites containing copper oxide and their gas sensing properties. *J. Sol–Gel Sci. Technol.* 26: 1085–1094.

9 Parashar, P. (2015). Preparation and characterization of subsurface silver particulate films on polymer blends of polystyrene/poly(2-vinyl pyridine)/poly(vinylpyrrolidone)/ poly(4-vinyl pyridine). *Soft Nanosci. Lett.* 5: 3–11.

10 Oliveira, M. and Machado, A.V. (2013). Preparation of polymer-based nanocomposites by different routes, pp. 1–22.

11 Picu, R.V. and Ozmusul, M.S. (2002). Enhanced Structure at the Interface between the Polymer Matrix and Spherical Nanoparticles in Polymer-Based Nanocomposites. *Tech. Proc. 2002 Inter. Conf. Comput. Nanosci. Nanotechnol.* 2: 399–402.

12 Ajayan, P.M. (2004). Bulk metal and ceramics nanocomposites. In: *Nanocomposite Science and Technology* (ed. P.M. Ajayan, L.S. Schadler and P.V. Braun), 1–75. Hoboken: Wiley-VCH.

13 Fornes, T.D. and Paul, D.R. (2003). Modeling properties of nylon 6/clay nanocomposites using composite theories. *Polymer* 44: 4993–5013.

14 Brune, D.A. and Bicerano, J. (2002). Micromechanics of nanocomposites: comparison of tensile and compressive elastic moduli, and prediction of effects of incomplete exfoliation and imperfect alignment on modulus. *Polymer* 43: 369–387.

15 Mittal, V. (2010). *Optimization of Polymer Nanocomposite Properties*, 1–19. Weinheim: Wiley-VCH.

16 Liu, T.X., Phang, I.Y., Shen, L. et al. (2004). Morphology and mechanical properties of multiwalled carbon nanotubes reinforced nylon-6 composites. *Macromolecules* 37: 7214–7222.

17 Pandey, P.P. (2020). Preparation and characterization of polymer nanocomposites. *Soft Nanosci. Lett.* 10: 1–15.

18 Ray, S.S. and Okamoto, M. (2003). Polymer/layered silicate nanocomposites: a review from preparation to processing. *Prog. Polym. Sci.* 28: 1539–1641.

19 Sambarkar, P.P., Patwekar, S.L., and Dudhgaonkar, B.M. (2012). Polymer nanocomposites: an overview. *Inter. J. Pharm. Pharm. Sci.* 4: 60–65.

20 Shi, H., Chen, J., Li, G. et al. (2013). Synthesis and characterization of novel plasmonic Ag/AgX-CNTs (X = Cl, Br, I) nanocomposite photocatalysts and synergetic degradation of an organic pollutant under visible light. *ACS Appl. Mater. Interfaces* 5: 6959–6967.

21 Armstrong, G. (2015). An introduction to polymer nanocomposites. *Eur. J. Phys.* 36: 063001.

22 Frimpong, R.A., Fraser, S., and Hilt, J.Z. (2007). Synthesis and temperature response analysis of magnetic-hydrogel nanocomposites. *J. Biomed. Mater. Res. A* 80: 1–6.

23 Charles, C.O. (2014). The benefits and applications of nanocomposites. *Inter. J. Adv. Eng. Technol.* 5: 12–18.

24 Kessler, M. and Goertzen, W. (2009). *Polymer Nanocomposites for Infrastructure Rehabilitation. Nanotechnology in Construction 3*, 241–250. Berlin, Heidelberg: Springer.

25 Okpala, C. (2013). Nanocomposites – an overview. *Inter. J. Eng. Res. Dev.* 8: 17–23.

26 Lyapina, M., Cekova, M., Krasteva, A. et al. (2016). Physical properties of nanocomposites in relation to their advantages. *J. IMAB* 22: 1056–1062.

7

Polymer Nanocomposites in Energy Storage System

7.1 Introduction

Energy storage systems provide a wide array of technological approaches to managing our power supply to create a more resilient energy infrastructure and bring cost savings to utilities and consumers. Energy storage systems are essential to the operation of power systems (https://energystorage.org/why-energy-storage/technologies). They ensure continuity of energy supply and improve the reliability of the system. Energy storage systems can be in many forms and sizes. The size, cost, and scalability of an energy storage system highly depend on the form of the stored energy. Energy can be stored as potential, kinetic, chemical, electromagnetic (EM), thermal, etc. Some energy storage forms are better suited for small-scale systems, and some are used only for large-scale storage systems [1].

7.2 Batteries

The range of electrochemical storage solutions include advanced chemistry batteries, flow batteries, and capacitors. For example, lithium-ion batteries have received a lot of thrust for their rapidly declining costs due to the growing popularity of electric vehicles. In a different type of battery, flow battery, energy is stored and is provided by two chemicals that are dissolved in liquids and stored in tanks [2]. These are well suited for longer-duration storage.

7.3 Thermal

Heat and cold are captured to create energy on-demand or offset energy needs. For example, molten salt stores solar-generated heat for use when there is no sunlight. Ice storage in buildings reduces the need to run compressors while still providing air-conditioning for several hours. Other systems use chilled water and dispatchable water heaters (https://www.nyserda.ny.gov/All-Programs/Programs/Energy-Storage/Energy Storage-for-Your-Business/Types-of-Energy-Storage). In all

Polymer Nanocomposites for Energy Applications, First Edition.
T. Daniel Thangadurai, Manjubaashini Nandhakumar, Sabu Thomas, and Ange Nzihou.

cases, excess energy charges the storage system (heat the molten salts, freeze the water, etc.) and is later released as needed.

7.4 Mechanical Storage

This is the innovative technology by harnesses kinetic or gravitational energy to store electricity. For example, flywheels store energy in a rapidly spinning mechanical rotor and are capable of absorbing and releasing high power for typically 15 minutes or less, although longer-duration systems are being developed. These systems can balance fluctuations in electricity supply and demand where they respond to a control signal adjusted every few seconds (https://www.eesi.org/papers/view/energy-storage-2019). They also recapture braking energy from electric trains in some installations or provide short-term power until a backup generation comes online during a grid outage, such as in a critical manufacturing process where the product would be lost by a momentary electric interruption.

7.5 Hydrogen

The excess electricity generation can be converted into hydrogen via electrolysis and stored. Hydrogen storage is a key enabling technology for the advancement of hydrogen and fuel cell technologies in applications, including stationary power, portable power, and transportation. Hydrogen has the highest energy per mass of any fuel; however, its low ambient temperature density results in a low energy per unit volume, therefore requiring the development of advanced storage methods that have the potential for higher energy density [3].

7.6 Pumped Hydropower

Pumped hydroelectric facilities are the most common form of energy storage on the grid and account for over 95% of the storage in use today. During off-peak hours, turbines pump water to an elevated reservoir using excess electricity. When electricity demand is high, the reservoir opens to allow the retained water to flow through turbines and produce electricity [4]. Siting these systems can be difficult because of the terrain needed and the large footprint.

7.7 Flywheels

The functionality of a flywheel system is quite simple, and you may have even played with it when you were a kid. A flywheel is a disk with a certain amount of mass that spins, holding kinetic energy. Modern high-tech flywheels are built with the disk attached to a rotor in an upright position to prevent gravity influence [5]. They are charged by a simple electric motor that simultaneously acts as a generator in the process of discharging. When dealing with efficiency, however, it gets more complicated, as stated by the rules of physics, they will eventually have to deal with friction during operation. Therefore, the challenge to increase that efficiency is to minimize friction.

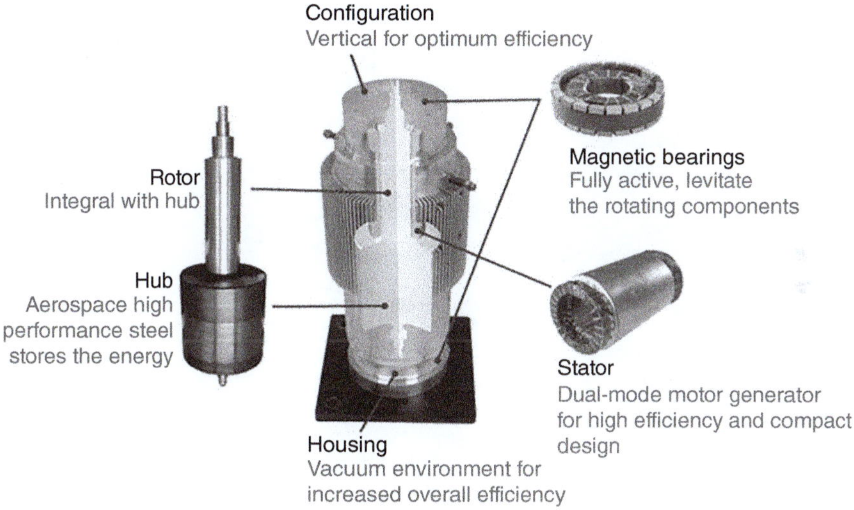

Figure 7.1 Pictorial representation of the inside view of a flywheel.

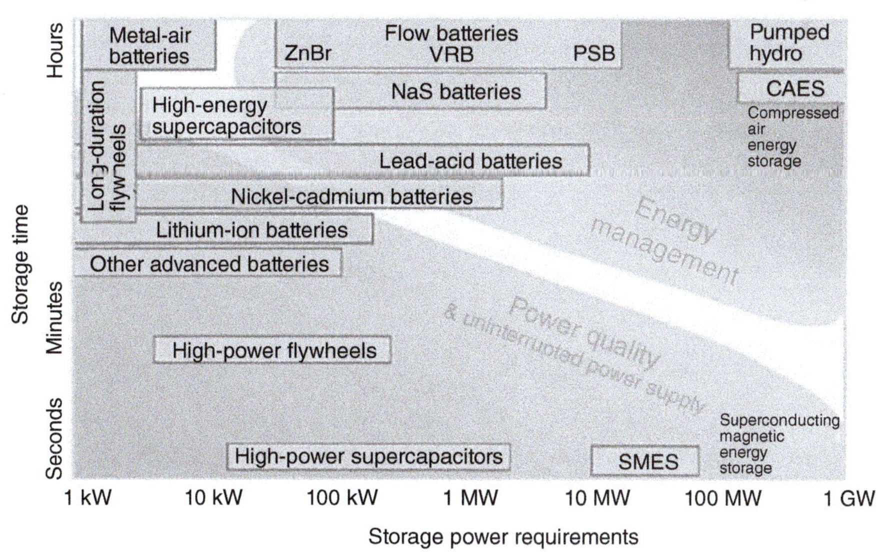

Figure 7.2 Chart representation of different types of energy storage systems.

This is mainly accomplished by two measures: the first one is to let the disk spin in a vacuum, so there will be no air friction, and the second one is to bear the spinning rotor on permanent and electromagnetic bearings so it floats. The spinning speed for a modern single flywheel reaches up to 16 000 rpm and offers a capacity of up to 25 kWh, which can be absorbed and injected almost instantly [6] (Figures 7.1–7.3, Table 7.1).

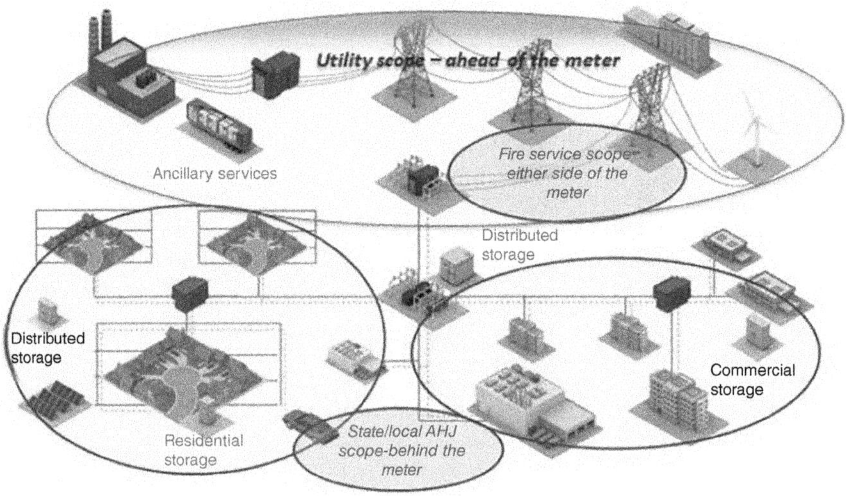

Figure 7.3 Applications of energy storage systems.

Table 7.1 The strengths and weaknesses of different energy storages.

Storage system	Max power rating (MW)	Discharge time	Max cycles or lifetime	Energy density (watt-hour per liter)	Efficiency
Pumped hydro	3 000	4–16 h	30–60 yr	0.2–2	70–85%
Compressed air	1000	2–30 h	20–40 yr	2–6	40–70%
Molten salt (thermal)	150	Hours	30 yr	70–210	80–90%
Li-ion battery	100	1 min–8 h	1,000–10 000 cycles	200–400	85–95%
Lead-acid battery	100	1 min–8 h	6–40 yr	50–80	80–90%
Flow battery	100	Hours	12 000–14 000 cycles	20–70	60–85%
Hydrogen	100	Minutes–week	5–30 years	600 (at 200 bar)	25–45%
Flywheel	20	Seconds–minutes	20 000–100 000 cycles	20–80	70–95%

7.8 Role of Polymer Nanocomposites in Energy Storage Applications

Polymer nanocomposites (PNCs) have attracted significant research interests due to their promising potential for versatile applications ranging from environmental remediation, energy storage, electromagnetic absorption, sensing, and actuation, transportation, and safety, defense system, information industry, to novel catalysts, etc. [7].

The incorporation of nanofillers into polymers can give rise to distinct properties from the synergistic effects of each component in the PNCs. PNCs can be synthesized by various methods that can be generally divided into four categories: melt intercalation, template synthesis, exfoliation adsorption, and *in situ* polymerization intercalation. The aspect ratio of the nanofiller, the nature of the nanofiller–matrix interface, and the percolation threshold are key aspects in the PNCs design and behavior [8]. The nanofillers are capable of imparting both structural and functional reinforcement to the PNCs. For example, significant enhancements in mechanical strength have been witnessed in the PNCs by the addition of nanofillers with an intrinsic high modulus. The uniform dispersion of the nanofillers and the strong interfacial interactions between the matrix and the nanofillers are mainly responsible for the significantly enhanced mechanical properties. The covalent bonding between the filler and matrix is considered the most effective mean to increase the interfacial shear stress for improving stress transfer. Moreover, attributed to the excellent thermal/electrical conductivity in the nanofillers (i.e. carbon nanostructures), thermally and electrically conductive polymers have been prepared with extended applications for sensors, EMI shielding materials, electrodes, and stimuli-responsive materials [9]. The bulk conductivity of the typical engineering polymers is reported to increase by 10^{10} to 10^{14} Sm^{-1} at very low filler concentrations (Figure 7.4).

The development of sustainable and renewable energy storage resources from renewable energy reservoirs such as wind and solar is in urgent demand due to the ever-increasing energy crisis arising from the depletion of conventional fossil

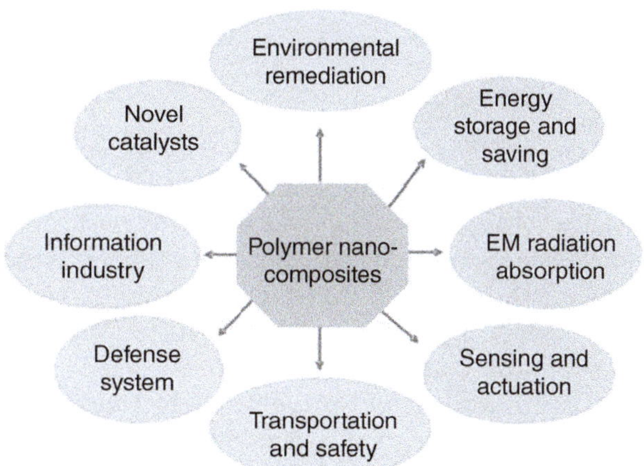

Figure 7.4 Different applications of polymer nanocomposites.

fuels [10]. Due to the intermittent nature of these renewable energy resources, reliable energy storage systems are in urgent need to store and stably output energy. Electrochemical energy storage systems (EESSs) that include electrochemical capacitors (ECs), batteries, and fuel cells have attracted remarkable research interest among various energy storage systems, for example, pumped hydro storage, thermal energy storage, compressed air energy storage, and flywheel energy storage [11].

7.9 Properties of Polymer Nanocomposites

PNC materials exhibit advantageous physical properties. These include improved thermal stability, flame retardancy, enhanced barrier properties, and improved mechanical properties. Commercially, the developments in PNCs are resulting in their increased usage [12]. The combination of improved properties, including weight reduction and low cost in the final product, provides important commercial applications in the transport sector.

7.9.1 Physical Properties

The physical properties such as density, crystallinity and solubility of the polymer are generally influenced marginally by the incorporation of nanomaterials, although in many cases these properties remain almost the same as those of pristine polymers. The lightweight characteristic of polymer is also found to be retained, as the increment of density for PNCs is negligible in almost all cases. This is one of the greatest achievements in PNCs by the incorporation of nanomaterials. The lightweight nature of such materials increases sophistication and aids in easy handling, as well as also reduces the cost per unit volume [13]. However, the effect of nanomaterials on the crystallinity of PNCs is inconsistent, though the effect is less significant in most cases. Again the solubility of PNC is generally difficult because of the interactions with nanomaterials, most of which are insoluble.

7.9.2 Rheological Properties

The study of the rheological behavior of PNCs is very important for their processing, interpretation of properties, and industrial applications. Rheological properties are related to the deformation and fluidity of such materials under applied external force. The variation in shear viscosity under different temperatures and stresses is studied. The trend of variation in viscosity helps to determine the dispersion quality of a nanomaterial within the polymer matrix [14]. In most cases, viscosity increases relative to the number of nanomaterials loading up to a certain level. The variations of the storage modulus correspond to the elastic response of the deformation, and the loss modulus corresponds to the plastic response of the deformation, and the loss factor corresponds to the ratio of the loss modulus to the storage modulus with the frequency of PNCs that indicate the strength and elasticity of the materials [15]. The higher the storage modulus value, the greater is the interaction between the nanomaterials and polymer matrices, and hence greater is the strength of the nanocomposite.

7.9.3 Mechanical Properties

Mechanical properties, especially the strength of a pristine polymer, increase significantly on the incorporation of suitable nanomaterials even at very low dose levels (<5 wt%) [16]. However, it is observed that the tensile strength usually increased along with the increasing loading of nanomaterial up to a certain dose level, beyond which it may deteriorate due to the agglomeration of nanomaterial in the polymer matrix. The higher the interaction of the nanomaterials with the polymeric matrix, the greater the increment of the mechanical strength. Thus, exfoliated nanocomposites show higher increments of strength compared to the intercalated ones, and the strength in both cases is always higher than conventional macro composites and pristine systems [17]. The higher surface activity of nanomaterials facilitates greater interactions, and thus the load-bearing capacity of the PNCs is enhanced due to the effective transfer of the load from mechanically weak polymer matrices to strong and rigid nanomaterials. Again, many reports show that flexibility or elongation at break decreases with the formation of nanocomposites, as compared to the pristine polymer matrices due to restriction of the mobility of polymeric chains by the rigid nanomaterials. However, it is noteworthy to mention that the choice of properly functionalized nanomaterials with suitable polymer matrices can also enhance the elongation at breaks in addition to an increase in strength, and therefore the resultant PNCs become a very tough material [18]. For example, *in situ* fabricated functionalized (reduced graphene oxide) RGO/hyperbranched polyurethane nanocomposites exhibited excellent toughness by increasing both tensile strength and elongation at break.

7.9.4 Thermal Properties

One of the most noteworthy shortcomings of polymers is their thermal stability at elevated temperatures, except for specially designed highly thermostable polymers. This poor thermal stability of pristine polymers can be improved to a considerable extent with the incorporation of suitable nanomaterials into them [19]. The incorporation of many nanomaterials augments thermal stability because nanomaterials generally serve as heat insulators, act as the mass transport barrier for volatiles generated during the decomposition process, and alter the pathway of their decomposition. This also increases glass transition temperature (T_g) due to the restriction of polymeric chain mobility by rigid nanomaterials. The formation of PNC may also enhance the crystalline melting temperature (T_m) as well as the total enthalpy of crystallization. In such cases, nanomaterials act as nucleating agents for the enhancement of crystallization of the crystalline polymer matrices [20].

7.9.5 Barrier and Chemical Resistance

Barrier property and chemical resistance of the pristine polymers also significantly increase upon the formation of suitable polymer nanocomposites. The nanomaterials with high aspect ratios provide very high surface areas, which hamper the diffusion pathways of different penetrating molecules. The rate of diffusion of a penetrating molecule further depends on the degree of dispersion of the nanomaterials [21]. The high aspect ratio provided by two-dimensional nanomaterials generally diminishes the gas permeability very effectively; thus, the dramatic improvement of barrier properties can be explained by using the concept of a tortuous or zigzag pathway. The tortuosity factor is taken as the ratio of the actual distance that the penetrants must travel to the shortest distance that they would travel in the absence of barriers. As the diffusion becomes difficult, so the chemical vapors or ions generated from different chemical environments also faced difficulties in their interaction with PNCs, and thereby chemical resistance is also enhanced [22].

7.9.6 Flame Retardancy

Nanomaterials are incorporated into polymeric matrices to increase the ignition temperature and to slow down decomposition. The flame-retardant PNCs are those that retard, inhibit, or suppress the heating, degradation, or decomposition, ignition of flammable gases, and combustion with heat generation during the burning process [23]. The formation of PNCs with clay, silica, carbon nanotube (CNT), carbon nanofiber, graphene, RGO, etc., increases the flame-retardant behavior of pristine polymers. These nanomaterials are collapsed or form nonflammable char and thus reduce the heat release rate during the heating and burning process. The char also prevents the fire from spreading by holding the structural integrity of the nanocomposites. PNCs are thus considered some of the most promising developments in the area of flame retardancy, as they offer significant advantages over conventional formulations [24]. These include regulatory concerns about the human and environmental contamination caused by toxic flame-retardant chemicals, which evolve during their combustion, high levels of loading to obtain the desired level of such property, additional costs, processing difficulties, and the deterioration of other valuable properties, etc., associated with a conventional system.

7.9.7 Optical Properties

In many cases, the transparency of a transparent polymer remains intact in the formation of nanocomposites. Other interesting optical properties, such as fluorescence, luminescence, and nonlinearity, are also exhibited by PNCs depending on the nanomaterials used. These days, different nanomaterials are explored to obtain fascinating optical properties of PNCs which have immense industrial applicability [25].

7.9.8 Electrical Properties

A material that allows the flow of current is a good conductor, while the one that opposes a current is regarded as an insulator. Most of the polymers in their pristine forms behave as insulators, mainly due to their covalent nature and lack of electronic or ionic pathways, except conducting polymers. However, the incorporation of many conducting nanomaterials confers interesting electrical properties to PNCs [26]. For example, the incorporation of 2 wt% of functionalized RGO in hyperbranched polyurethane resulted in a 10^{10} times increase in conductivity to the pristine polyurethane.

7.9.9 Dielectric Properties

Dielectric PNCs created by incorporating inorganic nanoparticles into polymer matrices show promising energy storing properties, which would make for higher-performance capacitors. Recently, it has been explored that the energy storage capability of polyvinylidene fluoride (PVDF)-based composites with variety of nanoparticles including $BaTiO_3$, $Ba_{1-x}Ca_xTiO_3$ ($X = 0.3 \pm 0.05$) [BCT], $BaZr_{1-x}Ti_xO_3$ ($X = 0.2 \pm 0.05$) [BZT], and $Ba_{0.5}Sr_{0.5}TiO_3$ [BST]. Colloidal nanoparticles were initially prepared in high yield by a solvothermal method and used as fillers in the PVDF polymer matrix [27]. The unique dielectric polymer films show enhanced dielectric constants, enhanced electrical breakdown strengths, and enhanced dielectric energy densities. The matrices exhibit higher dielectric permittivities than those of the polymer alone as well as higher breakdown strengths than the ceramic materials. These nanocomposite materials also allow for more versatility than traditional ceramics used in capacitors as they are lighter in weight than ceramics alone and their flexibility allows them to be molded within electronics.

7.9.10 Biological Properties

The most important biological properties such as biocompatibility, antimicrobial activity, and biodegradability of PNCs are necessary to study [28]. The preliminary *in vitro* biocompatibility of polymeric materials is assayed by calculating the cell survival rate in the presence of polymeric materials to be tested. Many PNCs are found to gain excellent biocompatibility by tailoring their architecture and functionalization. Polyurethane, polyamide, and polyester nanocomposites are the most explored biocompatible materials. Although many metals and metal oxide nanoparticles showed severe toxicity toward many cell lines as tested *in vitro*, greener route-mediated synthesized such nanoparticles exhibited promising cell compatibility. Furthermore, the toxicity of nanomaterials such as graphene and CNT is dramatically diminished after the functionalization and incorporation of these materials into polymer matrices [29]. Additionally, the adhesion, growth, and proliferation of cells of PNCs are found to be favorable for such biocompatible materials. Biomaterials are materials of synthetic or natural origin in contact with cells or biological fluids and projected to use for diagnostic, therapeutic, prosthetic, or storage applications without adversely affecting the living organisms. Various nanocomposites of polyurethane, polyglycolic acid, polylactic acid, polycaprolactone, chitosan-based polysaccharides, polyphosphazene, polyfumarate, polyorthoester, poly(glycerol sebacate), polypyrrole, polyarylates, poly(ether ester amide), poly(amidoamine), etc., are used for anchoring different types of cells [30].

References

1 De Andrade, L. and De Leao, T.P. (2012). A brief history of direct current in electrical power systems. *History of Electro-technology Conference (HISTELCON)* IEEE 1-6.

2 Masters, G.M. (2004). *Renewable and Efficient Electric Power Systems*. NJ: Wiley.

3 Sawin, J.L. and Sverrisson, F. (2014). Renewables: 2014 global status report. REN21 Secretariat, Paris, France.

4 Krause, P.C., Wasynczuk, O., and Sudhoff, S.D. (2002). *Analysis of Electric Machinery and Drive Systems*. NY: Wiley-IEEE Press.

5 Lienert, A. (2012). Mercedes-Benz fuel-cell car ready for market in 2014. Insideline.com.

6 Watts and Robert, G. (2002). *Innovative Energy Strategies for CO_2 Stabilization*. Cambridge University Press.

7 Mittal, V. and Aleman, C. (2015). *Synthesis Techniques for Polymer Nanocomposites*. Weinheim, Germany: Wiley-VCH.

8 Terrones, M., Martín, O., González, M. et al. (2011). Interphases in graphene polymer - based nanocomposites: achievements and challenges. *Adv. Mater.* 23: 5302–5310.

9 Mutiso, R.M. and Winey, K.I. (2015). Electrical properties of polymer nanocomposites containing rod-like nanofillers. *Prog. Polym. Sci.* 40: 63–84.

10 Yang, C., Wei, H., Guan, L. et al. (2015). Polymer nanocomposites for energy storage, energy-saving, and anticorrosion. *J. Mater. Chem. A* 3: 14929–14941.

11 Cao, H., Wang, X., Gu, H. et al. (2015). Carbon coated manganese monoxide octahedron negative-electrode for lithium-ion batteries with enhanced performance. *RSC Adv.* 5: 34566–34571.

12 Kasiri, A. and Brabazon, D. (2021). Materials used within polymer matrix composites (PMCs) and PCM production via additive manufacturing. *Ref. Module Mater. Sci. Mater. Eng.* https://doi.org/10.1016/B978-0-12-803581-8.11913-7.

13 Karak, N. (2019). *Fundamentals of Nanomaterials and Polymer Nanocomposites. Nanomaterials and Polymer Nanocomposites*, 1e, 1–45.

14 Duarah, R. and Karak, N. (2017). Facile and ultrafast green approach to synthesize bio-based luminescent reduced carbon nanodot: an efficient photocatalyst. *ACS Sustainable Chem. Eng.* 5: 9454–9466.

15 De, B., Gupta, K., and Karak, N. (2015). Biocide immobilized OMMT-carbon dot reduced nanohybrid/hyperbranched epoxy nanocomposites: mechanical, thermal, antimicrobial, and optical properties. *Mater. Sci. Eng. C* 56: 74–83.

16 Lagaly, G., Ogawa, M., and Dekany, I. (2006). Clay mineral organic interactions. In: *Handbook of Clay Science* (ed. F. Bergaya, B.K.G. Theng and G. Lagaly), 309–377. Amsterdam: Elsevier Ltd.

17 Thakur, S. and Karak, N. (2015). Alternative methods and nature-based reagents for the reduction of graphene oxide - a review. *Carbon* 94: 224–242.

18 Anandhan, S. and Bandyopadhyay, S. (2011). Polymer nanocomposites: from synthesis to applications. In: *Nano Composites and Polymers with Analytical Methods* (ed. J. Cuppoletti), 3–28. Croatia: In Tech Publishers.

19 Fahlman, B. (2011). *Materials Chemistry*. Netherlands: Springer https://doi.org/10.1007/978-94-007-0693-4.

20 Gilman, J.W. (1999). Flammability and thermal stability studies of polymer layered silicate (clay) nanocomposites. *Appl. Clay Sci.* 15: 31–49.

21 Zeng, Q.H., Yu, A., Lu, M., and Paul, D.R. (2005). Clay-based polymer nanocomposites: research and commercial development. *J. Nanosci. Nanotechnol.* 5: 1574–1592.

22 Zou, H., Wu, S., and Shen, J. (2008). Polymer/silica nanocomposites: preparation, characterization, properties, and applications. *Chem. Rev.* 108: 3893–3957.

23 Tran, Q.H., Nguyen, V.Q., and Le, A.T. (2013). Silver nanoparticles: synthesis, properties, toxicology, applications, and perspectives. *Adv. Nat. Sci.: Nanosci. Nanotechnol.* 4: 033001.

24 Vaia, R.A. and Giannelis, E.P. (1997). Lattice of polymer melt intercalation in organically modified layered silicates. *Macromolecules* 30: 7990–7999.

25 Ray, S.S. and Bousmina, M. (2006). *Polymer Nanocomposites and Their Applications*. New York: American Scientific Publishers.

26 Paul, D.R. and Robeson, L.M. (2008). Polymer nanotechnology: nanocomposites. *Polymer* 49: 3187–3204.

27 Riggsa, B.C., Adireddy, S., Rehm, C.H. et al. (2015). Polymer Nanocomposites for Energy Storage Applications. *Mater. Today: Proc.* 2: 3853–3863.

28 Kalita, S.J. (2008). *Nanostructured Biomaterials*. New York: Springer.

29 Gogoi, S., Maji, S., Mishra, D. et al. (2017). Nano-bio engineered carbon dot peptide-functionalized water-dispersible hyperbranched polyurethane for bone tissue regeneration. *Macromol. Biosci.* 17: 1600271.

30 Duarah, R., Singh, Y., Gupta, P. et al. (2016). High performance bio-based hyperbranched polyurethane/carbon dot-silver nanocomposite: a rapid self-expandable stent. *Biofabrication* 8: 045013.

8

Polymer Nanocomposites for Renewable Energy Storage System

8.1 Renewable Energy

Renewable energy resources (RESs) are forms of energy that are naturally replaced on our planet [1]. Examples of traditional renewable resources are hydropower and biomass (e.g. plant fuels such as wood traditionally have been used throughout history, mostly for heating). Modern renewable resources include wind, wave, tidal, solar, and geothermal.

Solar energy and wind energy have become important, and they started to be used as energy sources at considerable power levels. Besides, RES technologies allow energy generation at the locations at or near the energy consumed, and today this energy generation structure is called distributed generation. This causes changes in the conventional power network [2]. Variations in wind speed and solar irradiation lead to fluctuations in voltage and the frequency value of the power system, thus disturbs the power quality of the power system, especially in isolated networks. Energy storage systems with fast response and long life cycles are used to mitigate and delaminate these disturbances. An energy storage device is an apparatus used for storing electric energy when needed and releasing it when required [3]. As a measure to counter global warming, the role of energy storage device technology in fields such as renewable energy generation and hybrid automobile systems will become increasingly important. Grid-scale energy storage systems are introduced to improve power system stability [4].

8.2 Renewable Energy Storage

Energy storage systems with large capacity are used to shift generated energy from RESs which depend on meteorological conditions and are only accessible at certain periods to the desired periods [5]. Thus, more economical operating conditions can be achieved. Another significant advantage of these energy storage systems is their capability to increase the amount of energy exported to the grid during peak demand periods. Also, they have a great contribution to the fulfillment of the commitments that are given for the next day according to forecasting data [2]. Large energy storage systems such as 34 MW and 40 MWh have been installed for this purpose [6].

Polymer Nanocomposites for Energy Applications, First Edition.
T. Daniel Thangadurai, Manjubaashini Nandhakumar, Sabu Thomas, and Ange Nzihou.

Besides, generated energy from RESs can be stored at energy storage systems when the unit price of energy and energy demand is low, and then this stored energy can be exported to the grid when the unit price of energy and energy demand is high [7].

8.3 Polymers for Energy Storage

Polymers have wide applications in the areas of energy storage and conversion. Several recent advances in the control of the polymer molecular structure which allows the polymer properties to be more finely tuned have led to these advances and new applications [8]. Polymers for energy storage and conversion assimilate these advances in the form of a comprehensive text that includes the synthesis and properties of a large number of polymer systems for applications in areas such as lithium batteries, photovoltaics, and solar cells (Figure 8.1).

Nanocomposite polymer materials are commonly used in energy storage devices on account of their excellent dielectric performance [9]. Nevertheless, there is a long-term flaw that still exists between the dielectric constant and breakdown strength of nanocomposite (Figure 8.2).

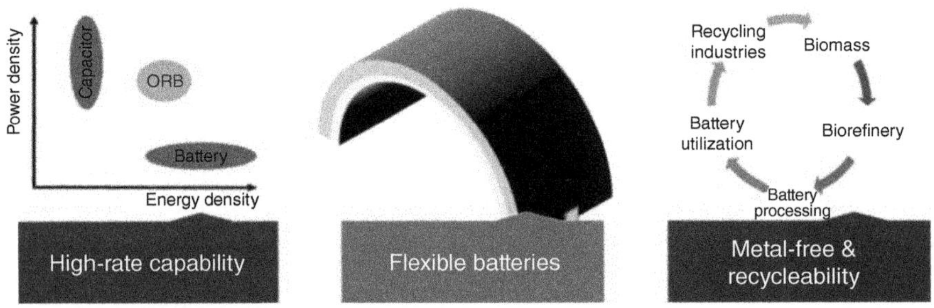

Figure 8.1 Advantages of polymer-based batteries.

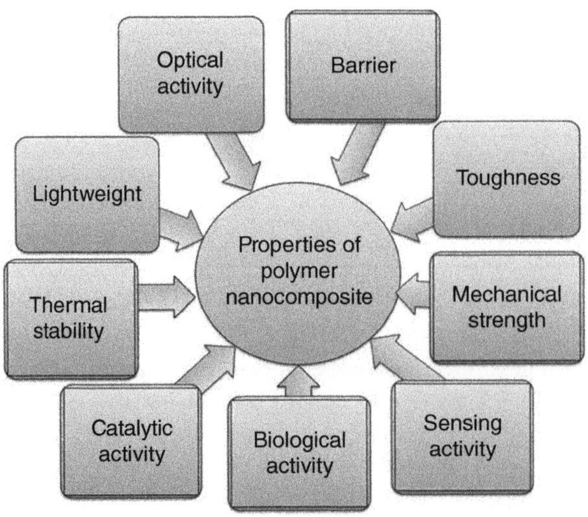

Figure 8.2 Significant properties of polymer nanocomposites.

Renewable sources like solar, wind, tidal, and bioenergy can satisfy the world's power requirements; also substitutes for petroleum-derived substances request a cause of carbon fragments [10]. The innovation of new, compact, and portable electronic devices has drawn much attention toward efficient energy storage. Since, more transportable portable devices such as smartphones, laptops, and smart health devices, have reduced suggestively in volume, and besides their abilities, storage potentials continue to raise intensely [11]. Therefore, the energy markets of these tools are abundant, which has led to an enormous improvement in the level of study toward high-performance applications [12].

8.4 Carbon-Based Storage Materials

Carbon-based materials and their composites hold encouraging employment in a broad array of fields, for example, energy storage devices, fuel cells, membranes sensors, actuators, and electromagnetic shielding. Carbon and its derivatives exhibit some remarkable features such as high conductivity, high surface area, excellent chemical endurance, and good mechanical durability [13]. On the other hand, characteristics such as docility, lower price, and high environmental resistance are some of the unique properties of conducting polymers (CPs). To enhance the properties and performance, polymeric electrode materials can be modified suitably by metal oxides and carbon materials resulting in a composite that helps in the collection and accumulation of charges due to the large surface area. The carbon–polymer nanocomposites (CPNCs) assist in overcoming the difficulties arising in achieving the high performance of polymeric compounds and delivering high-performance composites that can be used in electrochemical energy storage devices (Figure 8.3).

CPNCs have various applications in the energy accumulation, energy storage, packing, aerospace, and automotive areas. The important characteristics of these nanostructured substances are the comfort of processing, configuration adaptability, lightweight, and flexibility to requirements [14]. For energy storage, fuel cells and supercapacitors are supposed to be crucial components in updating the prospect of renewable energy schemes [15]. The demand for high energy and power density

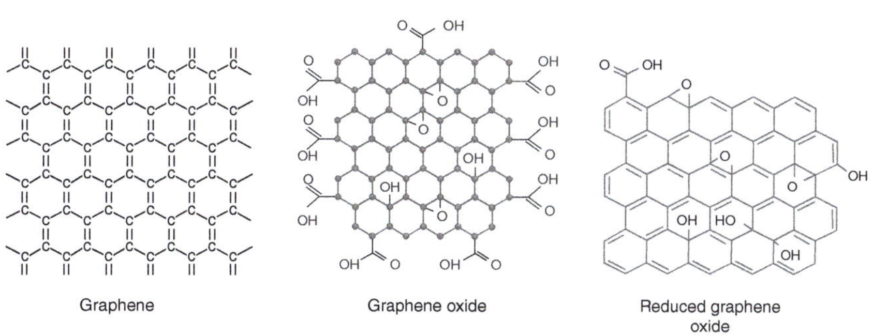

Figure 8.3 Chemical assembly of graphene, GO, and RGO.

devices at low cost leads to the discovery of novel nanocomposite materials for automotive and electric energy storage applications. Insulating polymers loaded by high-aspect-ratio conductive nanofillers, for example, carbon nanotube (CNT), as well as graphene nanoplatelets (GNPs), has been proved to be potential dielectric ingredients [16].

The energy storage methods require unique, paramount, and authentic approaches toward the storage of electric power by alternate renewable origins for assuring appropriate and dependable devices that can store a sufficient quantity of energy and later can be used for transport, electronic gadgets, electric-powered carriers, and distinct purposes. Electrochemical energy storage operations comprise supercapacitors, various kinds of batteries, and fuel cells. Energy exists in various forms in nature such as radiation, chemical, gravitational, potential, electricity, latent heat, and kinetic energy. Energy storage is the capture of produced energy so that it can be used later [17]. Battery and supercapacitors are the most commonly used energy storage devices, but the fuel cell does not store energy and instead it produces energy directly from the fuel and oxygen via electrochemical reactions in the cell (Figure 8.4).

Fuel cells are open systems, and fuels that provide energy are located outside, e.g. oxygen in the air and hydrogen outside the fuel storage tank which is supplied continuously to the reaction chamber (Figure 8.5, fuels are supplied from outside into the reaction chamber), while the battery is a closed system through the positive and negative electrode as the charge transfer intermediate [18]. Fuel cells have very high energy density even higher than batteries, but the lowest power density due to which they are not widely employed toward energy storage and renovation [19]. Fuel cells are still in the developing phase and in search of an effective application that can penetrate the energy market but still be competitive against batteries and supercapacitors.

Researchers prepared ferroelectric polymer nanocomposites using core–shell structured polymer@$BaTiO_3$ NPs as fillers. Three kinds of polymer@$BaTiO_3$ NPs

Figure 8.4 Types of electrochemical energy storage devices. Source: Maxwell Technologies, Inc./Wikimedia commons/CC by SA 3.0.

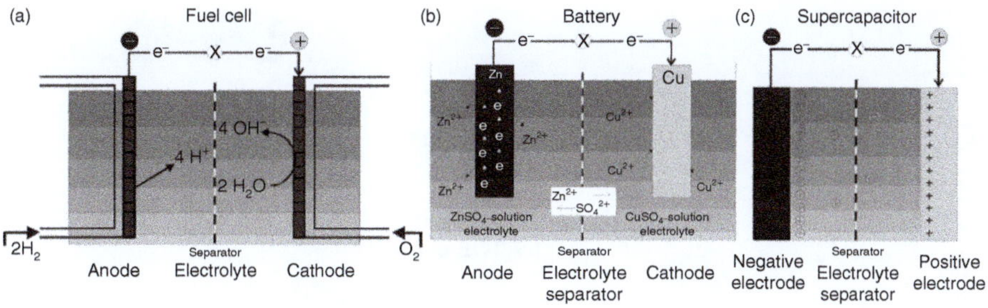

Figure 8.5 Representation of a fuel cell. (a) The constant source of ingredients and redox reactions in the cell, a battery, (b) the salient features of battery operation, and a supercapacitor, and (c) the energy storage at probe-conducting solution interface.

(i.e. PMMA@BT, PGMA@BT, and PHEMA@BT) with shells having different electrical properties were formulated by surface-initiated RAFT polymerization. The effects of the polymer shells on the crystal structure, morphology, breakdown strength, frequency-dependent dielectric properties, leakage currents, energy storage capability, and energy storage efficiency of the nanocomposites were investigated [20]. To get deep insights into the role of polymer shells in determining the electrical properties and energy storage of the nanocomposites, the polymers with the same molecular structure as the shells, for example, PMMA, PGMA, and PHEMA, were prepared by RAFT polymerization method, and their dielectric properties were also examined [21].

The core–shell nanoparticles with high dielectric constant shells can result in high electric displacement and thus high energy densities in the nanocomposites, but the high leakage currents cause high dielectric loss and low discharged energy densities in the nanocomposites [22]. Besides, the high dielectric constant polymer shells have high electrical conductivity, causing a significant decrease of breakdown strength and thus poor maximum-energy-storage capabilities. The nanocomposites filled with core–shell nanoparticles having low dielectric constant shell show low electric displacement and thus low energy densities, but the low leakage currents cause low dielectric loss and thus high discharged energy density [23]. Therefore, the optimum case is that the polymer shells have the combination of high dielectric constant and low leakage currents, which make the nanocomposites not only to have high discharged energy densities but also to have high energy efficiency. The cost associated with producing energy storage and conversion devices is driven by the relative abundance of materials, fabrication processes, and large energy costs of battery manufacturing and recycling.

The world's energy needs up to 2050 are depicted in Figure 8.6 and the forecast of the world's energy needs up to 2050 [25]. With the changing lifestyles of an increasing number of inhabitants, our energy rate demand will double from 14 TW (2010) to 28 TW (2050).

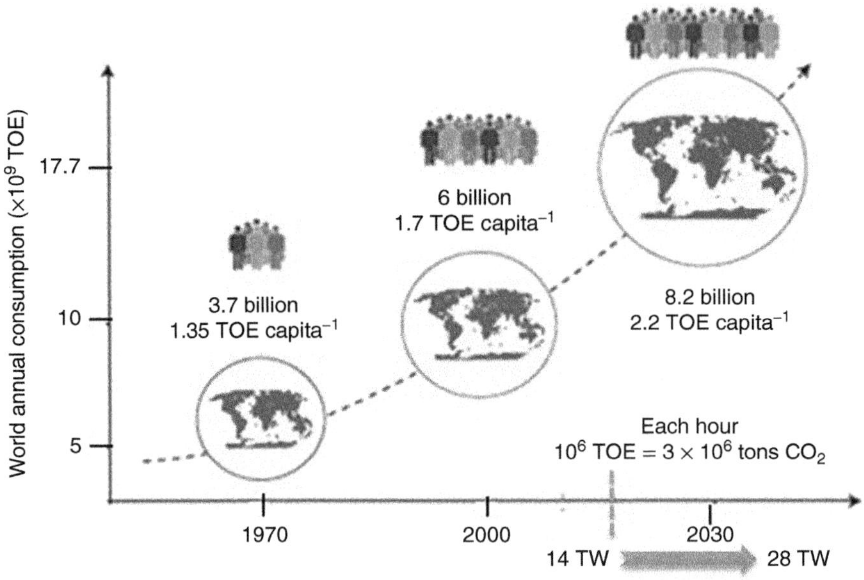

Figure 8.6 Graphical representation of estimation of the world's energy capacity for the future. Source: Larcher et al.[24]/Springer Nature.

8.5 Energy Storage Capability of Polymer Nanocomposites

Generally, the energy density (Ue) of dielectric materials is given by $Ue = \int E\, dD$, where E represents the applied electric field and D is the electric displacement. Thus, the energy storage capability of dielectric materials can be obtained from D–E loops by a modified Sawyer–Tower circuit. Figure 8.7 demonstrates the typical D–E loops of pristine polymer and *h*-DOPA@TiO_2 NWs/P(VDF-HFP) nanocomposites. The higher dielectric constant of TiO_2 NWs than the neat polymer gives rise to the increased electric displacement for the proposed nanocomposites as expected [26]. However, from those D–E loops, those fabricated nanocomposites possess higher remnant polarization than neat polymer due to the higher remnant polarization of the nanofillers than that of the poly(vinylidene fluoride-co-hexafluoropropylene) (P(VDF-HFP)). The high loadings of nanofillers usually bring about a remarkable reduction of discharged energy density because of the decreased integrated area of fatter D–E loops. Notably, attributing to adequate functionalization of nanowires, the loops of these thus-fabricated nanocomposites are depicted just as slim as that of pure polymer, especially at lower loadings of nanowires, indicating their high energy efficiencies (Figure 8.8).

Polymer nanocomposites are commonly used in energy storage devices on account of their excellent dielectric performance. Polyurea (PUA) is designed *in situ* to modify $BaTiO_3$ (BT) nanoparticles [27]. Based on the outstanding dispersity, promising compatibility, and exceptional insulating performance of

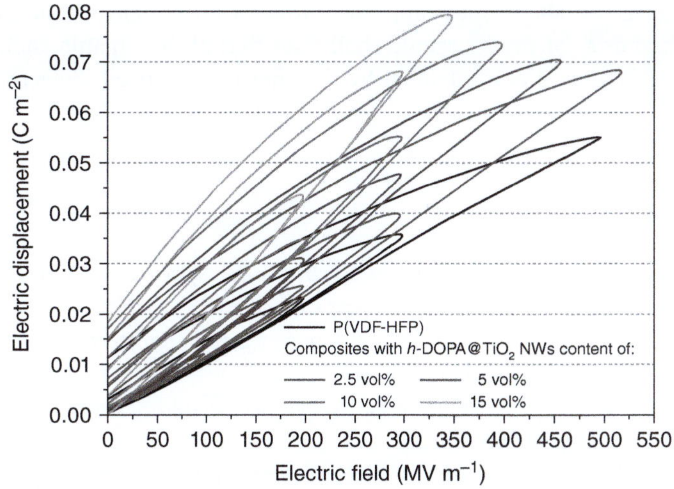

Figure 8.7 *D*–*E* loops under unipolar electric fields of 100 Hz for *h*-DOPA@TiO$_2$ NWs/P(VDF-HFP) nanocomposites with different filler concentrations.

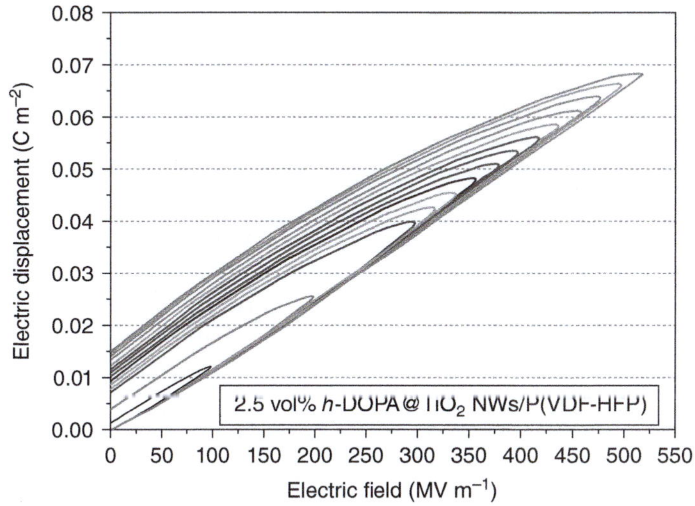

Figure 8.8 Pure P(VDF-HFP) and P(VDF-HFP)-based nanocomposites with 2.5 vol%.

BT@PUA particles, the P(VDF-CTFE)/BT@PUA nanocomposite displays improved breakdown strength and reliability. PVDF-based materials are commonly studied because of their high dielectric constant (~9), high breakdown strength as well as good stability, and are especially utilized in the energy storage and energy harvesting field [28]. BT nanoparticle attracts more attention to the preparation of the nanocomposite, because its preparation process is a mature technology and has been industrialization with relatively low cost. In general, the introduction of inorganic nanoparticles always generates more interfacial defects such as pore, crackle, or holes due to the poor compatibility with the polymer matrix.

For instance, the inorganic shell nanocomposite presents an enhanced energy density due to the higher ε_r of the inorganic shell than that of the organic shell, but the increase in ε_r is always accompanied by the sharply enhanced dielectric loss [29].

References

1 Masters, G.M. (2004). *Renewable and Efficient Electric Power Systems*. NJ: Wiley.

2 Li, X., Hui, D., and Lai, X. (2013). Battery energy storage station (BESS)-based smoothing control of photovoltaic (PV) and wind power generation fluctuations. *IEEE Trans. Sustainable Energy* 4: 464–473.

3 Masanori, S. and Masahiko, A. (2013). *Energy Storage Devices and Systems*. Hitachi Chemical Technical Report No. 55.

4 Zeng, J., Zhang, B., Mao, C. and Wang, Y. (2006). Use of battery energy storage system to improve the power quality and stability of wind farms. *International Conference on Power System Technology*, Chongqing, pp. 1–6.

5 Donnelly, M.K., Dagle, J.E., Trudnowski, D.J., and Rogers, G.J. (2002). Impacts of the distributed utility on transmission system stability. *IEEE Trans. Power Syst.* 11: 741–746.

6 Poullikkas, A. (2013). A comparative overview of large-scale battery systems for electricity storage. *Renewable Sustainable Energy Rev.* 27: 778–788.

7 Díaz-González, F., Sumper, A., Gomis-Bellmunt, O., and Villafáfila-Robles, R. (2012). A review of energy storage technologies for wind power applications. *Renewable and Sustainable Energy Rev.* 16: 2154–2171.

8 Isah, S. (2018). Advanced materials for energy storage devices. *Asian J. Nanosci. Mater.* 1: 90–103.

9 Zhou, Y., Liu, Q., Chen, F. et al. (2020). Significantly enhanced energy storage in core-shell structured poly (vinylidene fluoride-co-chlorotrifluoroethylene)/ $BaTiO_3$@polyurea nanocomposite films. *J. Mater. Sci.* 55: 11296–11309.

10 Meng, N., Ren, X., Santagiuliana, G. et al. (2019). Ultrahigh β-phase content poly(vinylidene fluoride) with relaxor-like ferroelectricity for high energy density capacitors. *Nat. Commun.* 10: 4535–4543.

11 Wang, Z., Zhu, M., Pei, Z. et al. (2020). Polymers for supercapacitors: boosting the development of the flexible and wearable energy storage. *Mater. Sci. Eng. R: Rep.* 139: 100520–100554.

12 Davis, S.E., Ide, M.S., Davis, R.J., and Bowman, S. (2013). Selective oxidation of alcohols and aldehydes over supported metal nanoparticles. *Green Chem.* 15: 17–45.

13 Siwal, S.S., Zhang, Q., Devi, N., and Thakur, V.K. (2020). Carbon-based polymer nanocomposite for high-performance energy storage applications. *Polymers* 12: 505–535.

14 Abbasi, H., Antunes, M., and Velasco, J.I. (2019). Recent advances in carbon-based polymer nanocomposites for electromagnetic interference shielding. *Prog. Mater. Sci.* 103: 319–373.

15 Devi, N., Ghosh, S.K., Perla, V.K. et al. (2019). Laboratory-based synthesis of the pure form of gananite (BiF_3) nanoparticles: a potential material for electrochemical supercapacitor application. *New J. Chem.* 43: 18369–18376.

16 Wang, F., Wang, H., and Mao, J. (2018). Aligned-graphene composites: a review. *J. Mater. Sci.* 54: 36–61.

17 Chauhan, A. and Saini, R. (2014). A review on integrated renewable energy system based power generation for standalone applications: configurations, storage options, sizing methodologies and control. *Renewable Sustainable Energy Rev.* 38: 99–120.

18 Winter, M. and Brodd, R.J. (2004). What are batteries, fuel cells, and supercapacitors? *Chem. Rev.* 104: 4245–4270.

19 Tang, H., Lin, Y., and Sodano, H.A. (2013). Synthesis of high aspect ratio $BaTiO_3$ nanowires for high energy density nanocomposite capacitors. *Adv. Energy Mater.* 3: 451–456.

20 Tang, H. and Sodano, H.A. (2013). Ultra high energy density nanocomposite capacitors with fast discharge using $Ba_{0.2}Sr_{0.8}TiO_3$ nanowires. *Nano Lett.* 13: 1373–1379.

21 Tomer, V., Polizos, G., Manias, E., and Randall, C.A. (2010). Epoxy-based nanocomposites for electrical energy storage. I: effects of montmorillonite and barium titanate nanofillers. *J. Appl. Phys.* 108: 074116–074125.

22 Wu, W., Huang, X.Y., Li, S.T. et al. (2012). Novel three-dimensional zinc oxide superstructures for high dielectric constant polymer composites capable of withstanding high electric field. *J. Phys. Chem. C* 116: 24887–24895.

23 Zhu, M., Huang, X., Yang, K. et al. (2014). Energy storage in ferroelectric polymer nanocomposites filled with core-shell structured polymer@$BaTiO_3$ nanoparticles: understanding the role of polymer shells in the interfacial regions. *ACS Appl. Mater. Interf.* 6: 19644–19654.

24 Larcher, D. and Tarascon, J.M. (2015). Towards greener and more sustainable batteries for electrical energy storage. *Nat. Chem.* 7: 19–29.

25 Vlad, A. and Balducci, A. (2017). Porous materials get energized. *Nat. Mater.* 16: 161–162.

26 Wang, G., Huang, X., and Jiang, P. (2017). Bio-inspired polydopamine coating as a facile approach to constructing polymer nanocomposites for energy storage. *J. Mater. Chem. C* 5: 3112–3120.

27 Zhou, Y., Liu, Q., Chen, F. et al. (2020). Significantly enhanced energy storage in core-shell structured poly (vinylidene fluoride-chlorotrifluoroethylene)/$BaTiO_3$@polyurea nanocomposite films. *J. Mater. Sci.* 55: 11296–11309.

28 Zhu, Y., Zhu, Y., Huang, X. et al. (2019). High energy density polymer dielectrics interlayered by assembled boron nitride nanosheets. *Adv. Energy Mater.* 9: 1901826.

29 Yang, Y., Gao, Z.S., Yang, M. et al. (2019). Enhanced energy conversion efficiency in the surface-modified $BaTiO_3$ nanoparticles/polyurethane nanocomposites for potential dielectric elastomer generators. *Nano Energy* 59: 363–371.

9

High-Performance Inorganic Polymer Nanocomposites-Based Solar Cells

9.1 Introduction

Photovoltaic devices based on nanocomposites composed of conjugated polymers and inorganic nanocrystals (NCs) show promise for the fabrication of low-cost third-generation thin-film photovoltaics [1]. The hybrid photovoltaic (HPVs) devices are expected to achieve a high power conversion efficiency (PCE), because they combine the advantageous characteristics of polymers and NCs, including the flexibility, lightweight, and low fabrication costs of polymer materials and the high electron mobility, size-dependent optical properties, and physical and chemical stability of inorganic nanocomposites [2]. Organic photovoltaic (OPVs) devices based on blends of conjugated polymers and fullerenes form interpenetrating donor–acceptor networks have been used in prototype bulk heterojunction (BHJ) geometry applications [3]. BHJs in polymer–inorganic hybrid solar cells may be formed by replacing the fullerenes, which act as organic nanoparticles, with inorganic semiconductors as electron acceptors for dispersal in the polymer matrix [4] (Figure 9.1).

9.2 Organic–Organic Composites

This composite, for example, includes the synthesis of carbon nanotubes (CNTs) with the *p*-type conducting polymers with the view to exploit the high conductivity of the nanotubes to assist the charge transport process in polymer photoactive medium. While synthesizing the CNT in *p*-type semiconducting polymers, several possible functionalizations are possible [5]. For example, Figure 9.2 shows typical possible functional groups in the synthesis of CNT-based composites. The purpose of synthesizing CNT-based nanocomposite is to enhance the charge carrier mobility in the bulk heterojunction organic medium where the CNT exhibited high carrier mobility and created an interconnected pathway that offers a long mean free path. The CNTs are not only expected to improve carrier mobility but also mechanical flexibility of the polymer materials which is compatible with roll-to-roll solution processing. The incorporation of CNT in organic photoactive medium encounters

Polymer Nanocomposites for Energy Applications, First Edition.
T. Daniel Thangadurai, Manjubaashini Nandhakumar, Sabu Thomas, and Ange Nzihou.

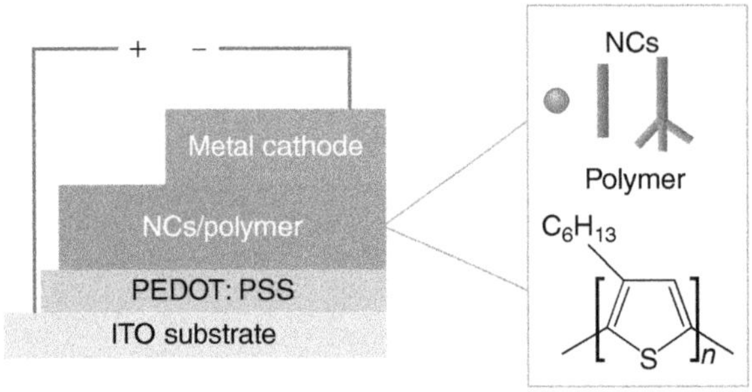

Figure 9.1 Schematic diagram showing the structure of a typical nanocomposites/polymer hybrid solar cell.

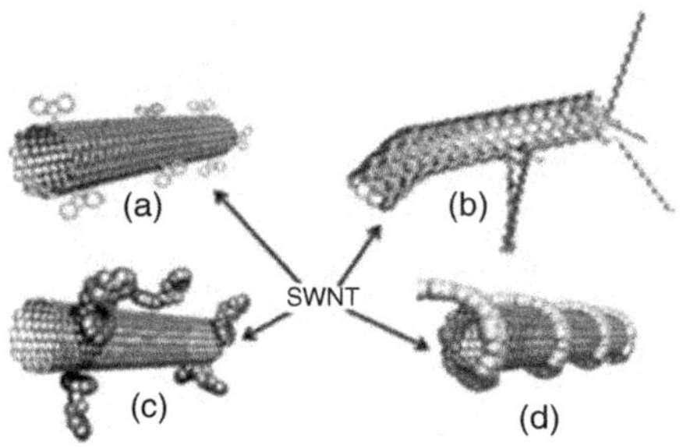

Figure 9.2 Various forms of functionalization of carbon nanotubes. (a) Defect functionalization, (b) sidewall of CNT, (c) noncovalent functionalization using noncovalent exohedral surfactant, and (d) noncovalent exohedral polymer rapping.

several challenges despite such attractive advantages of CNTs [6]. The formation of CNT defect clusters in the presence of a conducting polymer has resulted in poor performances compared to those devices without CNTs.

9.3 Inorganic Nanocomposites

The incorporation of nanocomposite in the photoactive layer of organic solar cells (OSCs) is expected to synergistically influence the performance of the device through additional exciton generation by localized surface plasmon resonance (LSPR) effect, exciton dissociation enhancement by nanocomposite–exciton coupling, improved charge transport assisted with reduced recombination, and the crystallinity enhancement of the polymer: fullerene BHJ blend for enhanced charge transport and collection [7]. The LSPR effect of metal nanoparticles enhances the electromagnetic field which expedites photoabsorption and generation of excess exciton in the photoactive layer of a photovoltaic device, while the light scattering effect increases the optical path length within the active layer which is also beneficial to exciton generation [8]. Moreover, exciton diffusion and charge dissociation are also encouraged when the electromagnetic field of the metal nanoparticles interacts with the generated excitons. However, despite being applicable in all functional layers of BHJ OSC, metal nanoparticles are more preferable when they are rightly incorporated in the active layer because the LSPR effect is restricted to a few nanometer scales [9]. Moreover, metal nanoparticles can facilitate the creation of hoping sites for holes, thus enhancing charge mobility via the introduction of dopant states within the bandgap of polymers which is more effective at relatively

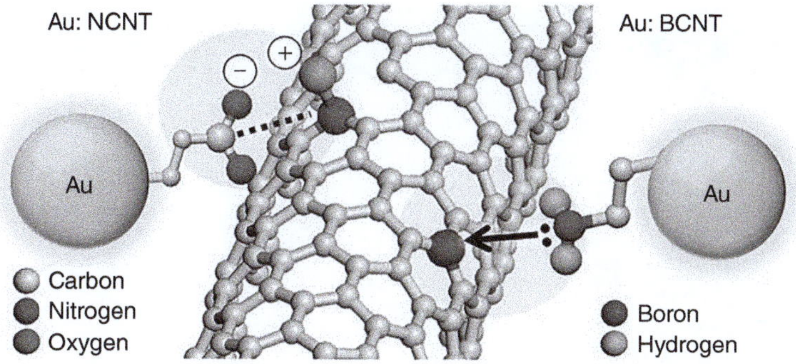

Figure 9.3 Schematic of chemical interactions between Au: NCNT and Au: BCNT nanocomposites.

low concentrations of nanoparticles to avoid the transfer of nonradiative energy which quenches excitons in the photoactive layer. Various reports involving the incorporation of nanocomposites in the photoactive layer have been demonstrated widely by researchers to improve the overall performance of OSCs [10] (Figure 9.3).

9.4 Nanocomposites in Perovskite Solar Cells

Nanoparticles incorporation as additives in the perovskite-based solar cell has been used in improving the morphology of perovskite films in terms of crystal growth, crystallization kinetics, and device performance. Due to their different chemical and physical properties, different nanoparticles have a unique effect on the formation of perovskite thin film. One of the most widely accepted methods is directly mixing metal nanoparticles into poly(3,4ethylenedioxylen-thiophene): poly(styrene sulfonic acid) (PEDOT: PSS), an active layer, or even as cathode interfacial materials [11]. Noble metallic nanoparticles are known to exhibit a strong absorption band in the UV–vis region, which lies within the optical absorption band of the conjugated polymers used in the active layer of OPV devices. The utilization of plasmonic metallic nanoparticles lies between interfaces or inside the buffer or the active layers of OPV devices to promote absorption, thereby increasing the optical thickness of OPV materials for light harvesting [12]. Plasmonic nanoparticles are expected to bring about a major reduction of both the perovskite film thickness and the amount of lead present in the device structure while at the same time ensuring a broad spectrum light absorption (Figure 9.4).

- ➢ Nanoparticles function as effective nucleation sites to promote the formation of perovskite lattice structures.
- ➢ Enhance the crystallinity of perovskite and help achieve large grain size.
- ➢ Enhances in light absorption through plasmonic effects.
- ➢ Nanoparticles act as distributes at the grain boundary, passivating the perovskite film minimizing the nonradiative recombination [13].
- ➢ Improve charge transport properties through metal nanoparticle doping.

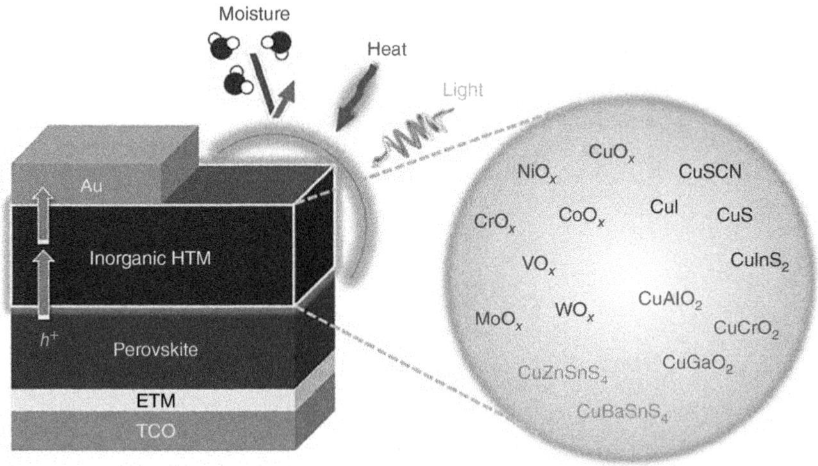

Figure 9.4 Schematic representation of perovskite solar cell.

9.5 Polymeric Nanocomposites in Dye-Sensitized Solar Cells (DSSCs)

The propensity to substitute certain components with polymers is to maximize performance and decrease costs for dye-sensitized solar cells (DSSCs). While 13% PCE has already been accomplished for regular rigid and liquid electrolyte-based DSSCs, a few DSSCs' key disadvantages, particularly when used in portable electronics, such as rigidity and high weight, frangibility of the glass electrode, leakage and instability are the difficulties of liquid electrolyte [14]. In comparison, according to the latest expense and life process study of DSSCs, transparent conductive oxides (TCOs)-coated glass compensate for greater than 60% of the expense and 55% of the global warming impact. Polymer nanocomposites, for instance, may be used to boost a cost/performance ratio of the product as part of different constituents of DSSC, such as flexible substrates, catalyst materials, and polymer electrolytes. Electrodes based on polyethylene terephthalate (PET) or polyethylene naphthalate (PEN) can replace glass electrodes, thereby enhancing the versatility and impact resistance of a DSSC. Also, polymer electrolytes can overcome liquid electrolyte leakage and volatility issues [15]. Because of highly expensive platinum, conducting polymers may be used as a catalyst layer too. Utilizing the plastic materials in DSSC allows flexible DSSC development. The use of polymeric structures in DSSCs also makes the manufacture of versatile solar cells that have gained considerable interest and found large areas of application in the solar industry due to their lightweight and cost-effectiveness [16].

Moumita Kar et al. show that hybrid solar cells including polymers and quantum dots (QDs) are extensively studied. In semiconductor polymers, the most widely studied polymer is regioregular poly(3-hexylthiophene) (P3HT) linearly consisting of thiophene units with pendant hexyl side chains which allow solubilization [17].

Also, Rajni Sharma et al. discussed hydrophobic nanoparticles into PCPDTBT: PCBM-based BHJ polymer solar cells resulted in significant improvement in solar cell *J–V* characteristics with enhancement in open-circuit voltage (VOC), short-circuit current density (JSC), and thereby overall improvement in cell efficiency [18]. Umair Baig et al. determined that the photovoltaic performance of the polypyrrole-titanium dioxide (PPy-TiO_2) nanocomposite has a 51.4% improvement with a photo-conversion efficiency of 8.07% as compared to pure TiO_2-based DSSC. By comparing the physical mixture of the PPy-TiO_2 nanocomposite and pristine TiO_2, the enhanced activity of the PPy-TiO_2 nanocomposite can be attributed to the reduced charge transfer resistance, the outstanding electrical conductance of the PPy, the nanosized structure of TiO_2, and their synergetic effect [19]. An innovative quasi-solid state (QSS) polymer electrolyte containing rare earth oxide nanoparticles and an ionic liquid was developed to enhance the efficiency and stability of dye-sensitized solar cells (DSSCs). It was found that DSSCs assembled with IL and Dy NPs incorporated polyvinylidene fluoride (PVDF) nanofibers had a higher conversion efficiency of 6.4%, and they had better stability than liquid electrolytes [20]. The nanocomposites that comprise isotactic polypropylene (iPP) and molybdenum disulfide nanotubes (MoS_2) are advanced nanomaterials that are simple, cost-effective, and eco-friendly processing [21].

The transparent conducting electrodes (TCEs) based on sulfonated graphene (SG)/PEDOT composites were prepared by *in situ* polymerization. Firstly, SG was prepared from graphene oxide (GO) in four steps: (i) reduction of GO with $NaBH_4$; (ii) sulfonation with the aryl diazonium salt of sulfanilic acid; (iii) post-reduction

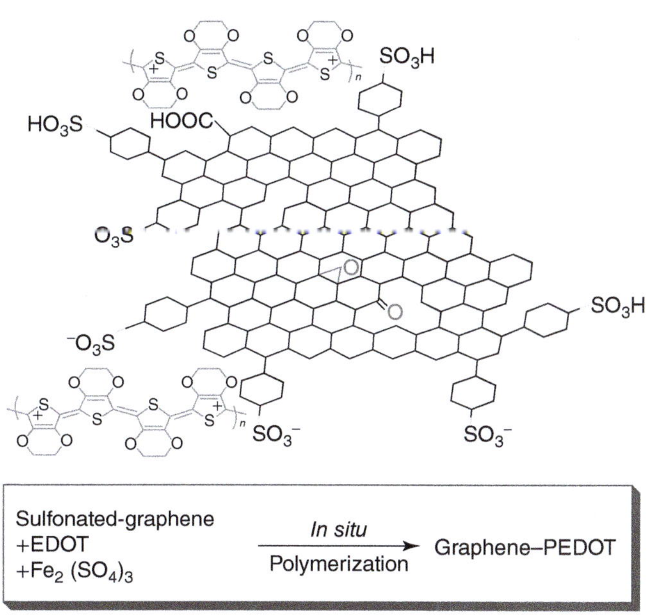

Figure 9.5 Schematic representation of sulfonated graphene (SG)/poly(3,4-ethylenedioxythiophene): poly(styrenesulfonate) (PEDOT) nanocomposites and its synthesis reaction conditions.

with N_2H_4; and (iv) functionalization with $NaNO_3$, sulfanilic acid, and azoisobutyronitride (AIBN). Subsequently, SG was dispersed in water followed by the addition of $Fe_2(SO_4)_3$ and the monomer ethylene dioxythiophene (EDOT) [22]. The mixture was stirred for 48 hours at 50 °C and then poured into methanol. The excess of EDOT and other impurities were removed through several washing cycles. The composites showed good processability both in water and organic solvents, superior transparency, high thermal conductivity, and thermal stability. A conductivity of $0.2\,S\,cm^{-1}$ and transmittances higher than 80% in the wavelength range of 400–1800 nm were observed for films with a thickness of a few nm [23]. This conductivity is much higher than that of a commercial poly(3,4-ethylene dioxythiophene): poly(styrene sulfonate) PEDOT: PSS product (10^{-6}–$10^{-5}\,S\,cm^{-1}$). Moreover, when a poly(methyl methacrylate) (PMMA) sheet coated with this composite was bent inward, it still retained high electrical conductivity ($0.18\,S\,cm^{-1}$) (Figure 9.5).

References

1 Nguyen, B.P., Kim, T., and Park, C.R. (2014). Nanocomposite-based bulk heterojunction hybrid solar cells. *J. Nanomater* 243041 20.

2 Gevorgyan, S.A., Medford, A.J., and Bundgaard, E. (2011). Aninterlaboratory stability study of roll-to-roll coated flexible polymer solar modules. *Sol. Energy Mater. Sol. Cells* 95: 1398–1416.

3 Tong, F., Kyusang, K., Daniel, M. et al. (2012). Flexible organic/inorganic hybrid solar cells based on conjugated polymer and ZnO nanorod array. *Semiconductor Sci. Technol.* 27: 105005–105010.

4 Reiss, P., Couderc, E., De Girolamo, J., and Pron, A. (2011). Conjugated polymers/semiconductor nanocrystals hybrid materials-preparation, electrical transport properties, and applications. *Nanoscale* 3: 446–489.

5 Chiechi, R.C., Havenith, R.W.A., Hummelen, J.C. et al. (2013). Modern plastic solar cells: materials, mechanisms, and modeling. *Mater. Today* 16: 281–289.

6 Yan, S., Lv, L., Ning, Y. et al. (2015). Effects of solvent additives on trap assisted recombination in P3HT: ICBA based polymer solar cells. *Phys. Status Solidi A* 212: 2169–2173.

7 Oseni, S.O. and Mola, G.T. (2017). Properties of functional layers in inverted thin-film organic solar cells. *Sol. Energy Mater. Sol. Cells* 160: 241–256.

8 Zhu, X., Zhang, F., An, Q. et al. (2015). Effect of solvent additive and ethanol treatment on the performance of PIDTDTQX: PC71BM polymer solar cells. *Sol. Energy Mater. Sol. Cells* 132: 528–534.

9 Lu, L. and Yu, L. (2014). Understanding low bandgap polymer PTB7 and optimizing polymer solar cells based on it. *Adv. Mater.* 26: 4413–4430.

10 Wang, M., Zhu, L., Zhou, M. et al. (2016). High efficiency organic bulk-heterojunction solar cells applying a new system of co-additives. *Mater. Lett.* 166: 227–230.

11 Thakur, U.K., Kisslinger, R., and Shankar, K. (2017). One-dimensional electron transport layers for perovskite solar cells. *Nanomater* 7: 95–121.

12 Liu, Y., Lang, F., Dittrich, T. et al. (2017). Enhancement of photocurrent in an ultra-thin perovskite solar cell by Ag nanoparticles deposited at low temperature. *RSC Adv.* 7: 1206–1214.

13 Hu, Z., Dong, S., Xue, Q. et al. (2015). In-situ synthesis of metal nanoparticle-polymer composites and their application as efficient interfacial materials for both polymer and planar heterojunctionperovskite solar cells. *Org. Electron.* 27: 46–52.

14 Mathew, S., Yella, A., Gao, P. et al. (2014). Dye-sensitized solar cells with 13% efficiency achieved through the molecular engineering of porphyrin sensitizers. *Nat. Chem.* 6: 242–247.

15 Brown, T.M., Rossi, F.D., Giacomo, F.D. et al. (2014). Progress in flexible dye solar cell materials, processes, and devices. *J. Mater. Chem. A* 2: 10788–10817.

16 Saleh, T.A., Shetti, N.P., Shanbhag, M.M. et al. (2020). Recent trends in functionalized nanoparticles loaded polymeric composites: an energy application. *Mater. Sci. Energy Technol.* 3: 515–525.

17 Kar, M., Sarkar, S., and Sarkar, P. (2020). Highly efficient inorganic-organic heterojunction solar cells based on polymer and CdX (X=Se, Te) quantum dots: an insight from a theoretical study. *J. Phys. Chem. C* 124: 11350–11357.

18 Sharma, R., Alam, F., Sharma, A.K. et al. (2014). ZnO anchored graphene hydrophobic nanocomposite-based bulk heterojunction solar cells showing enhanced short-circuit current. *J. Mater. Chem. C* 2: 8142–8151.

19 Baig, U., Gondal, M.A., Ilyas, A.M., and Sanagi, M.M. (2017). Bandgap engineered polymeric-inorganic nanocomposite catalysts: synthesis, isothermal stability, photocatalytic activity, and photovoltaic performance. *J. Mater. Sci. Technol.* 33: 547–557.

20 Thomas, M. and Rajiv, S. (2019). Dye-sensitized solar cells based on an electrospun polymer nanocomposite membrane as the electrolyte. *New J. Chem.* 43: 4444–4454.

21 Naffakh, M., Remskar, M., Marco, C. et al. (2011). Towards a new generation of polymer nanocomposites based on inorganic nanotubes. *J. Mater. Chem.* 21: 3574–3578.

22 Xu, Y., Wang, Y., Liang, J. et al. (2009). A hybrid material of graphene and poly(3, 4-ethyldioxythiophene) with high conductivity, flexibility, and transparency. *Nano Res.* 2: 343–348.

23 Diez-Pascual, A.M., Antonio Luceno Sanchez, J., Capilla, R.P., and Díaz, P.G. (2018). Recent developments in graphene/polymer nanocomposites for application in polymer solar cells. *Polymers* 10: 217–238.

10

Polymer Nanocomposites for Magnetic Energy and Thermal Energy Storage

10.1 Background of Polymer Nanocomposites for Energy Storage

Energy storage can be defined as the work done by the capacitor to move a unit charge from a negative plate to a positive plate [1]. Mathematically, the amount of energy stored is

$$dU_{\text{store}} = \frac{q}{C}dq \tag{10.1}$$

where dq denotes the unit charge. Besides, the voltage on a capacitor is proportional to the quantity of charge already stored on a capacitor [2]. If the amount of charge stored on a capacitor is Q, and the entire voltage of a battery appears on the capacitor, then the energy stored across the capacitor can be obtained as follows:

$$\begin{aligned} U_{\text{store}} &= \int_0^Q \frac{q}{C}dq = \frac{1}{C}\int_0^Q q\,dq \\ &= \frac{1}{C}\left(\frac{Q^2}{2} - \frac{0^2}{2}\right) = \frac{1Q^2}{2C} \end{aligned} \tag{10.2}$$

$$U_{\text{store}} = \frac{1Q^2}{2C} = \frac{1}{2}QV = \frac{1}{2}CV^2 \tag{10.3}$$

The energy storage expression in Eq. (10.2) can be written in other equivalent forms, as shown in Eq. (10.3), by using permutations based on the capacitance definition ($Q = CV$) [3]. From all the equivalent forms of U_{store}, it is obvious that the energy stored in a capacitor (i.e. $U_{\text{store}} = CV^2/2$) depends on the capacitance as well as voltage (V) bearability.

10.2 Energy Density

Energy density is generally defined as the total energy stored or discharged by the system per unit volume. Alternatively, it is also defined as the amount of energy either stored in or discharged out per unit mass [4]. However, the exact terminology for this definition is specific energy. Although both energy storage and

Polymer Nanocomposites for Energy Applications, First Edition.
T. Daniel Thangadurai, Manjubaashini Nandhakumar, Sabu Thomas, and Ange Nzihou.

energy discharge density are important in polymer nanocomposite film capacitors, recent research focuses on the latter one [5]. Mathematically, energy discharge density (U_e) is calculated by the definite integration of electric field (E) between remnant polarization (P_r) and maximum dielectric displacement (D_{max}), as shown in Eq. (10.4):

$$U_e = \int_{P_r}^{D_{max}} E \, dD \tag{10.4}$$

$$D = \varepsilon_0 \varepsilon_r E \tag{10.5}$$

Furthermore, for a linear dielectric material where the dielectric constant is independent of the electric field, such as polyurea and polythiourea, Eqs. (10.4) and (10.5) can be rewritten as Eqs. (10.6) and (10.7), respectively:

$$U_e = \frac{1}{2} DE \tag{10.6}$$

$$U_e = \varepsilon_0 \varepsilon_r E^2 \tag{10.7}$$

where ε_r is the relative dielectric constant of polymer-based nanocomposite material and ε_0 is the permittivity of free space. More importantly, Eqs. (10.6) and (10.7) illustrate that energy density is dependent on the polarization, the relative dielectric constant, and the square of the applied electric field that is bound to the dielectric breakdown strength (E_b) of a material [6]. Therefore, to increase the energy density of the material, it is critical to produce an electroactive material with high polarization, substantial dielectric constant, extraordinary breakdown strength, and low conductive and ferroelectric losses [7].

10.3 Superconducting Magnetic Energy Storage (SMES)

The superconducting magnetic energy storage (SMES) is an energy storage device that stores electrical energy in a magnetic field without change in chemical or mechanical forms. SMES is achieved by inducing DC into a coil made of superconducting cables of nearly zero resistance, generally made of niobiumtitane (NbTi) filaments that operate at very low temperatures (−270 °C). The current increases when charging and decreases during the discharge process and has to be converted for AC or DC voltage application [8]. When looking at the total system, however, it is clear that there is a considerable energy requirement for refrigeration. Also, the current has to flow through non-superconducting components and solid-state switches, which cause resistive losses. Despite this, the overall efficiency in commercial applications in the range of the MW is very high. Like a battery, a SMES system provides rapid response for either charge or discharge. Unlike a battery, the amount of energy available is independent of the discharge rating. The energy content of SMES in commercial use today is approximately 1 kWh, but the maximum power output is in the MW range and only limited by the rating of the power electronics [9]. Due to the complexity of the cooling system, SMES cannot be built cost-effectively for low-power outputs. The response time of a SMES is limited to a few milliseconds by the speed with which the need to release energy is detected and the speed of the subsequent switching operation of the power electronics. In practice, that is a few milliseconds [10]. The life time of the superconducting coil and the number of charges and discharge is very high and probably exceeds all competing technologies, although there is mechanical stress in the components leading to material fatigue [11]. Despite their good technological features, there are very few SMES systems built mainly due to their high cost.

A sandwich-structured polymer nanocomposite based on poly(vinylidene fluoride-co-hexafluoropropylene) (P[VDF-HFP]) was designed, which consists of a central layer with wide bandgap MgO nanowires and the outer layers with high-K $BaTiO_3$ nanofibers. MgO nanowires are utilized as the nanofillers in the sandwiched nanocomposites. It is found that the small amount of MgO nanowires in the central layer can effectively increase the breakdown strength of the sandwiched nanocomposites, while the loading of $BaTiO_3$ nanofibers (BT NFS) in the outer layers can enhance the electric displacement of the nanocomposites [12]. The improved dielectric and energy storage performance are attributed to synergistic effects of the spatial arrangement of multiple nano-inclusions in the trilayered structures.

Polyvinylidene fluoride (PVDF) and its copolymers-based nanocomposites have been extensively investigated because of their high dielectric constant and easy processing. Also, to enhance the energy storage capability of dielectric polymer nanocomposites, ultra-small platinum (Pt) nanoparticles are introduced. Hence, the air-stable ultra-small Pt nanoparticles (<2 nm) decorated polydopamine-encapsulated $BaTiO_3$ (Pt@PDA@BT) were used as nanofiller [13]. The core–satellite Pt@PDA@BTnanofiller effectively enhanced the breakdown strength and suppressed the leakage current of P(VDF-HFP) nanocomposites, since the energy storage capability and charge–discharge efficiency were significantly enhanced.

The structured polymer@$BaTiO_3$ core–shell nanoparticles with high dielectric constant shells can result in high electric displacement and thus high energy densities in the nanocomposites, but the high leakage currents cause high dielectric loss and low discharged energy densities in the nanocomposites. Also, the high dielectric constant polymer shells have high electrical conductivity, causing a significant decrease of breakdown strength and thus poor maximum energy storage capabilities [14]. The nanocomposites filled with core–shell nanoparticles having low dielectric constant shell show low electric displacement and thus low energy densities, but the low leakage currents cause low dielectric loss and thus high discharged energy density.

Electric double-layer capacitors (EDLCs) are also known as supercapacitors or ultra-capacitors. EDLCs store charge at an electrolyte–electrode interface leading to a promising power source. They have potential applications in mobile electronic equipment and electric vehicles. Supercapacitors are categorized into two types according to their charge storage mechanisms [15]. In the first type, EDLCs store charge by the electrostatic separation of charges that occurs at the electrode–electrolyte interface. In the second one, pseudo-capacitance arises from either superficial or multielectron transfer faradic reactions with fast charge–discharge properties [16].

Supercapacitors based on magnetic conductive polymer–graphene nanocomposites were investigated. The nanocomposite was obtained by mixing graphene with polypyrrole or magnetic polypyrrole. The conductive ink consisted of the nanocomposite, solvent, and binder. The supercapacitor electrodes were manufactured with conductive ink by using copper foil. Supercapacitor cells were composed of the separator (PTFE) and electrolyte (ionic liquid materials or acids). It is also suggested that the supercapacitors can be exposed to the magnetic field. The magnetic field can arrange the magnetic particles in the supercapacitor cell, and this can improve the uniformity inside of the cell resulting in better electronic transportation [17]. Therefore, this can give a chance to increase the specific capacitance.

10.4 Thermal Energy Storage (TES)

Among energy storage systems, particularly interesting for some applications is thermal energy storage (TES), defined as the temporary storage of excess heat that can be used where and when needed. TES technologies can reduce the mismatch between thermal energy demand and availability, thereby contributing to more efficient exploitation of intermittent energy sources [18]. Compared to other energy storage systems, the storage of energy in the form of heat (or cold) exhibits longer storage times and higher efficiency [19]. The cycle of a TES system, reported in Figure 10.1, comprises charging, storage, and discharging steps. TES technologies are currently employed for specific purposes in three main cases, namely: (i) to store waste/excess heat that can be released when needed, e.g. to recover waste industrial heat [20], or in solar thermal power plants during peak periods [21]; (ii) to keep the temperature in a specific range, e.g. in buildings to store excess energy during the day and release it at nighttime [22], or for body temperature regulation through smart thermoregulating garments [23]; and (iii) to temporarily store heat

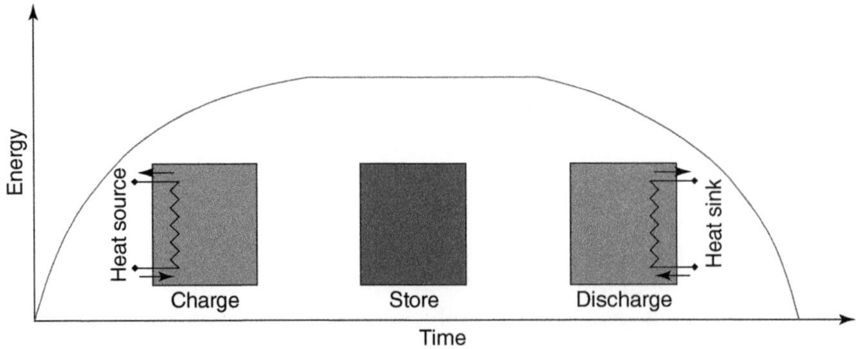

Figure 10.1 Typical working cycle of a TES system, showing the charging, storage, and discharging steps and the variation of stored energy over time.

and prevent a temperature rise that would otherwise damage a component, as in the thermal management (TM) of electronic devices [24].

In some applications, the desired product is then stored and released thermal energy, as in the cases (i) and (ii), these are generally referred to as examples of "thermal energy storage *properly said*," and they normally need energy storage systems with high thermal capacity to store as much energy as possible. In other applications, the excess heat is stored mainly to avoid a dangerous temperature rise, as in case (iii), these are examples of "TM," and their main requirement is usually a well-defined energy storage rate. It is not easy to distinguish between TES *properly said* and TM. Some other classifications categorize as TES properly said only the case (i), while the other cases are examples of TM as they must keep the temperature in an optimal range [25].

The most common classification of TES technologies is based on the way of varying the internal energy of the storage medium. Thermal energy can be stored and released through a temperature variation (*sensibleheat TES*, SH-TES), an endo/exothermic phase change (*latent heat TES*, LH-TES), or a thermochemical reaction (*thermochemical heat TES*, TH-TES) [26].

10.4.1 Sensible Heat Storage (SH-TES)

In sensible heat storage (SH-TES), energy is stored (and released) via an increase (or a decrease) in the temperature of the storage medium. The enthalpy variation is proportional to the temperature difference as represented in Figure 10.2a.

10.4.2 Latent Heat Storage (LH-TES)

Latent heat storage (LH-TES) involves the storage and release of heat through the phase transition of a phase change material (PCM). The vast majority of LH-TES systems are based on the melting–solidification transition, while evaporation-condensation phase changes are generally avoided, as the associated high volume variation increases the complexity of the confinement unit [27]. The variation of enthalpy in a solid-to-liquid phase transition is illustrated in Figure 10.2b.

10.4.3 Thermochemical Heat Storage (TH-TES)

Thermochemical heat storage refers to the techniques for storing and releasing heat through reversible endo/exothermic thermochemical reactions. The energy storage density of TH-TES materials is up to 10 times higher than that of SH-TES media and approximately 2 times higher than that of the most common LH-TES systems.

As explained above, among the applications of TES materials, (i) the heat storage for temperature control, for example in the buildings industry, or to produce smart textiles for body temperature regulation, and (ii) the temporary storage of heat to prevent overheating, as in the cooling systems for electronic devices. When a TES material is used for TM, it is usually only an extra component added to the main

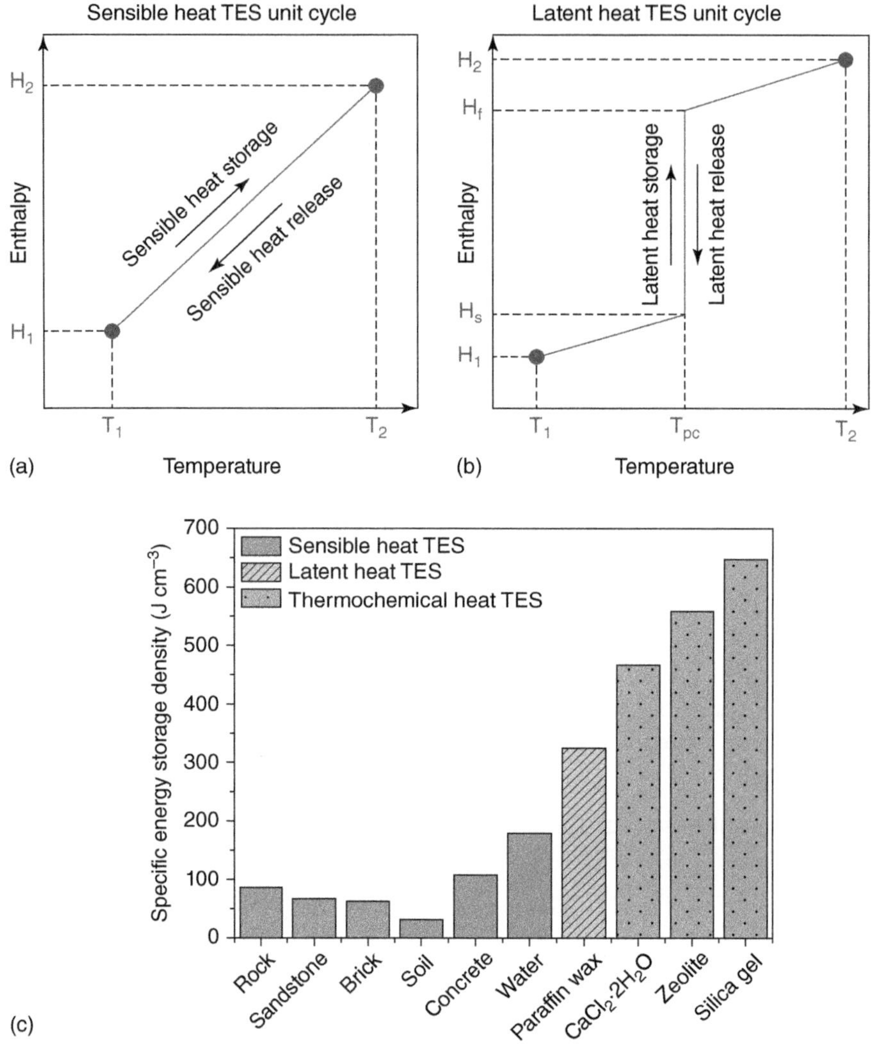

Figure 10.2 Variation of enthalpy of the storage medium as a function of its temperature in: (a) an SH-TES unit cycle; (b) LH-TES unit cycle; and (c) comparison of specific energy storage density values of different sensible, latent, and thermochemical TES media.

structure of a device. However, the resulting increase in weight and volume may be unacceptable for some applications where weight and volume reduction are crucial design parameters. In this case, it would be useful to have a multifunctional material that combines good mechanical properties and the heat storage/management function. With such materials, it would be possible to build part of the structure with the "thermal battery" material, or in other words, to design a structure that is part of the TM system. This approach is similar to that developed for an outstanding application of multifunctionality, i.e. the structural batteries, which are devices that can carry the mechanical load and store electrochemical energy simultaneously [28].

10.5 Thermoplastic Composites for TES

Thermoplastic composites have received considerable attention in the last decades due to some advantages over their thermosetting counterpart [29]. Even though the use of thermoplastic composites can be advantageous for some applications, a very exiguous number of examples have been reported in the literature that deals with thermoplastic structural or semi-structural composites containing organic PCMs [30] (Figure 10.3).

Three types of materials, namely (i) polyamide/(glass fiber) GF laminates containing paraffinic PCMs [31], (ii) PCM (microcapsules) MCs in semi-structural carbon–polyamide composites [32], (iii) reactive thermoplastic resin as a matrix for multifunctional carbon-fiber laminates [33] was developed. The prepared laminates were subjected to an in-depth characterization through dynamic–mechanical analysis (DMA). Three testing modes were investigated, namely single-frequency scans, multifrequency scans, and heating–cooling cycles. The trends of the storage modulus E' and the loss modulus E'' acquired during single-frequency scans are illustrated in Figure 10.4.

To facilitate the comparison, the values of E' of each laminate have been normalized to the value at 0 °C. The storage modulus of all laminates shows a marked decrease at the glass transition of the EL matrix, where E'' shows a peak. The laminates containing MC show an additional transition at the PCM melting temperature and the drop is more evident at high MC contents. Interestingly, the correlation between the drop amplitude and the MC weight fraction of the melting enthalpy is linear, with R^2 values higher than 0.98. Another DMA parameter that correlates linearly with the melting enthalpy is the area under the tan δ peak [34]. This implies that the DSC test allows one to predict, to a certain extent, the trend of the viscoelastic

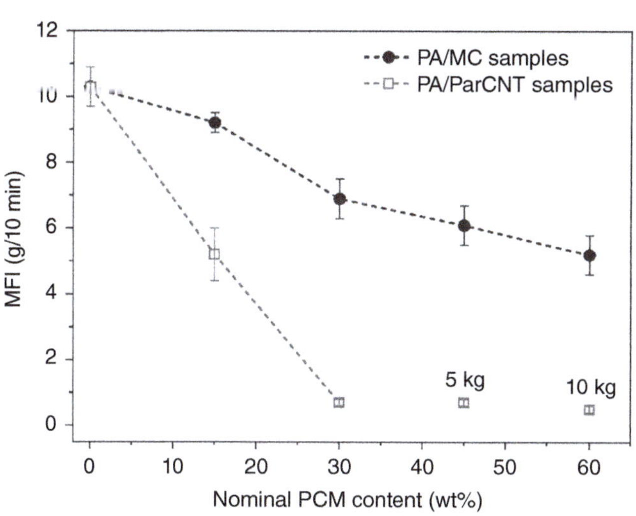

Figure 10.3 Melt flow index (MFI) of the PA/PCM blends as a function of the nominal PCM content (at 230 °C and 2.16 kg; single tests at 5 and 10 kg are also indicated).

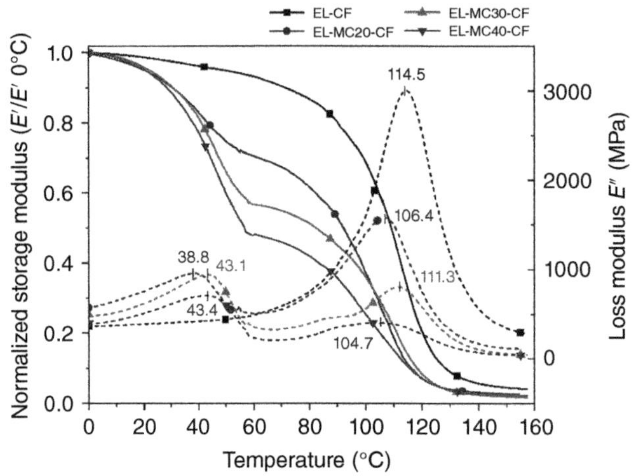

Figure 10.4 Results of the DMA tests on the samples EL-MCx-CF. The reported values represent the peak temperatures normalized storage modulus E'(solid lines) and loss modulus E'' (dashed lines) (single cantilever; 1 Hz; 3 °C min^{-1}).

properties of the composite in the temperature range around the PCM phase change. As structural TES composites must withstand repeated thermal cycles around the phase change temperature of the PCM, DMA tests were performed not only on heating but also on cooling, between−40 and 60 °C. This is not a common approach as no other studies can be found in the literature that analyzes the trend of viscoelastic parameters of PCM-enhanced polymers on cooling, to the best of the authors' knowledge. Figure 10.5a illustrates the trends of E' during a heating–cooling–heating scan performed at 3 °C min^{-1} on the three prepared MC-containing laminates. The absolute values of E' decrease with an increase in the MC weight fraction, which was expected from the decreasing values of the elastic modulus (Figure 10.5b).

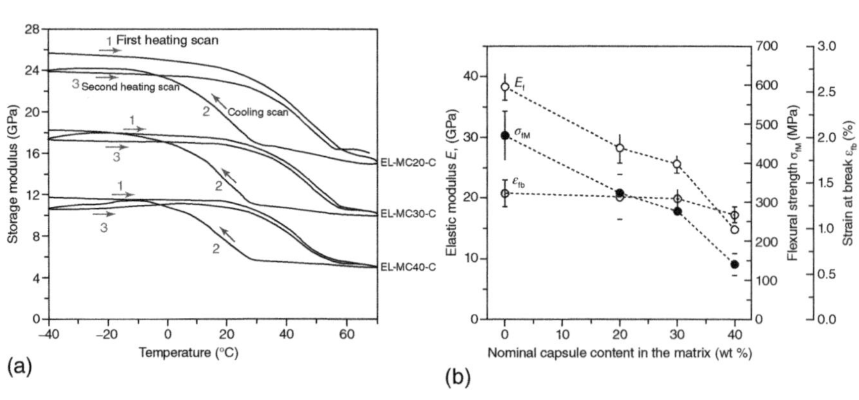

Figure 10.5 (a) DMA heating (1) cooling (2)–heating (3) scans on the samples EL-MCx-CF: values of E' in the temperature range around the phase change temperature of the PCM (single cantilever; 1 Hz; 3 °C min^{-1}); (b) results of the three-point bending tests on the samples EL-MCx-CF as a function of the nominal MC content in the matrix.

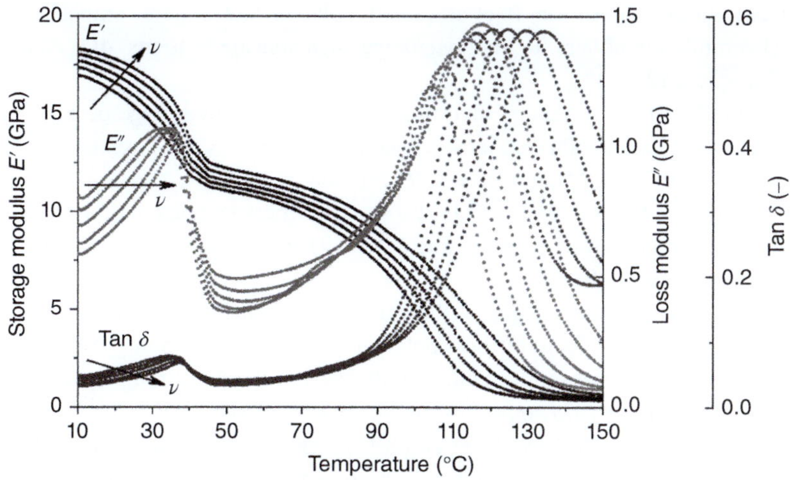

Figure 10.6 Representative DMA multifrequency thermograms of the sample EL-MC30-CF (single cantilever; 0.3-1-3-10-30 Hz; 0.3 °C min^{-1}).

The decreasing step of E' at the PCM melting is almost completely recovered on cooling, as it reaches 90–95% of the initial value. The recovery happens with a certain hysteresis, as the crystallization is found at lower temperatures than the melting, as also observed in DSC tests, due to reasons related to both thermal inertia and the thermodynamics of crystallization. The trends of E'' and tan δ also show that the peak on cooling (crystallization) is found approximately 30 °C lower than the melting peak. From these findings, the DMA test appears as a powerful technique to evaluate material stiffness during phase transition of PCM, especially combined with DSC analysis (for a better comparison, the same rate of heating–cooling is suggested). Finally, multifrequency DMA analysis was performed to assess the effect of frequency on the PCM melting and the glass transition of the EL resin. Figure 10.6 presents the results of the sample EL-MC30-CF.The shift of the signals toward higher temperature with increasing frequency is observable not only at the glass transition of the EL resin, as commonly observed in polymer composites, but also at the PCM melting. For this transition, the frequency sensitivity is higher below the peak temperature than above it, as after the peak temperature the curves are almost overlapped. This suggests that when the core is completely molten, the dependence of the signals on frequency weakens considerably.

References

1 Yang, L., Kong, X., Li, F. et al. (2019). Perovskite lead-free dielectrics for energy storage applications. *Prog. Mater Sci.* 102: 72–108.

2 Huang, X., Sun, B., Zhu, Y. et al. (2019). High- k polymer nanocomposites with 1D filler for dielectric and energy storage applications. *Prog. Mater Sci.* 100: 187–225.

3 Zhang, H., Marwat, M.A., Xie, B. et al. (2020). Polymer matrix nanocomposites with 1D ceramic nanofillers for energy storage capacitor applications. *ACS Appl. Mater. Interfaces* 12: 1–37.

4 Ma, W., Zhu, Y., Marwat, M.A. et al. (2019). Enhanced energy-storage performance with excellent stability under low electric fields in BNT- ST relaxor ferroelectric ceramics. *J. Mater. Chem. C* 7: 281–288.

5 Xie, B., Zhang, L., Marwat, M.A. et al. (2018). High energy storage performance for dielectric film capacitors by designing 1D $SrTiO_3@SiO_2$ nanofillers. *J. Adv. Dielectri.* 8: 1850039–1850048.

6 Prateek, B.R., Garg, A., and Gupta, R.K. (2019). Poly- (vinylpyrrolidone)/ poly(vinylidene fluoride) as guest/host polymer blends: understanding the role of compositional transformation on nanoscale dielectric behavior through a simple solution–process route. *ACS Appl. Energy Mater.* 2: 6146–6152.

7 Wang, Z., Nelson, J.K., Hillborg, H. et al. (2013). Dielectric constant and breakdown strength of polymer composites with high aspect ratio fillers studied by finite element models. *Compos. Sci. Technol.* 76: 29–36.

8 Chen, H.S., Cong, T.N., Yang, W. et al. (2009). Progress in electrical energy storage system: a critical review. *Prog. Nat. Sci.* 19: 291–312.

9 Nielsen, K.E. and Molinas, M. (2010). Superconducting Magnetic Energy Storage (SMES) in power systems with renewable energy sources. In: *IEEE International Symposium on Industrial Electronics*, 2487–2492.

10 Tan, X., Li, Q., and Wanga, H. (2013). Advances and trends of energy storage technology in Microgrid. *Electr. Power Energy Syst.* 44: 179–191.

11 Koshizuka, N., Ishikawa, F., and Nasu, H. (2003). Progress of superconducting bearing technologies for flywheel energy storage systems. *Physica C* 386: 444–450.

12 Lia, Z., Liu, F., Li, H. et al. (2019). Largely enhanced energy storage performance of sandwich-structured polymer nanocomposites with synergistic inorganic nanowires. *Ceram. Int.* 45: 8216–8221.

13 Wang, L., Huang, X., Zhu, Y., and Jiang, P. (2018). Enhancing electrical energy storage capability of dielectric polymer nanocomposites via the room temperature Coulomb blockade effect of ultra-small platinum nanoparticles. *Phys. Chem. Chem. Phys.* 20: 5001–5011.

14 Zhu, M., Huang, X., Yang, K. et al. (2014). Energy storage in ferroelectric polymer nanocomposites filled with core-shell structured polymer@$BaTiO_3$ nanoparticles: understanding the role of polymer shells in the interfacial regions. *ACS Appl. Mater. Interfaces* 6: 19644–19654.

15 Liew, C.W., Ramesh, S., and Arof, A.K. (2016). Enhanced capacitance of EDLCs (electrical double-layer capacitors) based on ionic liquid-added polymer electrolytes. *Energy* 109: 546–556.

16 Eng, A.Y.S., Chua, C.K., and Pumera, M. (2016). Facile labeling of graphene oxide for superior capacitive energy storage and fluorescence applications. *J. Phys. Chem.* 18: 9673–9681.

17 Yanik, M.O., Yigit, E.A., Akansu, Y.E., and Sahmetlioglu, E. (2017). Magnetic conductive polymer-graphene nanocomposites based supercapacitors for energy storage. *Energy* 138: 883–889.

18 Cabeza, L.F. (2014). *Advances in Thermal Energy Storage Systems: Methods and Applications*. Cambridge, UK: Woodhead Publishing.

19 Dincer, I. and Ezan, M.A. (2018). *Heat Storage: A Unique Solution For Energy Systems*. Cham, Switzerland: Springer.

20 Zhang, J., Zhang, H.H., He, Y.L., and Tao, W.Q. (2016). A comprehensive review on advances and applications of industrial heat pumps based on the practices in China. *Appl. Energy* 178: 800–825.

21 Sheng, N., Zhu, C., Sakai, H. et al. (2019). Synthesis of Al-25 wt% Si@Al_2O_3@Cu microcapsules as phase change materials for high-temperature thermal energy storage. *Sol. Energy Mater. Sol. Cells* 191: 141–147.

22 Ostry, M. and Charvat, P. (2013). Materials for advanced heat storage in buildings *11th Int. Conf. on Modern Building Materials, Structures and Techniques (MBMST 2013) (Vilnius: Procedia Engineering)*, pp. 837–843.

23 Cherif, C., Tran, N.H.A., Kirsten, M. et al. (2018). Environmentally friendly and highly productive bi-component melt spinning of thermoregulated smart polymer fibers with high latent heat capacity. *eXPRESS Polym. Lett.* 12: 203–214.

24 Kandasamy, R., Wang, X.Q., and Mujumdar, A.S. (2007). Application of phase change materials in thermal management of electronics. *Appl. Therm. Eng.* 27: 2822–2832.

25 Fredi, G., Dorigato, A., Fambri, L., and Pegoretti, A. (2020). Multifunctional structural composites for thermal energy storage. *Multifunct. Mater.* 3: 042001.

26 Pielichowska, K. and Pielichowski, K. (2014). Phase change materials for thermal energy storage. *Prog. Mater Sci.* 65: 67–12.

27 Campbell, F.C. (2010). *Structural Composite Materials (ASM International)*. Ohio: Materials Park.

28 Petrucci, R. and Torre, L. (2017). Filled polymer composites. In: *Modification of Polymer Properties* (ed. C.F. Jasso-Gastinel and J.M. Kenny), 23–46. Amsterdam: Elsevier.

29 Biron, M. (2013). *Thermoplastics and Thermoplastic Composites*. Oxford, UK: Elsevier, Ltd. https://doi.org/10.1016/B978-1-4557-7898-0.00001-9.

30 Fredi, G., Dorigato, A., and Pegoretti, A. (2018). Multifunctional glass fiber/polyamide composites with thermal energy storage/release capability. *Express Polym. Lett.* 12: 349–364.

31 Jamekhorshid, A., Sadrameli, S.M., and Farid, M. (2014). A review of microencapsulation methods of phase change materials (PCMs) as thermal energy storage (TES) medium. *Renewable Sustainable Energy Rev.* 31: 531–542.

32 Fredi, G., Dorigato, A., and Pegoretti, A. (2019). Novel reactive thermoplastic resin as a matrix for laminates containing phase change microcapsules. *Polym. Compos.* 40: 3711–3724.

33 Abdel Ghafaar, M., Mazen, A.A., and El-Mahallawy, N.A. (2006). Behavior of woven fabric reinforced epoxy composites under bending and compressive loads. *J. Eng. Sci.* 34: 453–469.

34 Fredi, G., Dorigato, A., and Pegoretti, A. (2020). Dynamic-mechanical response of carbon fiber laminates with a reactive thermoplastic resin containing phase change microcapsules. *Mech. Time-Depend. Mater.* 24: 395–418.

11

Polymer Nanocomposites for Triboelectricity and Hydrogen Storage

11.1 Polymer Nanocomposites for Triboelectricity

Energy harvesting systems for low-power devices are regularly being a prerequisite within the context of the internet of things and, in particular, for self-powered sensors in remote or inaccessible places. Triboelectric nanogenerators (TENGs) are an appropriate method for harvesting environmental mechanical energy otherwise unused in nature [1]. The benefit of the triboelectric phenomenon is the wide range of materials that can be used in the distinctive triboelectric mechanical modes: contact separation, lateral sliding, single electrode, and free-standing triboelectric layer mode. Being a process that can be carried out completely with polymers and the corresponding composites, the entire properties of the materials can be tailored for each specific application, including dimensions, geometry, and optical transparency [2]. The triboelectric power output can also be sturdily improved by tailoring the intrinsic properties of the polymers by synthesis and functionalization or by reinforcing with high-dielectric or other functional fillers. Further, the geometrical dimension and roughness of the materials can also be planned to maximize the generated energy [3]. The power density per area of triboelectric devices is the major among the aforementioned systems, reaching powers as high as 500 W m^{-2} and an energy conversion efficiency of 70% has been validated; further, they are lightweight and economical [4]. Compared with piezoelectric devices, triboelectricity can be more suitable for environmentally friendly energy production for portable devices, low-power devices in remote or inaccessible places, or for needed internet of things networks of sensors [5].

The assessment of the output power of different polymer and polymer composites using the triboelectric contact separation systems (10 N of force followed by 5 cm of separation per cycle) were clarified. Different materials were used as positive (mica, polyamide (PA66) and styrene/ethylene-butadiene/styrene (SEBS)) and negative (polyvinylidene fluoride (PVDF), polyurethane (PU), polypropylene (PP), and Kapton) charge materials [6].

Polymer Nanocomposites for Energy Applications, First Edition.
T. Daniel Thangadurai, Manjubaashini Nandhakumar, Sabu Thomas, and Ange Nzihou.

11.1.1 Energy Harvesting Application

A simple application was established by harvesting the triboelectric energy into a capacitor and later powering a LED. The two material pairs with the largest output powers, PA66:PP and Mica:30BT/PVDF, were applied. The triboelectric pairs were connected to an electrical circuit containing four diodes to transform the AC to DC voltage and charge a capacitor of 15 μF, as illustrated in Figure 11.1. The circuit follows a traditional DO-35 Schottky (D1–D4) rectifier bridge topology with an output electrolytic capacitor for energy storage, powering a load composed by a manual switch button, the LED, and the resistor [8]. This setup permits the energy to be stored and manually discharge over the load when the voltage level is appropriate. When the capacitor is charged, using a light switch, the LED was lighted on and the respective voltage drop at the capacitor ends was witnessed. By using Mica:30BT/PVDF or PA66:PP pairs, it is likely to charge the 15 μF (capacitor with 25 to 30 cycles, the capacitor being able to turn on the LED for a few seconds). It is noteworthy that this is attained with a small active area of 46.4×10^{-4} m^2 in each material. Thus, by increasing the active area of the materials to 0.14 m^2, the capacitor possibly will be charged in just one cycle [7]. Thus, implemented in an example (Figure 11.1), human walking can generate enough energy in a few steps to power the LEDs, taking into account the weight and area of the shoe.

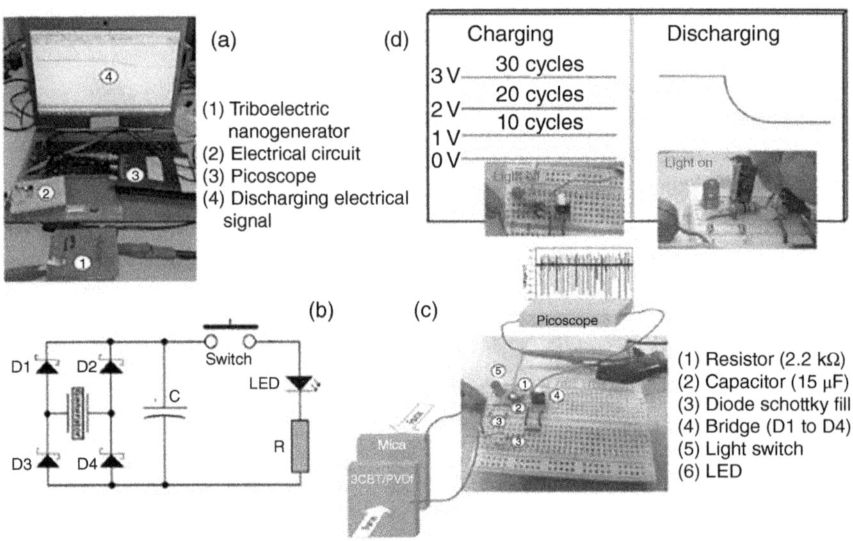

Figure 11.1 Illustration of the complete setup (a) with the pair of materials, detailed electronic circuit scheme (b), and picoscope connected to a laptop. (c) Electric circuit for powering the LED and (d) charge–discharge cycles using triboelectric materials (PA66:PP or Mica:30BT/PVDF pairs) as nanogenerators. Source: Rodrigues-Marinho et al. [7]/with permission from MDPI.

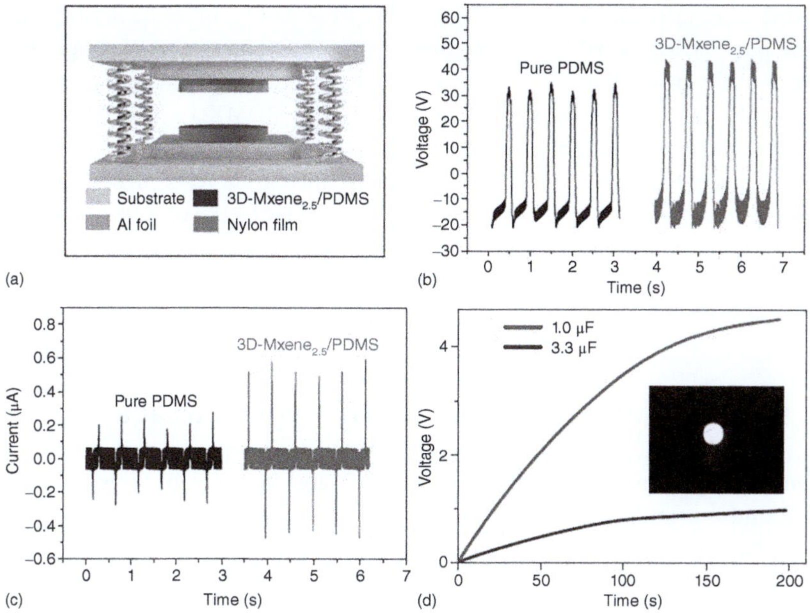

Figure 11.2 (a) Schematic diagram of 3D-MXene/PDMS-based TENG. The corresponding output (b) voltage and (c) current of the 3D-MXene$_{2.5}$/PDMS and pure PDMS-based TENG; (d) charging curve of two commercial capacitors (insert photograph shows a LED lit by the 3D MXene$_{2.5}$/PDMS-based TENG). Source: Wang et al. [10]/with permission from Elsevier.

TENGs based on coupled triboelectrification and electrostatic induction can convert mechanical energy from the environment to electrical energy. The 3D-MXene/PDMS composite can be also applied as a negative tribo-material in a TENG device [9]. Figure 11.2 shows the TENG structure based on the contact separate operation mode. The substrates and the springs were used to guarantee the normal operation of the TENG. Nylon film, acting as the positive tribo-material, is coupled with the 3D-MXene/PDMS to assemble the TENG [11]. The triboelectric performance was investigated using a homemade machine under an impact frequency of 2 Hz. The output voltages that are demonstrated in Figure 11.2 shows that the TENG assembled with 3D-MXene$_{2.5}$/PDMS exhibits an output voltage of 45 V, higher than that of the output voltage (i.e. 33 V) of the pure PDMS-based TENG [12]. The enhanced output performance can be ascribed to the micro-capacitors induced by the introduction of conductive MXene flakes in the dielectric PDMS. In other words, the increase of the dielectric constant can improve the capacitance and surface charge density of the nanocomposites, leading to high TENG performance. The enhanced dielectric constant has been verified by the dielectric measurement of the nanocomposites. The dielectric constant of the 3DMXene$_{2.5}$/PDMS nanocomposite is higher than 100 in the frequency range of 1 kHz to 1 MHz, which is much higher than that (i.e. 2.3) of pure PDMS [13]. The 3D-MXene$_{2.5}$/PDMS composite-based

TENG exhibits a high output current of 0.6 μA, approximately threefold higher than that of the pure PDMS-based TENG. As demonstrated above, this result originates from the significantly decreased electrical resistance of the 3DMXene$_{2.5}$/PDMS nanocomposite. A rectifier was integrated with the 3DMXene$_{2.5}$/PDMS-based TENG, and then two capacitors (1 and 3.3 μF) were serially connected to the circuit as electrical energy storage. The recorded energy-storing curve is presented that 80 seconds to charge the 1-μF capacitor to 3 V and takes 100 seconds to charge the 3.3-μF capacitor to 0.8 V. The 3D-MXene$_{2.5}$/PDMS-based TENG can light a LED, suggesting that the TENG can be used as a mechanical power source to drive portable electronic [10].

11.2 Polymer Nanocomposites for Hydrogen Storage

Hydrogen is one of the most effective, unpolluted, and lightweight fuels; though, it is not found naturally and must be generated from primary energy sources. It is expected to play a major role in future energy systems. Like electricity, it must be produced and transported. Although hydrogen has one additional advantage, it can be stored [14]. Electricity must be used as it is produced; it can be stored only if it is converted to another energy form. Currently, there are four main technologies for hydrogen storage out of which two are more mature and developed. These are hydrogen pressurization and hydrogen adsorption in metal hydrides. The remaining two technologies that are still in the research and technological development phase are the adsorption of hydrogen on carbon nanofibers and the liquefaction of the hydrogen. When hydrogen is produced, it can be stored to be used directly in fuel cells or transported to users to produce electricity. A hydrogen fuel cell uses hydrogen and oxygen to produce electricity and water, and a reversible hydrogen fuel cell could also use electricity and water to produce hydrogen and oxygen [15]. The essential elements of a hydrogen fuel cell comprise an electrolyze unit, to convert the electrical energy input into hydrogen, the hydrogen storage system itself, and a hydrogen energy conversion, to convert the stored chemical energy in the hydrogen back to electrical energy. Using these elements, water is electrolytically decomposed into hydrogen and oxygen. There are many concepts of hydrogen fuel cells; the principle technologies include proton exchange membrane fuel cells (PEM-FCs), alkaline fuel cells (AFCs), phosphoric acid fuel cells (PAFCs), and regenerative fuel cells (RFCs). Hydrogen fuel cells possess advantages, including high energy density, applicability at small and large scales, and simple modular use. In terms of their useful cycle life, they are estimated at more than 15 years and 20 000 charges and discharge cycles, respectively [16]. It is important to note that hydrogen-based storage technologies are considered as one of the promising technologies in load shifting applications (load shifting delays renewable energy delivery from nonpeak to peak utility demand) [17]. Several demo projects have been developed as a proof of concept concerning stand-alone systems with wind, photovoltaic generation, and hydrogen storage (Figure 11.3).

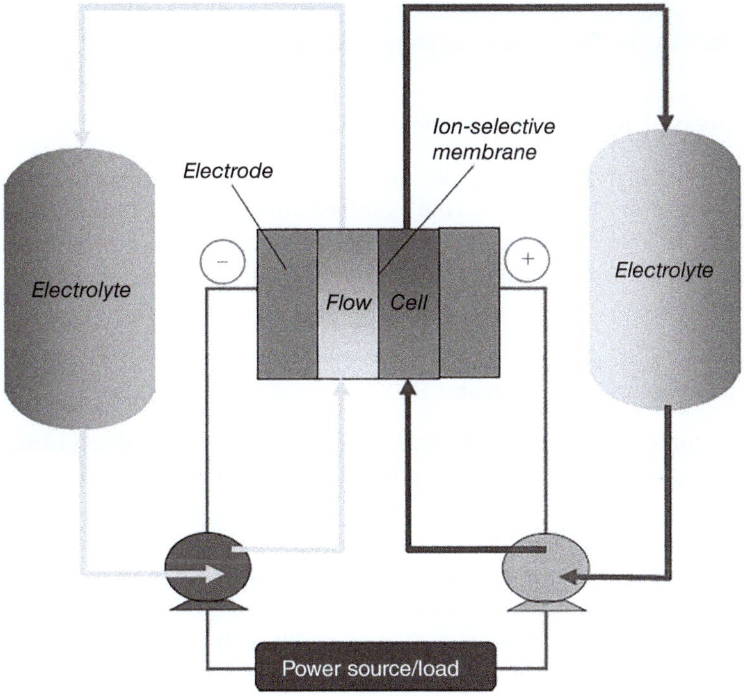

Figure 11.3 Schematic overview of a redox flow cell energy storage system.

11.3 Hydrogen-Based Energy Storage System

Hydrogen-based energy storage systems are one of the best electrical energy storage mediums. A hydrogen energy system can be easily integrated with renewable power sources like solar and wind. Even though the efficiency of hydrogen energy is not very high, their cost of storage capacity is very less when compared with the other energy storage techniques. Hydrogen energy storage is therefore economically best suited to situations where the total amount of energy stored is more valuable than efficiency [18]. Their essential elements comprise an electrolyte unit to convert the electrical energy into hydrogen, a reservoir, where the chemical energy is stored and a hydrogen energy conversion system such as a fuel cell convert the stored chemical energy into electrical energy. Hydrogen is produced, then compressed or liquefied, stored, and then converted back to electrical energy or heat energy. The major advantage of hydrogen is that it can replace all fossil fuel applications without the emission of harmful gases.

11.3.1 Liquid Hydrogen Storage

Hydrogen is cooled and liquefied and then kept cold in an insulated tank. Liquid hydrogen storage is one of the bulk storage methods. This method has a high safety record. The method does not hold good for large applications as the cost of liquefaction is very high because of the cryogenic storage technology [19].

11.3.2 Compressed and Stored in a Pressure Tank

Compressed gaseous hydrogen storage is suitable for both large- and small-scale applications as long-term storage where the hydrogen liquefaction method fails due to the high cost behind the liquefaction process. This method has some safety issues, and the initial cost of implementation is high due to the high cost of the compressors and pressure vessels.

11.3.3 Physical Adsorption in Carbon

The hydrogen is stored in gaseous form in the carbon through the adsorption process. The gaseous hydrogen can be adsorbed onto the surface of carbon to attain storage volumetric densities greater than liquid hydrogen. The carbon nanofiber is used as an adsorbent as it has improved hydrogen storage capability. Storage by absorption as chemical compounds or by adsorption on carbon materials has definite advantages from the safety perspective such that some form of conversion or energy input is required to release the hydrogen for use. Carbon materials such as activated carbons, carbon nanotubes, and carbon nanofiber have been the subject of intensive research.

11.3.4 Complex Compounds-Microsphere Hydrogen Storage

Microsphere hydrogen storage is well suitable for vehicular hydrogen storage. This system consists of hollow glass spheres that are charged with hydrogen for an hour and discharged by heating and reducing pressure. The microspheres can be easily transferred from one tank to another. The disadvantages of this system are immature system characterization and a lower storage period.

11.3.5 Metal Hydrides

Metal hydrides can be formed by the reaction of metal or alloy with hydrogen. The metal hydride formation is an exothermic process. When sufficient heat is supplied, the hydrogen can be released from the metal hydride [20]. Metal hydrides have (i) good safety characteristics, (ii) high storage density, and (iii) divided as low and high dissociation temperature hydrides. The low-temperature hydrides suffer from low hydrogen fraction and have low hydrogen loading capability. In the high-temperature hydrides, a heat source is required to generate a high temperature of dissociation and offers high heat storage capabilities.

Hydrogen (H_2) produced from renewable resources has been considered as an important energy storage vector to fully exploit the benefits of renewable energies. Hydrogen has a high gravimetric energy density of 142 MJ kg^{-1} (higher heating value, HHV) which is three times higher than that of gasoline or diesel (~ 45 MJ kg^{-1}, HHV), and it can be used to generate energy in internal combustion engines or in fuel cells, where water is the only by-product. However, hydrogen has a low volumetric energy density, making its storage a key challenge in the development of any future hydrogen-based economy [21]. Conventionally, H_2 is stored in tanks either as a compressed gas or a liquid. However, the safe and practical use of these H_2 storage methods is constrained by the large size of the storage tank, the cost required to liquefy H_2, low volumetric density of H_2, loss of liquid H_2 due to "boil-off," and critical safety concerns due to high pressures involved. Innovations in material development have demonstrated that H_2 can be stored by physisorption on large surface area adsorbents, and in that context, carbon-based nanostructures, metal–organic frameworks (MOFs), zeolites, and polymers with intrinsic microporosity have all been studied extensively as physisorbed [22].

In the field of energy storage, recently investigated nanocomposites show promise in terms of high hydrogen uptake and release with enhancement in the reaction kinetics [23]. Polymeric nanocomposite materials consist primarily of polyaniline, a conducting polymer, which has conductivity on the order of 1 Siemens cm^{-1} [24]. The chemical makeup of polyaniline in its emeraldine form has anions, which allow hydrogen ions to bond to the material very well. Due to its porosity, the surface area of the conducting polymeric matrix material will allow for additional hydrogen bonding in the form of physisorption [25]. The polymer nanostructures are widely studied for their interesting physical and chemical properties and are deployed for practical applications. Conducting polyaniline is a type of polymer that could be prepared in powder, gel, or other suitable forms and has been employed for storing hydrogen at varying densities [26].

The PIM-1/AX21 composite with 60 wt% AX21 provides enhanced hydrogen adsorption kinetics and a total hydrogen storage capacity of up to 9.35 wt% at 77 K; this is superior to the US Department of Energy hydrogen storage target. Tensile testing indicates that the ultimate stress and strain of PIM-1/AX21 are higher than those of the MIL-101 or PAF-1 containing composites and are sufficient for use in hydrogen storage tanks. The data presented provides new insights into both the design and characterization methods of polymer-based composite membranes [27]. The nanoporous polymer-based composites offer advantages over powders in terms of safety, handling, and practical manufacturing, with potential for hydrogen storage applications either as means of increasing storage or decreasing operating pressures in high-pressure hydrogen storage tanks.

The development of materials based on polymer nanocomposites for hydrogen storage with a lower temperature of desorption might contribute to the consolidation of the use of hydrogen as sustainable energy. The incorporation of hydrides into polymers can be an alternative to obtain materials with excellent hydrogen storage properties. Combining their hydrogen sorption abilities, superior protection against oxidation, hydrides morphology stability, and the low density of the polymers allows them to be further explored for mobile hydrogen storage applications. Nanocomposite material consisting of a polyaniline matrix, which can be functionalized by either catalytic doping or incorporation of nano variant, is considered to be a potential promise for hydrogen storage [28].

The porosity, hydrogen uptake, and mechanical properties of a processable microporous polymer (PIM-1) can be tuned by the addition of a highly porous filler (PAF-1). The surface area, total pore volume, and hydrogen uptake of the composites were found to follow rule-of-mixtures trends. PIM-1-based composites with large accessible surface areas maintained microporosity to efficiently adsorb hydrogen molecules. This was achieved by adding various amounts (0–37.5 wt%) of high-surface-area porous aromatic framework PAF-1 to a PIM-1 matrix. With high surface area additives, such as PAF-1, with PIM-1 we have been able to form composite materials possessing enhanced hydrogen storage capacity associated with the advantageous mechanical properties and processability of the polymer [29]. High-pressure hydrogen uptake measurements were performed to confirm the potential of this approach to generate composite materials that are potentially able to meet the DoE gravimetric target under useful pressure conditions.

References

1 Luo, J. and Wang, Z.L. (2019). Recent advances in triboelectric nano-generator based self-charging power systems. *Energy Storage Mater.* 23: 617–628.

2 Dharmasena, G.R.D.I. and Silva, S.R.P. (2019). Towards optimized triboelectric nanogenerators. *Nano Energy* 62: 530–549.

3 Saxon, D.J., Luke, A.M., Sajjad, H. et al. (2020). Next-generation polymers: isosorbide as a renewable alternative. *Prog. Polym. Sci.* 101: 101196.

4 Chen, J. and Wang, Z.L. (2017). Reviving vibration energy harvesting and self-powered sensing by a triboelectric nanogenerator. *Joule* 1: 480–521.

5 Wang, Z.L. (2017). On Maxwell's displacement current for energy and sensors: the origin of nanogenerators. *Mater. Today* 20: 74–82.

6 Pan, S. and Zhang, Z. (2019). Fundamental theories and basic principles of triboelectric effect: a review. *Friction* 7: 2–17.

7 Rodrigues-Marinho, T., Castro, N., Correia, V. et al. (2020). Triboelectric energy harvesting response of different polymer-based materials. *Materials* 13: 4980.

8 Zou, H., Zhang, Y., Guo, L. et al. (2019). Quantifying the triboelectric series. *Nat. Commun.* 10: 1427.

9 Yang, J., Chen, J., Liu, Y. et al. (2014). Triboelectrification-based organic film nano-generator for acoustic energy harvesting and self-powered active acoustic sensing. *ACS Nano* 8: 2649–2657.

10 Wang, D., Lin, Y., Hu, D. et al. (2020). Multifunctional 3D-MXene/PDMS nanocomposites for electrical, thermal, and triboelectric applications. *Composites, Part A* 130: 105754.

11 Zheng, Q., Shi, B., Fan, F. et al. (2014). In vivo powering of pacemaker by breathing-driven implanted triboelectric nano-generator. *Adv. Mater.* 26: 5851–5856.

12 Zhang, L., Zhang, B., Chen, J. et al. (2016). Lawn structured triboelectric nanogenerators for scavenging sweeping wind energy on rooftops. *Adv. Mater.* 28: 1650–1656.

13 Zu, G., Kanamori, K., Nakanishi, K., and Huang, J. (2019). Superhydrophobic ultra-flexible triple network graphene/polyorganosiloxane aerogels for a high-performance multifunctional temperature/strain/pressure sensing array. *Chem. Mater.* 31: 6276–6285.

14 Winter, C.J. (2009). Hydrogen energy – abundant, efficient, clean: a debate over the energy system of change. *Int. J. Hydrogen Energy* 34: 1–52.

15 Sherif, S.A., Barbir, F., and Veziroglu, T.N. (2005). Wind energy and the hydrogen economy – review of the technology. *Sol. Energy* 78: 647–660.

16 Paster, M.D., Ahluwalia, R.K., Berry, G. et al. (2011). Hydrogen storage technologies options for fuel cell vehicles: well-to-wheel costs, energy efficiencies, and greenhouse gas emissions. *Int. J. Hydrogen Energy* 36: 14534–14551.

17 Rabiee, A., Khorramdel, H., and Aghaei, J. (2013). A review of energy storage systems in micro-grids with wind turbines. *Renewable Sustainable Energy Rev.* 18: 316–326.

18 Smith, W. (2000). The role of fuel cells in energy storage. *J. Power Sources* 86: 74–83.

19 Bahadur, J., Contescu, C.I., Ramirez-Cuesta, A.J. et al. (2017). Properties of immobile hydrogen confined in microporous carbon. *Carbon* 117: 383–392.

20 Bimbo, N., Ting, V.P., Sharpe, J.E., and Mays, T.J. (2013). Analysis of optimal conditions for adsorptive hydrogen storage in microporous solids. *Colloids Surf., A* 437: 113–119.

21 Srinivasan, S., Emre Demirocak, D., Kaushik, A. et al. (2020). Reversible hydrogen storage using nanocomposites. *Appl. Sci.* 10: 4618.

22 Bardhan, R., Ruminski, A.M., Brand, A., and Urban, J.J. (2011). Magnesium nanocrystal-polymer composites: a new platform for designer hydrogen storage materials. *Energy Environ. Sci.* 4: 4882–4895.

23 Stejskal, J. (2002). Preparation of a conducting polymer (IUPAC Technical Report). *Int. Union Pure Appl. Chem.* 75: 857.

24 Virji, S., Kaner, R.B., and Weiller, B.H. (2006). Hydrogen sensors based on conductivity changes in polyaniline nanofibers. *J. Phys. Chem.* B110: 22266.

25 Fusalba, F., Gouerec, P., Vellers, D., and Belanger, D.J. (2001). Electrochemical characterization of polyaniline in nonaqueous electrolyte and its evaluation as electrode material for electrochemical supercapacitors. *Electrochem. Soc.* 148: A1–A6.

26 Tian, M., Rochat, S., Polak-Kraśna, K. et al. (2019). Nanoporous polymer-based composites for enhanced hydrogen storage. *Adsorption* 25: 889–901.

27 Gonc Beatrice, C.A., Rodrigues Moreira, B., Dantas de Oliveira, A. et al. (2020). Development of polymer nanocomposites with sodium alanate for hydrogen storage. *Int. J. Hyd. Energy* 45: 5337–5534.

28 Liemann, M.U., Srinivasan, S.S., Phani, A.R. et al. (2008). Nanomaterials for hydrogen storage applications. *A Rev. J. Nanomater.* 2008: Article ID: 950967.

29 Rochat, S., Polak-Krasna, K., Tian, M. et al. (2017). Hydrogen storage in polymer-based processable microporous composites. *J. Mater. Chem.* A5: 18752–18761.

12

Polymer Nanocomposites for Supercapacitors and Battery Application

12.1 Battery-Based Energy Storage System

An electrical battery is made of a stacked electrochemical cell, and the operation is based on electrochemical energy storage. Rechargeable batteries have a higher initial cost but can be recharged inexpensively and reused many times. Batteries are modular and nonpolluting and have fewer environmental impacts. The energy conversion in a secondary battery is reversible. The energy flow in the battery energy storage system is based on the charging and discharging process. In the grid, energy storage photovoltaic arrays applications of rechargeable batteries are used for load leveling. The unused excess electric energy during the low-demand period is stored in the batteries and is supplied to the grid during peak load periods such as storing power generated during the day to be used at night.

Some of the available battery technologies include lead-acid batteries, nickel-cadmium, sodium-ion batteries, sodium-sulfur batteries, lithium-ion batteries, and flow batteries like Hydrogen Vanadium redox and Regenesys redox [1]. The Li-ion batteries are suitable for portable devices and applications requiring high energy density and high overall efficiency. Due to low investment and maintenance costs, Li-ion batteries are currently the predominant technology. The vanadium and the flow batteries can be used for applications where high power is required for a long duration.

The efficiency of the rechargeable batteries can be 60–80%, and it varies with the factors like recharge cycle, depth of discharge, and temperature effect. Batteries are sensitive to the environment. During an electrical charge and discharge cycle, the temperature change in the battery must be controlled or it can cause adverse effects on the battery's life expectancy. The type of battery being used will determine how resistant it is to life degradation due to temperature. The battery's life cycle can be defined as the number of charge–discharge cycles that a battery can supply depending on the depth of discharge. The battery cycle applications require the battery energy storage system to charge and discharge multiple times a day. The battery's life cycle varies with the depth of charge [2]. The battery's life cycle will be high until the depth of discharge is relatively low. However, if the depth of discharge is large, then the battery's life cycle can be degraded.

Polymer Nanocomposites for Energy Applications, First Edition.
T. Daniel Thangadurai, Manjubaashini Nandhakumar, Sabu Thomas, and Ange Nzihou.

A battery is an electrochemical device that can deliver, in the form of electric energy, the chemical energy generated by electrochemical reactions. These reactions are set in train inside a basic cell, between two electrodes plunged into an electrolyte, when a load is connected to the cell's terminals [3]. The reaction involves the transfer of electrons from one electrode to the other through an external electric circuit. A battery consists of single or multiple cells, connected in series or parallel or both depending on the desired output voltage and capacity. Each cell consists of: (i) an anode or negative electrode which provides electrons to the load and is oxidized during the electrochemical reaction; (ii) the cathode or positive electrode which accepts electrons and is reduced during the reaction; (iii) the electrolyte which provides the medium for transfer of electrons between the anode and the cathode; and (iv) the separators between positive and negative electrodes for electrical insulation [4]. During discharge, electrochemical reactions at the two electrodes generate a flow of electrons through an external circuit. During the charging process, the electrochemical reactions are reversed via the application of the external voltage across the electrodes. The main difference between different battery systems is the materials used as electrodes and electrolytes, which determine the specific characteristics of the batteries. Separators are made of polymeric materials, paper, or paper board. The external case is composed of steel, polymeric materials, or paper board. Electrodes and electrolytes change as a function of the different applications of batteries [5]. The potentially hazardous components of batteries include mercury, lead, copper, zinc, cadmium, manganese, nickel, and lithium. The development of improved battery technology is critical for advancements in a variety of applications ranging from hybrid electric vehicles to consumer electronics, and improved battery performance depends on the development of materials for the various battery components [6].

12.2 Types of Battery

Batteries can be classified as primary batteries, which are not rechargeable, or secondary batteries, which can be recharged. Secondary batteries can be considered, since primary batteries are not viable for bulk energy storage. The construction of a secondary battery is facilitated by the short lead times, potentially convenient sitting, and the technology's modularity [7]. However, large-scale utility battery storage has been rare up until fairly recently because of low energy densities, small power capacity, high maintenance costs, cycle life, and a limited discharge capability. Batteries are potentially suitable for utility-scale battery energy storage applications including lead-acid battery, nickel-based battery, sodium-sulfur battery, and lithium-based battery [8].

12.2.1 Lead-Acid Battery

A lead-acid battery is the most mature and the cheapest energy storage device of all the battery technologies available. The lead-acid batteries are based on chemical reactions involving lead dioxide (cathode electrode), lead (anode electrode), and sulfuric acid which act as the electrolyte. There are two major types of lead-acid batteries: (i) flooded batteries (commonly used) and (ii) valve-regulated batteries (used in research and development) [9]. The lead-acid battery has a low cost (US$ 300–600 kWh^{-1}) and high reliability and efficiency (70–90%). In addition to the relatively poor performance of the battery at low and high ambient temperatures, and its relatively short lifetime, the main disadvantages of the lead-acid battery are the necessity for periodic water maintenance and its low specific energy and power [10].

12.2.2 Nickel-Based Battery

The nickel-based batteries are mainly nickel-cadmium (NiCd), nickel-metal hydride (NiMH), and nickel-zinc (NiZn) batteries. All three types use the same material for the positive electrode and the electrolyte which is nickel hydroxide, and an aqueous solution of potassium hydroxide with some lithium hydroxide, respectively [11]. As for the negative electrode, the NiCd type uses cadmium hydroxide, the NiMH uses a metal alloy, and the NiZn uses zinc hydroxide. Nickel-cadmium batteries compete with lead-acid batteries because they have a higher energy density, a longer life cycle (more than 3500 cycles), and lower maintenance requirements [12]. Despite the above advantages of the nickel-cadmium battery over the lead-acid battery, nickel-cadmium was not commercially successful, mainly due to its considerable costs at more than 10 times of lead-acid and it contains toxic components [13].

12.2.3 Sodium–Sulfur Battery (NaS)

NaS batteries are employed in power systems for more than 20 projects in Japan. The NaS consists of liquid-sulfur as the positive electrode, liquid-sodium as the negative electrode, and solid beta alumina ceramic as electrolyte [14]. NaS battery cells are usually designed in a tubular manner where the sodium is normally contained in an interior cavity formed by the electrolyte [15]. Compared with the other leading battery technologies, NaS shows a much more attractive energy density and has a long cycle capability and a millisecond response for full charging and discharging operations, which presents good potentials to be applied in micro-grid applications for power regulations [16]. The energy density and the energy efficiency of this type of battery are very high of 151 $kWh\ m^{-3}$ and 85%, respectively. Additional important features of NaS batteries are no self-discharge, low maintenance, and 99% recyclability. NaS battery can be widely used in aggregated energy storage [17].

12.2.4 Lithium-Based Battery

Lithium-polymer battery lifetime can only reach about 600 cycles. Regarding its self-discharge, this is much dependent on temperature, but it has been reported to be around 5% per month. Compared to the Li-ion battery, the lithium-polymer battery's operational specifications dictate a much narrower temperature range, avoiding lower temperatures [18]. However, lithium-polymer batteries are lighter and safer with minimum self-inflammability.

12.2.5 Flow Battery Energy Storage (FBES)

Flow batteries (redox flow batteries) are a relatively new system. In a flow battery, the battery is charged and discharged by a reversible chemical reaction between the two liquid electrolytes of the battery. Unlike conventional batteries, the liquid electrolytes are contained in separate tanks [19]. During operation, these electrolytes are pumped through the electrochemical reactor, in which a chemical redox reaction takes place and electricity is produced. Due to this storage of the electrolytes outside the reactor, the specifications of the battery are flexible; the power and the energy content of the system can be specified separately. It is very easy to increase the number of electrolytes or to replace the electrolytes [20]. Moreover, the design of the power cell can be optimized for the power rating needed, as this is independent of the amount of electrolyte used. Flow batteries can release energy continuously at a high rate of discharge for up to 10 hours. Flow batteries are distinguished from fuel cells by the fact that the chemical reaction involved is often reversible, i.e. they are generally of the secondary battery type and so they can be recharged without replacing the electroactive material [21]. One of the major advantages of flow batteries is that their energy capacity is easily scalable, since it depends on the volume of the stored electrolyte. Three types of flow batteries were developed up to the stage of commercialization and demonstration. These types are vanadium redox batteries (VRB), polysulphide bromide batteries (PSB), and zinc-bromine batteries (ZnBr). Finally, the efficiency of VRB systems is the most competitive advantage in terms of operating cost, system life, maintenance, and safety [22].

12.3 Conducting Polymer Nanocomposites

Conducting polymer (CP) matrix-based nanocomposites has generated considerable attention recently. CP provides tremendous scope for tuning their electrical conductivity from semiconducting to metallic regimes [23]. CP conductivity can be tuned by electrical manipulation of the polymer backbone by the nature of the dopant, by the degree of doping, by blending with other polymers, and by creating composites with inorganic materials. The feasibility of CP as materials for supercapacitors was demonstrated first by Rudge and coworkers [24].

Subsequently, a variety of such polymers has been investigated and, among them, polyaniline (PANI), polypyrrole (Ppy), polythiophene (PTH), and their substituted monomers are important [25]. Prerequisites for CP-based supercapacitors are environmental stability, facile doping, development, and ease of fabrication. In the chemical synthesis of PANI and Ppy about their feasibility as supercapacitors, various dopants, initiators, templates, and oxidants have been employed. Soft templates include various surfactants, seeded growth, and bio-templates [26]. For stability and suitable adhesion upon electrodes, binders such as polytetrafluoroethylene, polyvinylidene fluoride, N-methyl pyrrolidone (NMP), and Nafion have been employed. A novel synthesis of PANI nanowires using dilute polymerization on gold/chromium layers thermally etched on polyethylene terephthalate films leads to the design of micro-supercapacitors [27].

PANI and Ppy nanocomposites with diverse forms of carbon and oxides have been synthesized using various oxidants such as iron (III) chloride, ammonium persulfate, potassium permanganate, potassium dichromate, iron (II) chloride, and iron (III) nitrate, under acidic conditions [28]. The synthesized material can then be pressed into pellets or coated onto electrodes using appropriate binders. Several interesting morphologies arise in chemical syntheses of CP nanocomposites, if soft templates such as cetyltrimethylammonium bromide, dodecylbenzene sulfonic acid, dodecyl (trimethyl) azanium bromide, *n*-octadecyl (trimethyl) ammonium, or sodium dodecyl sulfate are employed [29]. Apart from carbon-based substrates, other electrodes commonly employed are nickel, stainless steel (SS), titanium, indium tin oxide, and tantalum, glassy carbon, gold, platinum, and silver [25].

Conducting PANI is a promising CP due to its high conductivity, ease of preparation, good environmental stability, and a large variety of applications [30]. These properties make this polymer suitable for gas sensors, functional hybrids, as a pH-switching electrical-conducting biopolymer hybrid for sensor applications, as an electrically active redox biomaterial for sensor applications, and as a matrix for preparation of CP nanocomposites [31]. Consequently, there has been rising attentiveness in the preparation of nanocomposites based on PANI. A literature review on CP nanocomposites shows that PANI has been used for nanocomposite preparation [32]. If nanostructure filler particles are magnetic in a polymer matrix, then such nanocomposites show enhanced magnetic and transport properties of the polymer [33].

12.4 Fuel Cells

Fuel cells have the merit of converting chemical energy into electrical energy. Among the several varieties of available fuel cells (solid oxide fuel cell, molten carbonate fuel cell, alkaline fuel cell, polymer electrolyte membrane fuel cell [PEMFC], and direct methanol fuel cell [DMFC]), polymer membranes are being used only in PEMFC and DMFC. The schematic depiction of PEMFC and DMFC is shown in Figure 12.1. Both PEMFC and DMFC are comprised of an anode and a cathode separated by a proton-conducting polymer membrane [34]. The fuel (hydrogen in the case of PEMFC and methanol in the case of DMFC) gets dissociated at the catalyst layer present in the anode. This generates protons and electrons at the anode. The proton passes through the polymer membrane, whereas the electron is being dragged through an external electrical circuit. The proton and the electron combine with the reduced oxygen at the cathode surface and generate water. The electron transmitted through the external electrical circuit generates electricity [35].

The objective of the polymer membrane in PEMFCs and DMFCs is to provide a path for proton conduction from anode to cathode, and hence, a limited number of membranes are suitable for this application. The membranes also face a harsh environment during the energy conversion process [36]. Hence, the main requirements of polymer membranes suitable for fuel cell applications are excellent

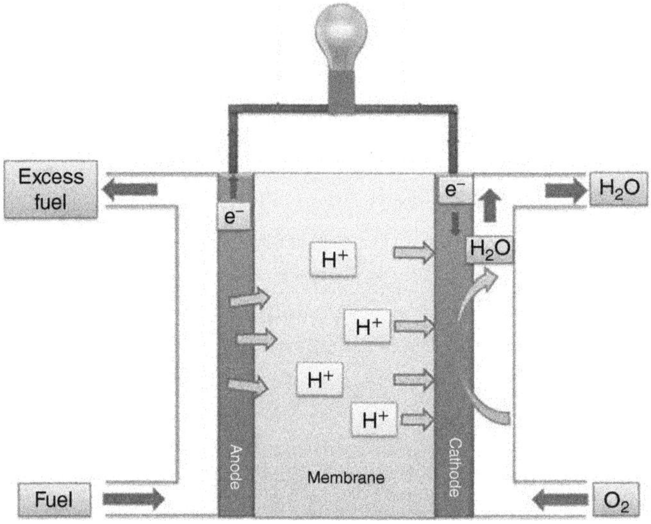

Figure 12.1 Schematic depiction of PEMFC and DMF.

proton conductivity, good electrical insulation, high thermo-mechanical, and chemical stability, cost-effectiveness, good barrier property, low swelling stresses, and capability for membrane electrode assembly fabrication.

There are two varieties of nanoclays, namely, montmorillonite (MMT) and Laponite (synthetic hectorite clay) that are being used for the fabrication of polymer nanocomposite membranes [37]. Surfaces of the nanoclays are modified by ionic, covalent, or plasma modification techniques taking into account the presence of replaceable Na^+ ions in the interlayer gallery spacing and the –OH groups present on the edges of the nanoclays. Ionic modifications of nanoclays are performed in two ways (e.g. acid activation and ion exchange with alkyl ammonium ions) by the replacement of exchangeable Na^+ ions present in the interlayer gallery spacing of nanoclays. The covalent modification of nanoclays involves the reaction between the –OH groups present on the surface of the nanoclays with alkyl silanes [38]. However, for plasma treatment, the end of the modifier capped with vinyl groups is covalently attached to the clay surface. In most cases, surface modifiers containing $-SO_3H$ and $-PO_3H$ groups are being used to further enhance the proton-conducting nature of nanoclays. Mishra et al. have reported the presence of H_3PO_4 on the surface of Laponite XLS. These H_3PO_4 are generated *in situ* due to the hydrolysis of peptizer ($Na_4P_2O_7$) adsorbed on the surface of the clay during its acid activation [39].

Electrostatic capacitors, electrolytic capacitors, electrochemical capacitors, batteries, and fuel cells are examples of energy storage devices [40]. These energy storage devices find applications depending on the energy and power that they can store or the cost of fabrication and installation in systems. The most widely used energy storage devices are summarized in Table 12.1 with their performance comparison [41].

Table 12.1 Optimal parameters of electrolytic capacitors, supercapacitors, and batteries.

Battery parameters	Electrolytic	Capacitor	Supercapacitor
Charge time	0.3–3 h	10^{-6}–10^{-3} sec	1–30 s
Discharge time	1–5 h	10^{6}–10^{-3} s	1–30 s
Energy density (Wh kg^{-1})	20–100	<0.1	1–10
Power density (W kg^{-1})	~50–200	>10 000	~1000–2000
Cycle life	~500–2000	>500 000	>100 000
Charge–discharge efficiency	~0.7–0.85	~1.0	~0.9–0.95
Operating voltage (V)	1.25–4.2	6–800	2.5–3 per cell
Operating temperature (°C)	−20 to 65	−20 to 100	−40 to 85

Polymer nanocomposites based on nematic liquid crystals and $CaCu_3Ti_4O_{12}$ (CCTO) nanoparticles have been prepared [42]. The host matrix is polymer-dispersed liquid crystals (PDLC) in which LC (E7) droplets are dispersed in different polymer blends ratios of polyvinyl chloride/polyaniline (PVC/PANI) [43]. A pseudo-perovskite structure material, CCTO, has established considerable interest among researchers, because it offers a new opportunity to generate ceramic/polymer dielectric composites. The CCTO has no piezoelectric property, although CCTO exhibits a colossal and stable dielectric constant (~104) in a wide temperature range (100–600 °C) [44]. Also, CCTO is lead-free and/or other hazardous metals-free such as cadmium which is a heavy metal used in the battery industry makes it a suitable choice for fabricating eco-friendly energy storage devices [45]. Therefore CP based on nonmaterial has been recognized to be a facilitating technology for building up high-performance energy storage and conversion devices. Composites based on PANI have recently gained extensive importance in energy storage applications owing to their distinct electrical conduction, environmental stability, and redox characteristics [46].

Polymer nanocomposite membranes are generally prepared by organic–inorganic nanofillers, which show the advantages of organic parts, such as flexibility, chemical stability, and processability, and the inorganic part of nanofillers exhibits the advantage of thermal and mechanical stabilities [47]. The crystallinity of the polymer structure is reduced due to the incorporation of hydrophilic inorganic nanofillers of different sizes, shapes, concentrations, and interactions, which results in increasing the amorphous phase and ionic conductivity. Fillers are generally categorized as (i) solid nonporous fillers such as titanium oxide, silica nanoparticles, and (ii) solid porous fillers such as zeolite, porous metal oxides, and so on. Praseodymium oxide (PSO, Pr_6O_{11}) is a rare earth metal oxide (dielectric constant k ~ 26–30, bandgap ~3.9 eV) with high k values, and the composites prepared by using PSO find an application in storage devices. Owing to its high dielectric nature, praseodymium finds many potential applications in nanodevices. It shows high electrical conductivity due to electron hopping between mixed metal ion

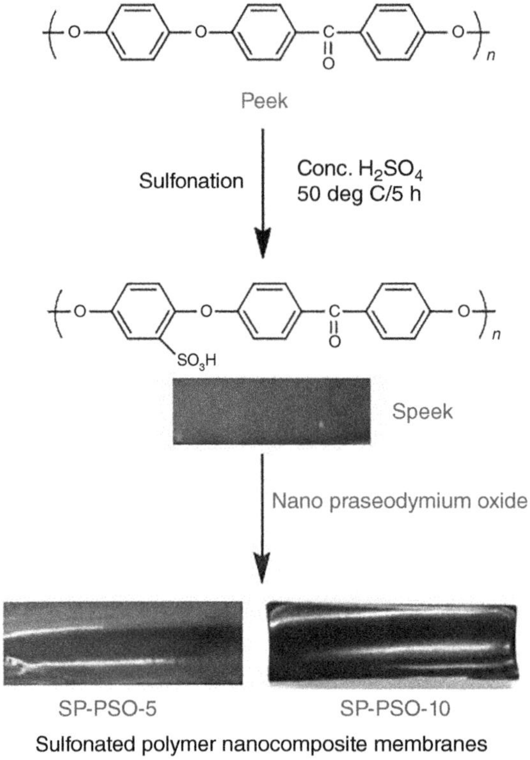

Figure 12.2 Schematic representation of the preparation of sulfonated polymer nanocomposite membranes.

valence states of lattice [48]. Malleable organic–inorganic polymer nanocomposite membranes with evenly dispersed metal oxide nanoparticles were prepared using sulfonated poly (ether ether ketone) (SPEEK) as a base material and PSO as an inorganic additive proton conductivity. PEEK is an aromatic, semicrystalline polymer and thus shows mild solubility in organic solvents because of its crystallinity. The crystallinity can be decreased and solubility can be increased by the sulfonation of PEEK [49] (Figure 12.2).

12.5 Capacitor and Supercapacitor Energy Storage

Capacitors store energy as an electric charge between two plates metal or conductive separated by an insulating material known as a dielectric when a voltage differential is applied across the plates. When one plate is charged with electricity from a direct current source, the other plate will have induced the opposite sign. The factors that determine the capacitance are the size of the plates, the separation of the plates, and the type of material used for the dielectric [50]. The energy stored in the capacitors is directly proportional to their capacity and the square of the voltage between the terminals of the electrochemical cell, while the capacity is proportional to the distance between the electrodes. Capacitors are already used in many utility power

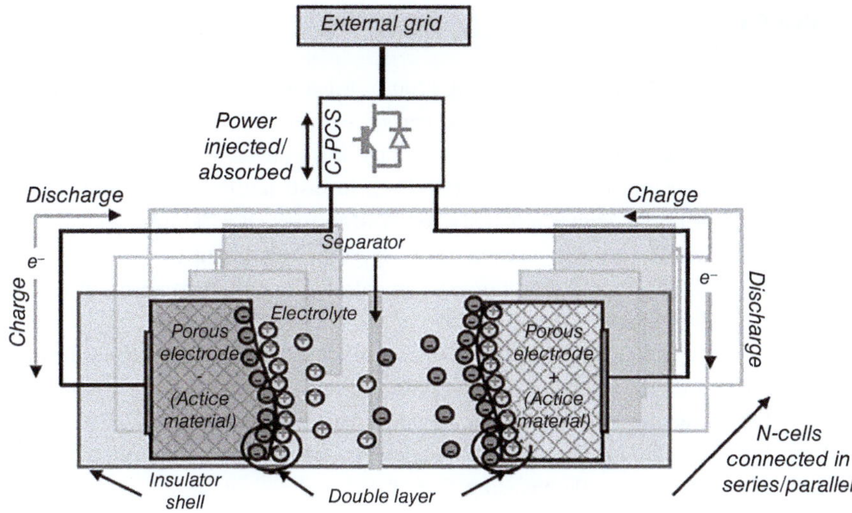

Figure 12.3 Schematic of a capacitor storage system.

control applications. The advantages of capacitors for small energy storage and short discharge are long cycle life and immediate recharge capability [21]. However, the main problem presented by capacitors is low energy density. If a large capacity is required, the area of the dielectric must be very large.

Supercapacitors have the same principle as capacitors except that the insulating material is replaced by an electrolyte ionic conductor in which ion movement is made along a conducting electrode with a very large specific surface providing higher energy density to the system. The design of the electrodes and the choice of electrolyte allow a very high charge density on the electrode surfaces but limit the voltage to approximately [22]. Despite the low voltage, the energy content is much higher than in conventional capacitors and can reach the scale of a few Wh for some of the largest supercapacitors which are now commercially available. Supercapacitors are connected to form larger modules with upto 1 kWh energy content and can be further joined together for larger energy storage units (Figure 12.3).

Supercapacitors have very high power output and energy storage systems now under trial reach approximately 50–100 kW. In most applications, the energy stored will supply the load only for a few seconds to minutes. The number of charges and discharge cycles is for all practical purposes nearly unlimited, but the energy put in fast cyclic operation is limited [51]. A control circuit to balance the individual voltages of each supercapacitor is necessary for safe and reliable operation if supercapacitors are connected in series to achieve a high output voltage. The lifetime of supercapacitors will probably be in the range of large conventional capacitors, e.g. 10 years. The in–out efficiency is very high, but the self-discharge rate is considerable compared with batteries [52]. Finally, it can be noted that the most important drawback of supercapacitors is their high cost estimated at five times that of lead-acid battery cost.

References

1 Smith, S.C., Sen, P.K., and Kroposki, B. (2008). Advancement of energy storage devices and applications in electrical power system. *Proc. IEEE* .

2 Nigim, K. and Reiser, H. (2009). Energy storage for renewable energy combined heat, power and hydrogen fuel (CHPH2) infrastructure. *Proceedings of the Electrical Power and Energy Conference IEEE*, Montreal, QC, Canada.

3 Huggins, R.A. (2010). *Energy Storage*. USA: Springer Science LLC.

4 Fergus, F.W. (2010). Recent developments in cathode materials for lithium-ion batteries. *J. Power Sources* 195: 939–954.

5 Anouti, M., Dougassa, Y.R., Tessier, C. et al. (2012). Low-pressure carbon dioxide solubility in pure electrolyte solvents for lithium-ion batteries as a function of temperature measurement and prediction. *J. Chem. Thermodyn.* 50: 71–79.

6 Soloveichik, G.L. (2011). Battery technologies for large-scale stationary energy storage. *Ann. Rev. Chem. Biomol. Eng.* 2: 503–527.

7 Chen, H.S., Cong, T.N., Yang, W. et al. (2009). Progress in electrical energy storage system: a critical review. *Prog. Nat. Sci.* 19: 291–312.

8 Bradford, R. (2009). Capturing grid power. *IEEE Trans. Power Energy Mag.* 7: 32–41.

9 Beaudin, M., Zareipour, H., Schellenberglabe, A., and Rosehart, W. (2010). Energy storage for mitigating the variability of renewable electricity sources: an updated review. *Energy Sustainable Dev.* 14: 302–314.

10 Nair, N.K.C. and Garimella, N. (2010). Battery energy storage systems: assessment for small-scale renewable energy integration. *Energy Build.* 42: 2124–2130.

11 Morioka, Y., Narukawa, S., and Itou, T. (2001). State of the art of alkaline rechargeable batteries. *J. Power Sources* 100: 107–116.

12 Broussely, M. and Pistoia, G. (2007). *Industrial Applications of Batteries. From Cars to Aerospace and Energy Storage*. Amsterdam: Elsevier B.V.

13 Hall, P.J. and Bain, E.J. (2008). Energy storage technologies and electricity generation. *Energy Policy* 36: 4352–4355.

14 Wen, Z., Cao, J., Gu, Z. et al. (2008). Research on sodium-sulfur battery for energy storage. *Solid State Ionics* 17: 1697–1701.

15 NGK Insulators Ltd. Website. http://www.ngk.co.jp/english/.

16 Bito, A. (2005). Overview of the sodium-sulfur battery for the IEEE Stationary Battery Committee. In: *IEEE Power Engineering Society General Meeting*, vol. 2, 1232–1235. IEEE.

17 Ritchie, A.G. (2001). Recent developments and future prospects for lithium rechargeable batteries. *J. Power Sources* 96: 1–4.

18 Kousksou, T., Bruel, P., Jamil, A. et al. (2014). Energy storage: applications and challenges. *Sol. Energy Mater. Sol. Cells* 120: 59–80.

19 Goswami, D.Y. and Kreith, F. (2007). *Energy Conversion*. Florida: CRC Press Taylor & Francis Group.

20 Nguyen, T. and Savinell, R.F. (2010). Flow batteries. *Electrochem. Soc. Interface* 19: 54–56.

21 Kazempour, S.J., Parsa Moghaddam, M., Haghifam, M.R., and Yousefi, G.R. (2009). Electric energy storage systems in a market-based economy: comparison of emerging and traditional technologies. *Renewable Energy* 34: 2630–2639.

22 Yamamura, T.T., Wu, X., Ohta, S. et al. (2011). Vanadium solid-salt battery: solid-state with two redox couples. *J. Power Sources* 196: 4003–4073.

23 Dhawale, D.S., Vinu, A., and Lokhande, C.D. (2011). Stable nanostructured polyaniline electrode for super-capacitor application. *Electrochim. Acta* 56: 9482–9487.

24 Palaniappan, S. and Devi, S.L. (2008). Novel chemically synthesized polyaniline electrodes containing a fluoroboric acid dopant for supercapacitors. *J. Appl. Polym. Sci.* 107: 1887–1892.

25 Shi, L., Wu, X., Lu, L. et al. (2010). Intercalated polyaniline nanosheets prepared from lyotropic liquid crystalline solutions and their capacitive performance. *Synth. Met.* 160: 989–995.

26 Li, Y., Zhao, X., Xu, Q. et al. (2011). Facile preparation and enhanced capacitance of the polyaniline/sodium alginate nanofiber network for supercapacitors. *Langmuir* 27: 6458–6463.

27 Wang, K., Zou, W., Quan, B. et al. (2011). An all-solid-state flexible micro-supercapacitor on a chip. *Adv. Energy Mater.* 1: 1068–1072.

28 Yan, Y., Cheng, Q., Wang, G., and Li, C. (2011). Growth of polyaniline nanowhiskers on mesoporous carbon for super-capacitor application. *J. Power Sources* 196: 7835–7840.

29 Mi, H., Zhang, X., Yang, S. et al. (2008). Polyaniline nanofibers as the electrode material for supercapacitors. *Mater. Chem. Phys.* 112: 127–131.

30 Kondawar, S.B., Thakare, S.R., Bompilwar, S., and Khati, V. (2009). Nanostructure titania reinforced conducting polymer composites. *Int. J. Modern Phys. B* 23: 3297–3304.

31 Tiwari, A., Prabaharan, M., Pandey, R., and Li, S. (2010). Vacuum-deposited thin film of Aniline–Formaldehyde condensate/$WO_3 \cdot nH_2O$ nanocomposite for NO_2 gas sensor. *J. Inorg. Organomet. Polym. Mater.* 20: 380–386.

32 Sharma, A.K., Sharma, Y., Malhotra, R., and Sharma, J.K. (2012). Solvent tuned PANI-CNT composites as advanced electrode materials for supercapacitor application. *Adv. Mater. Lett.* 3: 82–86.

33 Zhou, W., Xiao, J., Chen, Y. et al. (2011). A review on composite papers of graphene oxide, carbon nanotube, polymer/GO and polymer/CNT: processing strategies, properties, and relevance. *Polym. Adv. Technol.* 22: 1747.

34 Bose, S., Kuila, T., Mishra, A.K. et al. (2012). Carbon-based nanostructured materials and their composites as supercapacitor electrodes. *J. Mater. Chem.* 22: 767–784.

35 Choi, B.G., Huh, Y.S., Park, Y.C. et al. (2012). Enhanced transport properties in polymer electrolyte composite membranes with graphene oxide sheets. *Carbon* 50: 5395–5402.

36 Zheng, Y.Z., Zhang, M.L., and Gao, P. (2007). Preparation and electrochemical properties of multiwalled carbon nanotubes–nickel oxide porous composite for supercapacitors. *Mater. Res. Bull.* 42: 1740–1727.

37 Biswas, S. and Drzal, L.T. (2010). Multilayered nanoarchitecture of graphene nanosheets and polypyrrole nanowires for high-performance super-capacitor electrodes. *Chem. Mater.* 22: 5667–5671.

38 Mishra, A.K. and Valodkar, M.C. (2017). Polymer nanocomposites for energy and fuel cell applications. In: *Properties and Applications of Polymer Nanocomposites* (ed. D. Tripathy and B. Sahoo). Berlin, Heidelberg: Springer https://doi.org/10.1007/978-3-662-53517-2_6.

39 Zhang, Y., Feng, H., Wu, X. et al. (2009). Progress of electrochemical capacitor electrode materials: a review. *Int. J. Hydrogen Energy* 34: 4889–4899.

40 Basnayaka, P.A. (2013). Development of nanostructured graphene/conducting polymer composite materials for super-capacitor applications. Graduate Theses and Dissertations, University of South Florida, US., https://digitalcommons.usf.edu/etd/4864.

41 Labeeb, A.M., Ibrahim, S.A., Ward, A.A., and Abd-El-Messieh, S.L. (2020). Polymer/liquid crystal nanocomposites for energy storage applications. *Polym. Eng. Sci.* 60: 2529–2540.

42 Homes, C.C., Vogt, T., Shapiro, S.M. et al. (2001). Optical response of high-dielectric-constant perovskite-related oxide. *Science* 293: 673–676.

43 Morsi, M., Amr, S., and Gurpreet, S. (2016). *Processing, Properties and Design of Advanced Ceramics and Composites VI*. Hoboken, New Jersey: Wiley.

44 Lu, X., Hu, Y., Wang, L. et al. (2016). Macroporous carbon/nitrogen-doped carbon nanotubes/polyaniline nanocomposites and their application in supercapacitors. *Electrochim. Acta* 189: 158–195.

45 Rezakazemi, M., Sadrzadeh, M., and Matsuura, T. (2018). Thermally stable polymers for advanced high-performance gas separation membranes. *Prog. Energy Combust. Sci.* 66: 1–41.

46 Aslan, A. and Bozkurt, A. (2014). Nanocomposite membranes based on sulfonated polysulfone and sulfated nano-titania/NMPA for proton exchange membrane fuel cells. *Solid State Ionics* 255: 89–95.

47 Kausar, A. (2017). Overview on conducting polymer in energy storage and energy conversion system. *J. Macromol. Sci. A* 54: 640–653.

48 Muthumeenal, A., John Rethinam, A., and Nagendran, A. (2016). Sulfonated polyethersulfone based composite membranes containing heteropolyacids laminated with polypyrrole for electrochemical energy conversion devices. *Solid State Ionics* 296: 106–113.

49 Paster, M.D., Ahluwalia, R.K., Berry, G. et al. (2011). Hydrogen storage technologies options for fuel cell vehicles: well-to-wheel costs, energy efficiencies, and greenhouse gas emissions. *Int. J. Hydrogen Energy* 36: 14534–14551.

50 Rabiee, A., Khorramdel, H., and Aghaei, J. (2013). A review of energy storage systems in micro-grids with wind turbines. *Renewable Sustainable Energy Rev.* 18: 316–326.

51 Rafik, F., Gualous, H., Gallay, R. et al. (2007). Frequency, thermal and voltage super-capacitor characterization and modeling. *J. Power Sources* 165: 928–934.

52 Gualous, H., Bouquain, D., Berthon, A., and Kauffmann, J.M. (2003). Experimental study of super-capacitor serial resistance and capacitance variations with temperature. *J. Power Sources* 123: 86–93.

13

Electrochemical Energy Storage System

13.1 Introduction

Due to the growing energy crisis caused by the depletion of conventional fossil fuels, the production of sustainable and renewable storage resources in reservoirs for renewable energy such as the wind and the solar energy sector is urgently needed [1]. Because renewable energy supplies are intermittent, efficient energy storage systems need to store and generate renewable energy rapidly and stably. The research interest of the electrochemical energy storage systems (EESSs) with electrochemical capacitors (ECs), batteries [2], and fuel cells among different storage systems has been important such as hydro-pumped storage, thermal energy, compressed air energy storage, and flywheel storage [3]. Lithium-ion batteries (LIBs) are the relevant application of polymer nanocomposites. LIBs have become the most popular and frequently used rechargeable battery, with the benefits of high working voltage, huge power, lower toxicity, and long life cycle. Graphite, silicon, and metal oxide are commonly used in LIB as anode materials and $LiCoO_2$, $LiFePO_4$, Li[NiCoAl]O_2, Li [MnNiCo]O_2, and $LiMn_2O_4$ [4] as cathode materials.

EESS launch an essential element in the advance of sustainable energy technologies. The electrical energy created from renewable resources, for example, solar radiation or wind, provides great potential to meet our energy requirements sustainably. Though, these renewable energy technologies generate electricity irregularly and thus need proficient and dependable electrical energy storage methods. For commercial and residential end use, electricity must be constantly available at any time of the day. The second-to-second fluctuations can cause major interruptions with costs estimated to be in the tens of billions of dollars yearly [5]. Therefore, the development of new electrochemical energy storage (EES) systems will be vital in the use of large-scale solar or wind-based electricity generation. Moreover, significantly enhanced EES systems are essential to permit the widespread use of hybrid electrical vehicles (HEVs), plug-in hybrids, and all-electric vehicles. Progresses in ESSS performance, reliability, and efficacy are desirable in the development of modern portable electronic devices such as laptops and smartphones (http://www.seas.ucla.edu/~pilon/EES.html).

ECs, also recognized as supercapacitors or ultra-capacitors, are normally classified into two classes based on their diverse energy storage mechanisms, i.e. electric

Polymer Nanocomposites for Energy Applications, First Edition.
T. Daniel Thangadurai, Manjubaashini Nandhakumar, Sabu Thomas, and Ange Nzihou.

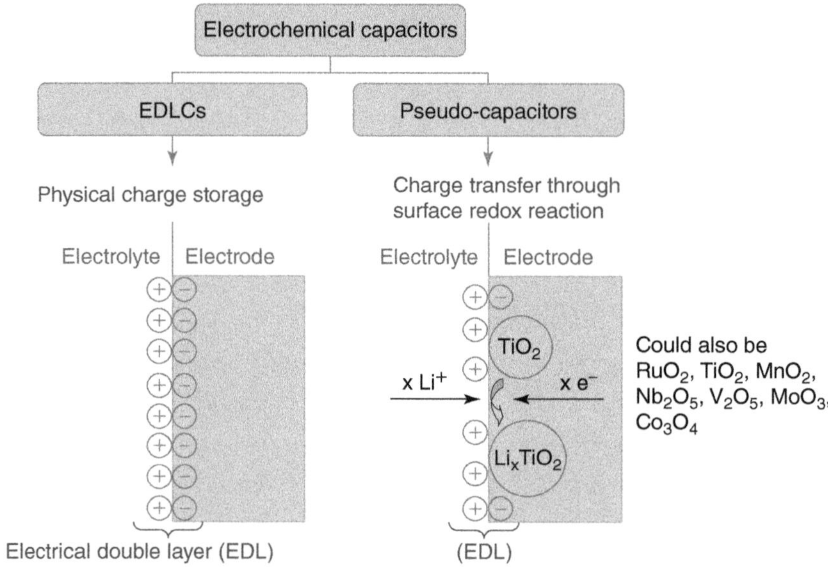

Figure 13.1 Classification of electrochemical capacitors into electrical double-layer capacitors and pseudo-capacitors.

double-layer capacitors (EDLCs) and pseudo-capacitors. First, EDLCs store charges physically in electric double layers developing near the electrode–electrolyte interfaces. Thus, the course is extremely reversible and the cycle of life is primarily infinite. In contrast, pseudo-capacitors store energy via not only electric double layer such as that in EDLCs but also fast surface oxidation–reduction (redox) reactions as well as potential ion intercalation in the electrode. The performance of ECs is measured by the combination of the electrode material and electrolyte used. There are three main categories of electrode materials used for ECs, namely (i) carbon-based materials, (ii) transition metal oxides, and (iii) conductive polymers. Likewise, three types of electrolyte materials are used for ECs, including (i) aqueous electrolytes, (ii) organic electrolytes, and (iii) ionic liquids (Figure 13.1).

The performances of ECs can be likened to the Ragone chart plotting their respective energy and power densities as illustrated in Figure 13.2 for different electrical energy storage devices. Owing to their physical charge storage, capacitors feature very large power densities equated with batteries and fuel cells but low energy densities. Instead, batteries and fuel cells have large energy densities but low power densities due to their slow reaction kinetics. ECs bridge the gap between capacitors and batteries/fuel cells [6]. They afford the prospect of keeping the high energy density of batteries without compromising the high power density of capacitors.

Polymer composites play a fundamental role in EESSs. They are widely used as both packaging and active materials in these devices. For example, conductive composites are used to efficiently conduct and distribute charge effectively through electrodes of supercapacitors, batteries, and fuel cells. In that case, the electronic conductive pathways formed in the composites are critical to delivering and

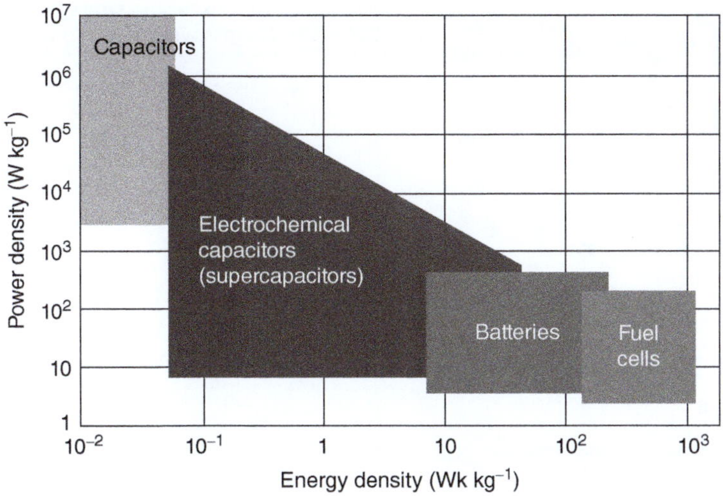

Figure 13.2 Power density vs. energy density of various energy storage systems.

extracting current out of the device. The capacity of electrical energy storage devices is largely determined by the electrode material as well as by its ability to store and release energy at the electrochemical interface between an electrode and an electrolyte [7]. The various materials studied as active nanofiller, graphene, carbon nanotubes, and several metal oxides have shown good results because of their high conductivity, surface area, and pseudo-capacitive properties.

However, many challenges related to the complete understanding of the role of these fillers and their stability and efficiency in charge storage applications still exist. This research activity is devoted to the issues associated with the development, synthesis, characterization, and use of new advanced composites materials for EESS (http://www.ipcb.cnr.it/index.php/en/staff/permanent/25-research/advanced-materials). Such systems include high-performance polymer materials, biodegradable materials, conductive polymers, as well as new and advanced nanofillers.

Nonaqueous redox flow batteries (NRFBs) provide high energy density compared to aqueous flow batteries. However, NRFBs struggle from active material crossover and low ionic conductivity, which are the major problems to be overcome for their commercialization. Accordingly, rapid and certain transport of the charge balancing ions and negligible crossover of the redox-active material is extremely preferred. The proof-of-concept demonstration for redox-active violates polymers in which coulombic efficiency of flowable electrodes is maintained by the size-exclusion-based mechanism of separation is provided. The polymer materials reveal high solubility in propylene carbonate (one molar) to permit high storage capacity. The size exclusion strategy will permit the use of commercial off-the-shelf (COTS) membranes in flow batteries, thus helping to bring down the price and upsurge the ionic conductivity and coulombic efficiency of flow batteries. For the first time, a redox-active polymer (RAP) that encounters the requirements for use

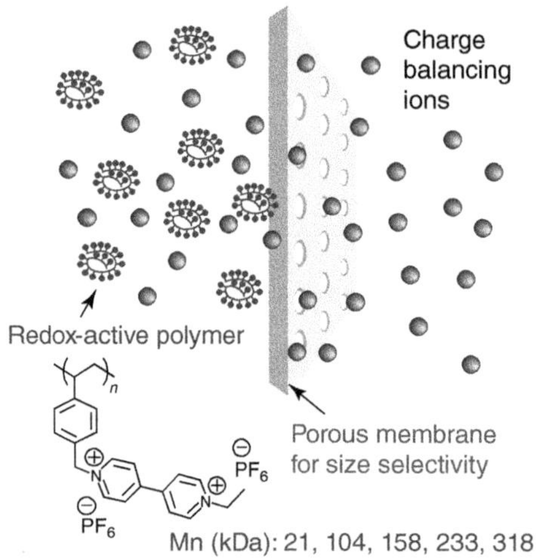

Figure 13.3 Size exclusion strategy for active material crossover prevention.

as an analyte in NRFBs is established. Molecular design principles are valuable to develop redox-active nanostructures and colloidal particles for use in nonaqueous colloidal flow batteries (Figure 13.3).

The design of chemically stable and electrochemically reversible RAPs is of great attention for energy storage machinery. The tether length and position of pendants have a pronounced character in their reactivity, electrochemical stability, and long-distance charge transfer kinetics. The systematic molecular design approaches to investigate the impact of pendant–backbone electronic interactions on the performance of redox-active pendant units were utilized. The study of viologen oligomers and polymers showed that a shorter or more rigid tether has positive shifts in the reduction potential and displays enhanced levels of electrochemical kinetics. The investigation of cyclopropenium appended polymers points out that polymers with

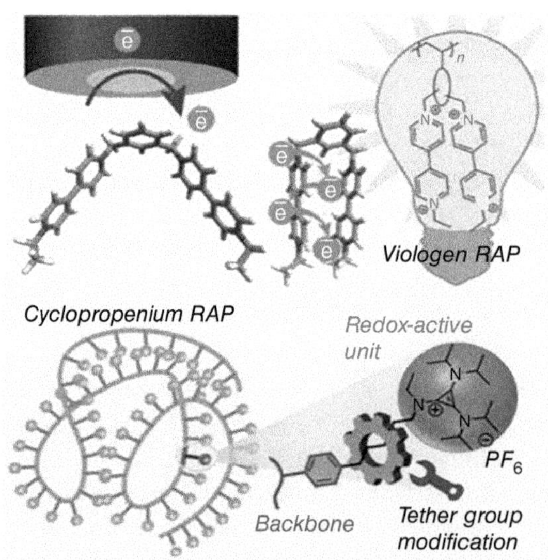

Figure 13.4 Polymer design for superior electrochemical performances.

extended tether groups show enhanced reversibility in cyclic voltammetry. The performance is mirrored in the stability of the charged state tested in galvanostatic half-cells. The capacity decays for the polymers were structure-dependent, providing empirical understanding into the design of the next generation of RAP materials (https://mooregroup.beckman.illinois.edu/energy-storage-materials) (Figure 13.4).

13.2 Need for Energy Storage System

13.2.1 Energy Reality and Increasing Renewable Penetration

The current worldwide electric generation capacity is estimated to be about 20 TW hours ($\times 10^{12}$ watts). Approximately 68% of today's electrical energy is supplied from fossil fuels: coal (42%), natural gas (21%), oil (5%), nuclear (14%), hydro (15%), and the remaining 3% from renewable energy technologies. Even with aggressive conservation and development of new, efficient technologies, the worldwide electricity demand is predicted to double by the middle of the century and triple by the end of the century [8]. Electricity is the dominant form of energy used (e.g. 40% of all energy consumption in the United States by 2002), and the electricity demand is increasing at a faster pace than overall energy consumption. At the same time, oil and natural gas production is predicted to peak over the next few decades. Coal has been the dominant source of electricity generation in the world; abundant coal reserves may maintain current consumption levels longer than oil and gas. However, every kWh of electricity generated by burning coal coproduces an average 1000 g of life cycle CO_2 emission, a greenhouse gas, that is widely considered as the primary contributor to global warming. In the United States alone, coal power plants emit 1.5 billion tons of CO_2 per year, and emissions from developing countries are accelerating [9]. To reduce greenhouse gas emissions, many countries are adopting emission regulations (cap-and-trade or variants) and carbon trading which benefits industries with a small carbon footprint and requires those producing higher emissions to purchase carbon allowances.

The environmental concerns over the use of fossil fuels and their resource constraints, combined with energy security concerns, have spurred great interest in generating electric energy from renewable sources. Solar and wind energy are among the most abundant and potentially readily available. The solar radiation energy, the earth receives in 1 hour, is enough to meet worldwide energy requirements for a year [10]. Capturing a small percentage of potential wind energy could also contribute significantly to meeting the world's electrical energy requirements. While advances in technology are still needed to harvest renewable energy economically, solar and wind power technologies have grown quickly. Globally, the total electricity from installed wind power reached 74.3 GW in 2006 and 94 GW in 2007. The World Energy Council (WEC) estimates that new wind capacity worldwide will total up to 474 GW by 2020. The output from photovoltaic (PV) module installations is currently growing at 40% per year worldwide. The United States targets 100 GW of solar power by 2020 [11] (Figures 13.5 and 13.6).

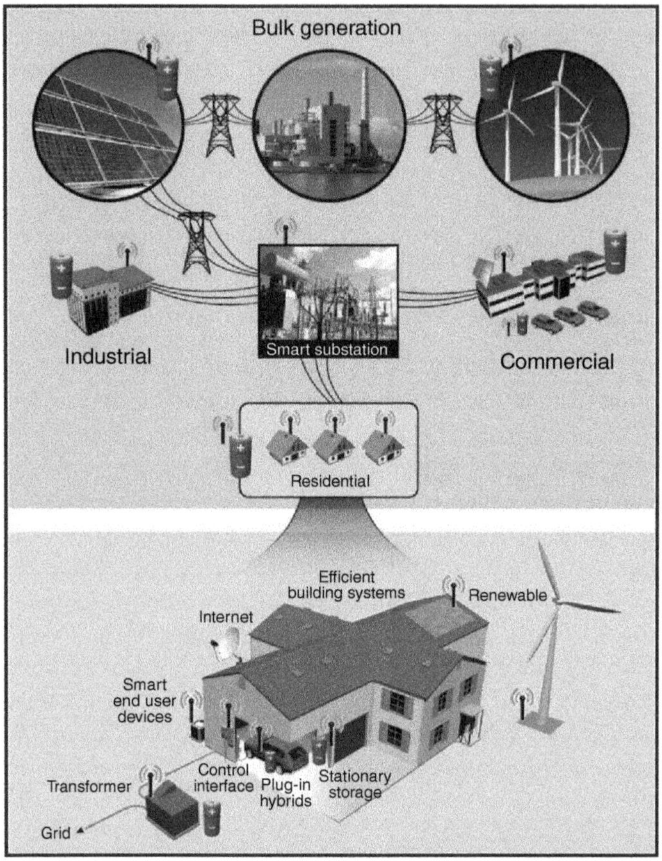

Figure 13.5 Schematic of applications of electricity storage for generation, transmission, distribution, and end customers and a future smart grid that integrates with intermittent renewables and plug-in hybrid vehicles through two-way digital communications between loads and generation or distribution grids. Source: Yang et al. [5]/with permission from U.S. National Library of Medicine.

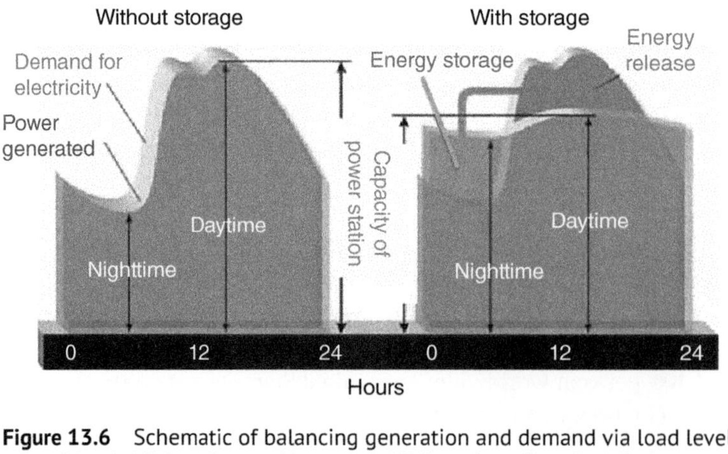

Figure 13.6 Schematic of balancing generation and demand via load leveling, a typical case of load shifting. Source: Yang et al. [5]/American Chemical Society.

References

1 Cook, T.R., Dogutan, D.K., Reece, S.Y. et al. (2010). Solar energy supply and storage for the legacy and nonlegacy worlds. *Chem. Rev.* 110: 6474–6502.

2 Mishra, A., Shetti, N.P., Basu, S. et al. (2019). Carbon cloth-based hybrid materials as flexible electrochemical supercapacitors. *ChemElectroChem* 6: 5771–5786.

3 Ding, K., Gu, H., Zheng, C. et al. (2014). A novel material $Li_2NiFe_2O_4$: preparation and performance as anode of lithium-ion battery. *Electrochim. Acta* 146: 585–590.

4 Etacheri, V., Marom, R., Elazari, R. et al. (2011). Challenges in the development of advanced Li-ion batteries: a review. *Energy Environ. Sci.* 4: 3243–3262.

5 Yang, Z., Zhang, J., Kintner-Meyer, M.C.W. et al. (2011). Electrochemical energy storage for green grid. *Chem. Rev.* 111: 3577–3613.

6 Huggins, R.A. (2016). Introduction to electrochemical energy storage. In: *Energy Storage*. Springer Cham: https://doi.org/10.1007/978-3-319-21239-5-9.

7 Wang, X., Salari, M., Jiang, D. et al. (2020). Electrode material–ionic liquid coupling for electrochemical energy storage. *Nat. Rev. Mater.* 5: 787–808.

8 (2010). *World Energy Outlook*. Paris: International Energy Agency.

9 Powell, C. and Morreale, B. (2008). Materials challenges in advanced coal conversion technologies. *MRS Bull.* 33: 309–315.

10 Hayman, B., Wedel-Heinen, J., and Brøndsted, P. (2008). Materials challenges in present and future wind energy. *MRS Bull.* 33: 343–353.

11 Meier, P.J., Wilson, P.P.H., Kulcinski, G.L., and Denholm, P.L. (2005). US electric industry response to carbon constraint: a life-cycle assessment of supply-side alternatives. *Energy Policy* 33: 1099–1108.

14

Electrical Energy Storage System

14.1 Introduction

Dielectric materials are the basis of a fundamental electric circuit element, a dielectric capacitor, which can be found in almost all electric circuits. Dielectric capacitors are used to control and store electric charge and electrical energy in electrical and electronic devices, such as electric power converters, pulse power systems, and electric power systems [1]. For example, dielectric capacitors convert the DC electricity from sources such as batteries to AC electricity in the power inverters of hybrid electric vehicles (HEVs), in which motors are driven by AC power [2]. Dielectric polymers are critical to meet the increasing demands for high-energy-density capacitors operating in harsh environments, such as aerospace power conditioning, underground oil and gas exploration, electrified transportation, and pulse power systems. A crosslinked fluoropolymer poly (chlorotrifluoroethylene-co-vinylidene fluoride) (P(CTFE-VDF)) is prepared by the melting processing of the mixture of P(CTFE-VDF), dicumyl peroxide (DCP) as the initiator, and triallyl isocyanurate (TAIC) as the co-agent [3]. The crosslinked P(CTFE-VDF) exhibits much improved discharged energy densities and greater charge–discharge efficiencies, along with excellent breakdown strength and cyclic stability, at elevated temperatures when compared with the current dielectric polymers. Further, the optimized crosslinked P(CTFE-VDF) can deliver an energy density of $2.67\,J/cm^3$ with a charge–discharge efficiency of >90% at $400\,MV\,m^{-1}$ and 150 °C (Figure 14.1), which exceeds most high-temperature polymer dielectrics including c-BCB, PI, PEI, FPE, SO_2-PPO, and SO_2-PIM, and polymer nanocomposites [5]. The comprehensive characterization of conduction current, thermally stimulated current, and embedded charges indicate that the crosslinked structure efficiently functions as molecular deep trapping sites to inhabit charge carrier transport and reduce conduction current and thus leads to significantly improved capacitive performance [4].

The ferroelectric polymers are promising dielectric energy storage media for film capacitors due to their superiority in excellent dielectric properties, high breakdown strength, and flexibility [6]. Compared with ceramic dielectrics, polymeric ones possess superiorities in high breakdown strength, mechanical flexibility, lightweight, and easy processing, which are preferable materials for capacitors. However, the intrinsic low dielectric constant of polymers seriously

Polymer Nanocomposites for Energy Applications, First Edition.
T. Daniel Thangadurai, Manjubaashini Nandhakumar, Sabu Thomas, and Ange Nzihou.

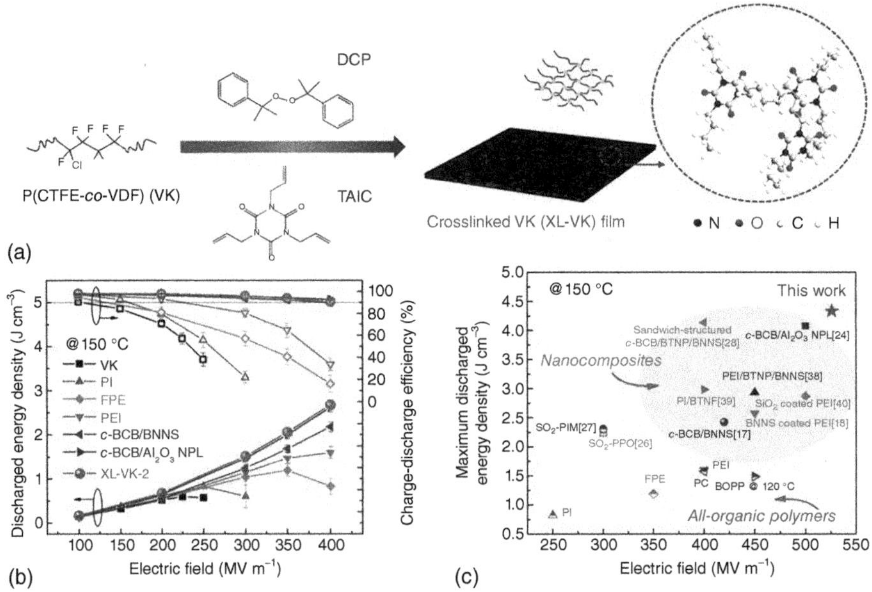

Figure 14.1 (a) Schematic of the preparation of crosslinked P(CTFE-VDF)(XL-VK), (b) discharged energy densities and charge–discharge efficiencies of XL-VK and high-temperature dielectrics at 150 °C, and (c) comparison of the discharged energy density of high-temperature dielectric polymers and polymer composites at 150 °C. Source: Li et al. [4]/Royal Society of Chemistry.

limits the energy density [7]. For instance, the dielectric constant of the commercial biaxial-oriented polypropylene (BOPP) is only 2.2, resulting in an energy density of fewer than $2\,J\,cm^{-3}$. Ferroelectric polymers such as poly(vinylidene fluoride) (PVDF) exhibit higher dielectric constants of ~10, which have been receiving substantial attention in the aspect of energy storage applications. Furthermore, the nanocomposites by incorporating ceramic nanofillers with even higher constants (>100) into the ferroelectric polymer matrix are expected to further enhance the energy storage performances [8]. The ferroelectricity of polymers lies in the dipoles arranged in the same direction, producing spontaneous displacement. Polymeric ferroelectricity has been discovered in odd nylon, polylactic acid (PLLA), polylactic glycolic acid (PLGA), as well as PVDF and its copolymers, among which PVDF and its copolymers are the best-known and most widely used. PVDF is a semicrystalline polymer with multiple crystalline phases, which usually can be divided into four polymorphs according to the molecular chain configurations, including the α-phase (in trans-gauche conformation TGTG), β-phase (in all-trans planar zigzag conformation TTTT), γ-phase (in a conformation of three trans-linked to a gauche TTTG), and δ-phase (a polar version of the α-phase) Figure 14.2 [9].

The exploration of high-energy-density electrostatic capacitors capable of operating both efficiently and reliably at elevated temperatures is of great significance to meet advanced power electronic applications. The energy density of a capacitor is strongly dependent on the dielectric constant and breakdown strength of a dielectric

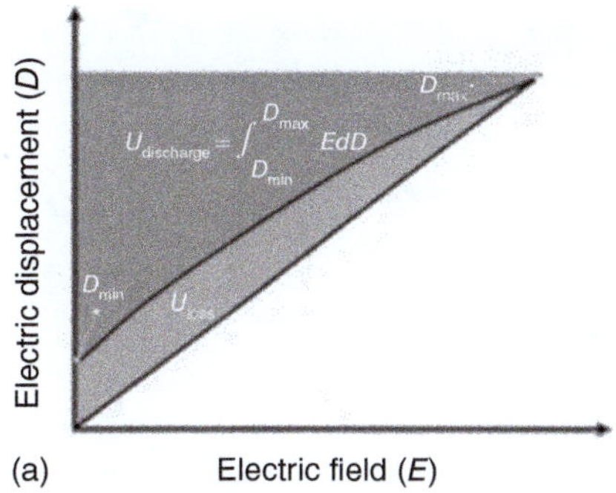

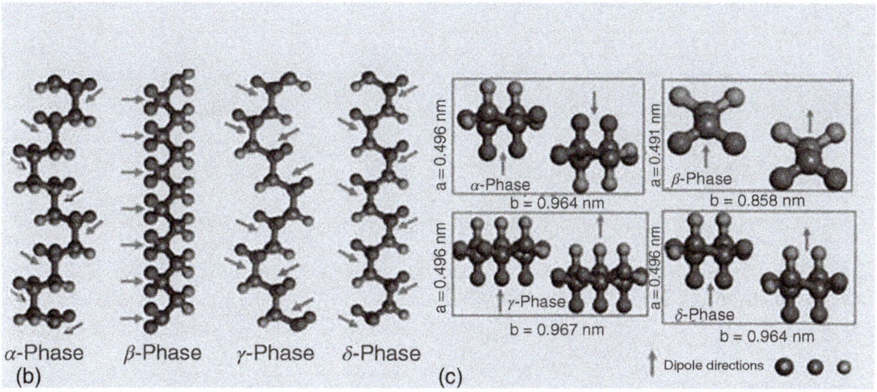

Figure 14.2 (a) Graphical representation of D–E loops used for the calculation of energy density, (b) schematic chain conformations of PVD, (c) unit cells of α, β, γ, and δ forms of PVDF crystals viewed along the c-axes. Source: Jiang et al. [9]/American Institute of Physics/CC BY-4.0.

material. BNNSs were fabricated via a modified liquid-phase exfoliation method [10]. Binary PEI/BTNP and PEI/BNNS and ternary PEI/BTNP/BNNS nanocomposite films were fabricated through solution casting processes (Figure 14.3).

In the transmission electron microscopy (TEM) image of the exfoliated BNNS (Figure 14.4A), the thickness of BNNS is around 2.2 nm with two to six layers. The width of the nanosheets is around 400 to 600 nm as demonstrated [12]. Figure 14.4B, C show the cross-sectional images of the poly(ether imide) (PEI) nanocomposite films filled with BTNPs and BNNSs, respectively, indicating that the inorganic fillers are well dispersed in the PEI matrix. Stable ternary suspensions in nitroxide mediated polymerization (NMP) at a relatively high concentration of 25 : 1.5 : 2.5 mg ml^{-1} of PEI to BTNPs to BNNSs are shown in Figure 14.4D [13]. It can be seen from Figure 14.4E that the nanocomposite film displays excellent mechanical flexibility even at a relatively high filler concentration, for example, 1.27 vol% BTNPs and 6.05 vol% BNNSs. The cross-sectional morphology of the ternary nanocomposite is exhibited in Figure 14.4F, in which the homogeneous filler dispersion is verified by the EDS mapping of different element distributions (Figure 14.4G) [14].

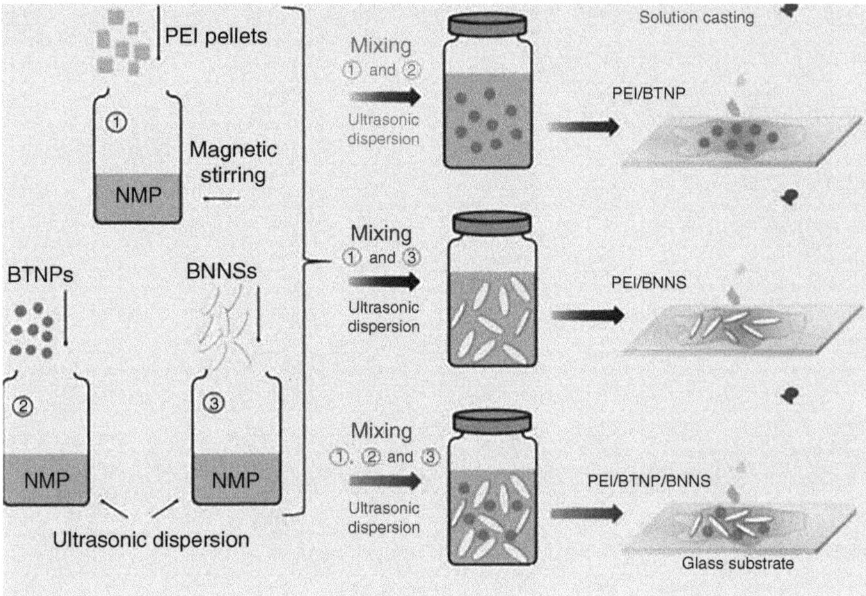

Figure 14.3 Schematic of the preparation of the poly(ether imide) (PEI)-based nanocomposite films. Source: Reprinted with permission from Li et al. [11]/John Wiley & Sons/CC BY-4.0.

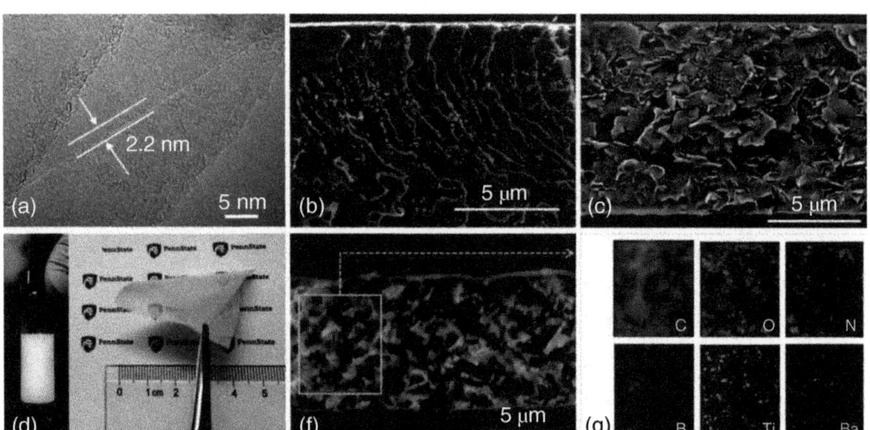

Figure 14.4 (A) TEM image of BNNS with exposed edge showing the layer structure, (B) cross-sectional SEM image of the PEI nanocomposite with 1.27 vol% BTNPs, (C) cross-sectional SEM image of the PEI nanocomposite with 6.05 vol% BNNSs, (D) photograph of PEI pellet, BTNP, and BNNS dissolved in NMP, (E) photograph of a bent ternary PEI nanocomposite film containing 1.27 vol% BTNPs and 6.05 vol% BNNSs, (F) cross-sectional SEM image of the ternary PEI nanocomposite with 1.27 vol% BTNPs and 6.05 vol% BNNSs, and (G) element mapping of carbon (C), oxygen (O), nitrogen (N), boron (B), titanium (Ti), and barium (Ba) on the orange-marked area in F established using EDS. Source: Li et al. [11]/with permission from John Wiley & Sons, Inc.

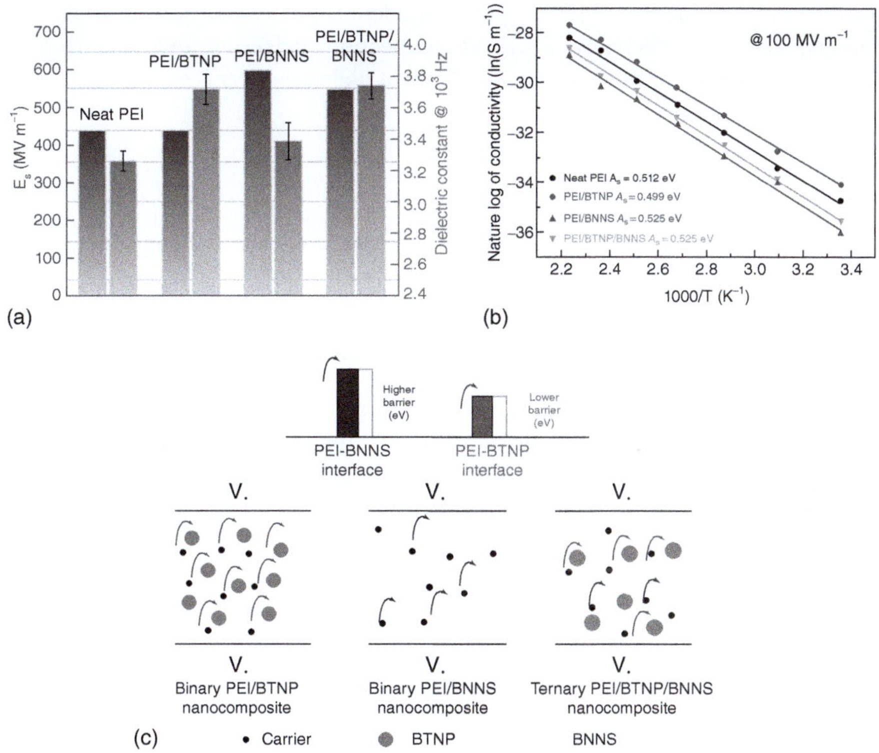

Figure 14.5 (a) Comparison of the Weibull breakdown strength at 150 °C and the dielectric constant at 103 Hz and room temperature of neat PEI, the binary PEI/BTNP nanocomposite with 1.27 vol% BTNP, the binary PEI/BNNS nanocomposite with 6.05 vol% BNNS, and the ternary PEI nanocomposite with 1.27 vol% BTNPs and 6.05 vol% BNNSs, (b) Arrhenius plots of electrical conductivity of neat PEI, the binary PEI/BTNP nanocomposite with 1.27 vol% BTNPs, the binary PEI/BNNS nanocomposite with 6.05 vol% BNNSs, and the ternary PEI nanocomposite with 1.27 vol% BTNPs and 6.05 vol% BNNSs under 100 MV m^{-1}, and (c) schematics of the interfacial barrier and the carrier transport process of the PEI-based nanocomposites. Source: Li et al. [11]/John Wiley & Sons/CC BY-4.0.

Figure 14.5A compares K and Eb of neat PEI, binary 1.27 vol% BTNP-filled PEI, 6.05 vol% BNNS-filled PEI nanocomposite, and the ternary PEI nanocomposite with 1.27 vol% BTNPs and 6.05 vol% BNNSs. A high K of 3.74 along with a low tan δ of 0.00715 at 103 Hz and room temperature as well as an outstanding Eb of 547 MV m^{-1} measured at 150 °C are obtained in the ternary PEI nanocomposite, significantly outperforming pristine PEI. It is known that electrical conduction plays a critical role in dielectrics operating at elevated temperatures and high electric fields. Therefore, investigation on the conduction behaviors of the composites at a relatively high electric field, for example, 100 MV m^{-1}, at varying temperatures is shown in Figure 14.5B. In contrast, the addition of BNNSs increases the activation energy to 0.525 eV, which is attributed to a higher barrier at the interface between polymer matrix and BNNS than that of BTNP (Figure 14.5C) [11]. In the rationally designed ternary composites, BNNSs not only give rise to a higher energy barrier for the carrier to overcome during the conduction process because of their wide bandgap but also effectively block the connection of BTNPs, thus ensuring the ternary polymer composites maintain excellent dielectric strength.

References

1 Bell, A.J. (2008). Ferroelectrics: the role of ceramic science and engineering. *J. Eur. Ceram. Soc.* 28: 1307.

2 Montanari, D., Saarinen, K., Scagliarini, F. et al. (2009). Film capacitors for automotive and industrial applications. In: *Proceedings of CARTS USA 2009*, 23–38. Electronic Components Industry Association.

3 Zhou, Y. and Wang, Q. (2020). Advanced polymer dielectrics for high-temperature capacitive energy storage. *J. Appl. Phys.* 127: 240902.

4 Li, H., Gadinski, M.R., Huang, Y. et al. (2020). Crosslinked fluoropolymers exhibiting superior high-temperature energy density and charge–discharge efficiency. *Energy Environ. Sci.* 13: 1279.

5 Chen, S., Meng, G., Kong, B. et al. (2020). Asymmetric alicyclic amine-polyether amine molecular chain structure for improved energy storage density of high-temperature crosslinked polymer capacitor. *Chem. Eng. J.* 387: 123662.

6 Tan, D.Q. (2020). Review of polymer-based nanodielectric exploration and film scale-up for advanced capacitors. *Adv. Funct. Mater.* 30: 1808567.

7 Michalczyk, P. and Bramoulle, M. (2003). Ultimate properties of the polypropylene film for energy storage capacitors. *IEEE Trans. Magn.* 39: 362.

8 Rabuffi, M. and Picci, G. (2002). Status quo and future prospects for metallized polypropylene energy storage capacitors. *IEEE Trans. Plasma Sci.* 30: 1939.

9 Jiang, Y., Zhou, M., Shen, Z. et al. (2021). Ferroelectric polymers and their nanocomposites for dielectric energy storage applications. *APL Mater.* 9: 020905.

10 Wang, Y., Zhou, X., Chen, Q. et al. (2010). Recent development of high energy density polymers for dielectric capacitors. *IEEE Trans. Dielectr. Electr. Insul.* 17: 1036–1042.

11 Li, H., Ren, L., Ai, D. et al. (2019). Ternary polymer nanocomposites with concurrently enhanced dielectric constant and breakdown strength for high-temperature electrostatic capacitors. *InfoMat* 1–12.

12 Zhang, G., Fan, B., Zhao, P. et al. (2018). Ferroelectric polymer nanocomposites with complementary nanostructured fillers for electrocaloric cooling with high power density and great efficiency. *ACS Appl. Energy Mater.* 1: 1344–1354.

13 Li, H., Xie, Z., Liu, L. et al. (2019). High-performance insulation materials from poly(ether imide)/boron nitride nanosheets with enhanced DC breakdown strength and thermal stability. *IEEE Trans. Dielectr. Electr. Insul.* 26: 722–729.

14 Niu, Y. and Wang, H. (2019). Dielectric nanomaterials for power energy storage: surface modification and characterization. *ACS Appl. Nano Mater.* 2: 627–642.

15

Real-Time Applications of Polymer Nanocomposites

15.1 Introduction

The need of the hour is to create a large-scale energy storage system to integrate renewable energy sources, such as solar and wind, to be effectively used in the grid and to avoid output fluctuation. Redox flow batteries (RFBs) have received tremendous attention for this kind of energy storage. Vanadium-based RFBs received much attention; however, their low volumetric energy storage capacity, limited solubility, and need for ion exchange membranes are some of the major drawbacks. Recently, redox organic polymers have been proposed as charge storage materials, which are potentially low cost, soluble in a range of solvents, along with the possibility to tune their redox properties, solubility, and crossover by introducing different substituents. Recent developments toward developing water-soluble redox-active polymeric materials have opened a new avenue for developing safer and greener RFBs. However, water-soluble polymers and membranes with very high selectivity are crucial for this kind of RFBs. New polymers with enhanced solubility and redox activity along with fabrication of novel Janus membranes for better selectivity and low crossover of active polymers are synthesized. The project is intended to develop a novel aqueous organic polymer-based redox flow battery (ORFB) system based on Janus nanoporous and charged membranes as well as tailor-made electro/redox-active organic oligomers and polymers. Two crucial components, i.e. Janus membrane separator and redox-active polymers, for an aqueous organic polymer-based ORFB system are developed (Figure 15.1a, b).

To realize the full potential of ORFB, novel tailor-made redox-active polymers and Janus membranes, which are critical components for this type of battery, are developed. The focus is water-soluble redox-active polymers based on monomers with redox-active moieties and Janus membranes to achieve high cell performance by limiting the crossover of materials and long-term application. The functionalization of these polymers is also be undertaken by introducing aqueous solubilizing units such as cationic and anionic functional groups. The second part of the work plan is to fabricate charged and ion conductive nanoporous Janus membranes by chemical modification or layer-by-layer deposition method with well-defined nanopores to control the ionic transport as well as selectivity concerning anolyte and catholyte. We are analyzing the Janus membranes regarding molecular weight cutoff (monomer to

Polymer Nanocomposites for Energy Applications, First Edition.
T. Daniel Thangadurai, Manjubaashini Nandhakumar, Sabu Thomas, and Ange Nzihou.

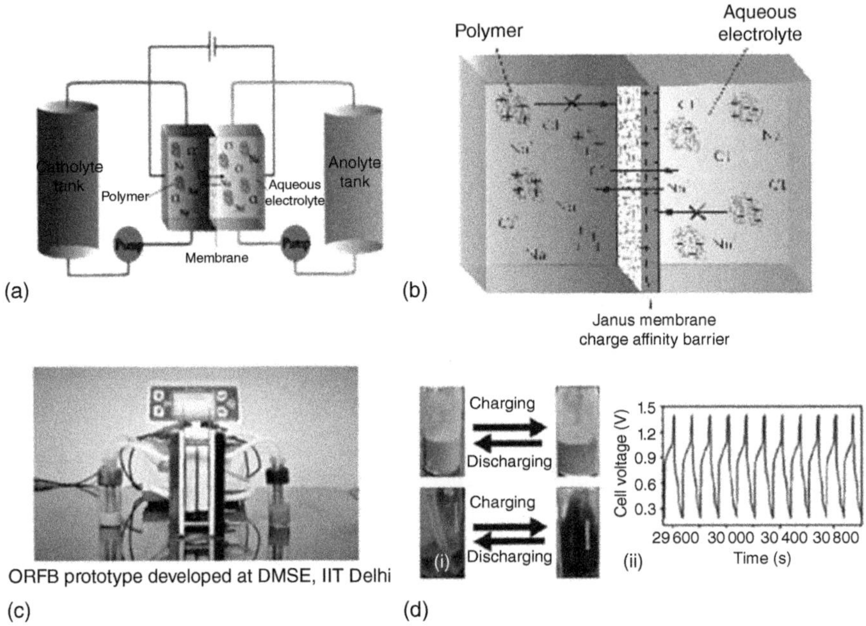

Figure 15.1 (a) and (b) Demonstration of an electrochemical cell with a Janus membrane, (c) prototype model, and (d) (i) color change of electrolyte and anolyte polymers and (ii) charge–discharge curves. Source: Tripathi [1]/Department of science and technology/ Public Domain.

oligomer and polymer), surface charges, ion conductivity, diffusion and crossover, and their electrochemical behavior. Finally, a lab-scale prototype RFB consisting of Teflon support and flow chambers, current collectors, EPDM gaskets, graphite felt electrodes, and reservoirs for catholyte and anolyte connected to a peristaltic pump is fabricated. The galvanostatic charging–discharging of the developed batteries and their life cycle were assessed [1]. A lab-scale prototype organic polymer-based RFB using synthesized polymers and Janus membranes is displayed in Figure 15.1c. Characterization and performance assessment of ORFB under neutral pH and aqueous conditions were shown in Figure 15.2.

The goal of the project is to improve the tensile strength, stress rupture, fatigue life, and fracture toughness of T1000 carbon/epoxy composite flywheel material by at least 30% so that it can rotate faster. The specific objectives are to fabricate ZnO nanowire/T1000 carbon/epoxy hybrid composites and study their mechanical properties. The properties of interest are interfacial strength, tensile strength and stiffness (normal and transverse), tensile stress rupture, fatigue life, and Mode-I fracture toughness to fabricate prototypes of composite flywheels (25 cm length × 25 cm OD × 2.5 mm thickness) and perform spin test to determine the burst speed and energy density.

The approach is to integrate ZnO nanowires at the fiber–matrix interface (by hydrothermal method) in T1000 carbon/epoxy composites to increase the multidirectional properties and durability of flywheel composites so that they rotate faster

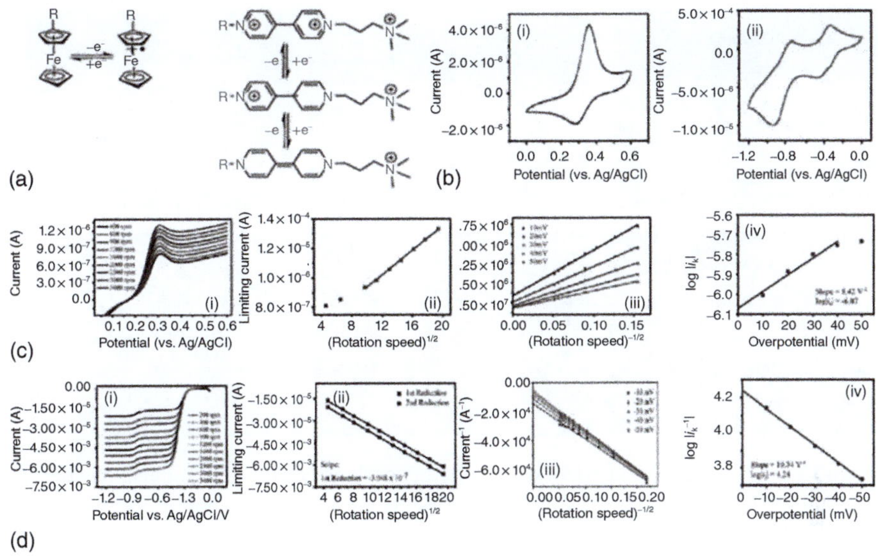

Figure 15.2 (a) Redox equilibrium of ferrocene and viologen polymers, (b) cyclic voltameter of (i) ferrocene and (ii) viologen polymers, (c) RDE voltameter of ferrocene polymer (i) current vs. potential, (ii) Levish, (iii) Koutecky-Levich, and (iv) Tofel plot, and (d) RDE voltammeter of viologen polymer (i) current vs. potential, (ii) Levish, (iii) Koutecky-Levich, and (iv) Tofel plot. Source: Tripathi [1]/Department of science and technology/Public Domain.

and thereby gain energy density. Laboratory scale coupon tests will be conducted to study the strengthening, durability, and delamination failure of composites. Thin cylindrical composite flywheel prototypes (25 cm length × 25 cm OD × 2.5 mm thickness) will be fabricated by resin transfer molding (RTM) and subjected to spin tests to determine their burst speeds and energy density. This approach is a novel attempt to increase the strength and hence the energy density of flywheel composites, and such studies have not been conducted at the international as well as national levels [2].

Novel ZnO nanowire/T1000 carbon/epoxy hybrid composite flywheel materials show at least 30% improvement in tensile strength, stress rupture, fatigue life, and fracture toughness. Database on the properties and guidelines on strengthening, durability, and damage tolerance of ZnO nanowire/T1000 carbon/epoxy composites are useful for flywheel designs. Thin cylindrical flywheel prototypes are made up of proposed hybrid composite materials (Figure 15.3).

Design and Development of new generation ion exchange membranes have excellent stability in an acidic and alkaline environment and have mechanical and thermal stability and electrochemical performance. The membranes have equivalent properties to that of Nafion® and Neosepta®. The comparative performance evaluation in RFB, specifically, zinc and vanadium RFB, the development of an alternative organic redox couple will be attempted [3].

There are three different approaches to develop ion exchange membrane with a common polymer backbone and structural resemblance to perfluorinated

Figure 15.3 (a) Porsche 911 GT3R with carbon fiber flywheel, (b) carbon fibers flywheel for grid regulation, (c) and (d) fibers coated with ZnO nanowires. Source: Reprinted with permission from Gowthaman [2].

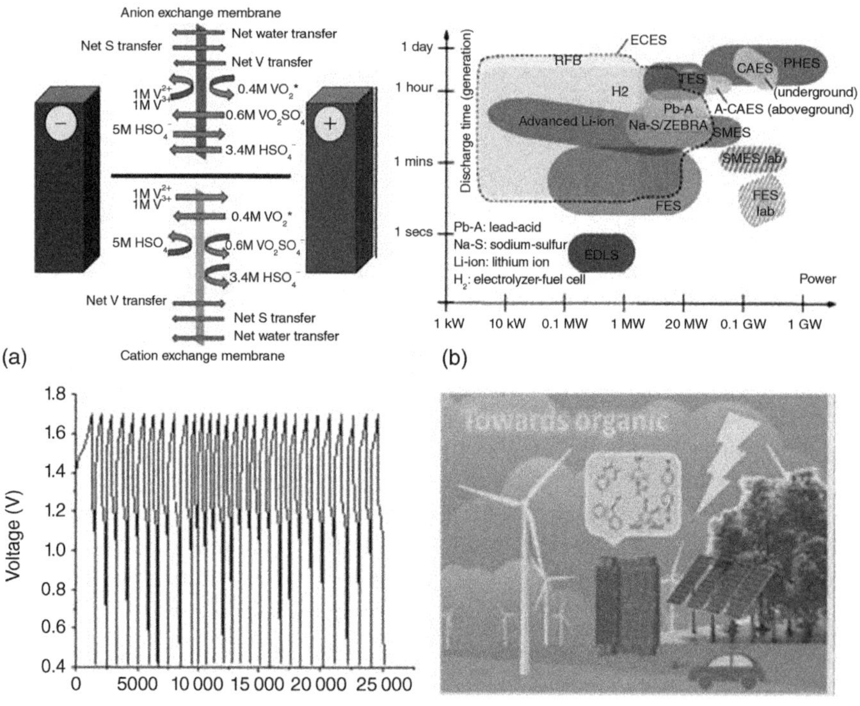

Figure 15.4 (a) Ions and water transfer directions in typical vanadium redox flow battery with anion and cation exchange membranes, *Electrochim. Acta* 101 (2013) 27–40, (b) ions and water transfer directions in typical vanadium redox flow battery with anion and cation exchange membranes [4], (c) all iron alkaline redox flow battery performance with polyethylene-based interpolymer cation exchange, and (d) cartoon showing the utility of redox flow battery in greed applications with organic redox molecules for sustainable development. Source: Modified from Ke et al. [5].

polyelectrolyte membranes, Nafion developed by DuPont, and Aquivion® developed by Solvay specialty polymers. The first approach will be an interpenetrating polymer network, where functional monomers will be copolymerized with perfluorinated polymers. The second and third approaches will be based on a functional modification of perfluorinated polymers with suitable organic molecules to dock the ionic cluster on the polymer backbone. The developed membranes will be thoroughly evaluated for their electrochemical and physicochemical properties (Figure 15.4). The developed membranes have a plethora of applications in energy devices for the generation and utilization of green and clean energy.

15.2 Polymer–Graphene/Carbon Nanotube

Rechargeable lithium batteries are considered as one of the most promising energy storage technologies, especially as power sources for emerging applications, such as plug-in hybrid electric vehicles and electric vehicles [6]. In these applications, besides high energy density, high power density is also essential [7]. In traditional lithium-ion batteries, cathode materials are usually lithium transition metal oxides or phosphates (such as $LiCOO_2$ or $LiFePO_4$), which can be reversibly de-/re-embedded with Li^+ ions [8]. However, the diffusion kinetics of lithium-ion inorganic materials is slow, which will impair the charge and discharge performance of lithium batteries [9]. In addition to the traditional inorganic materials, organic cathode materials including small molecules and polymers have become the research hot spot of the new generation of green lithium battery electrodes in recent years due to their sustainability and environmental friendliness [10]. Nevertheless, the poor conductivity of organic materials limits the rate performance of electrodes [11].

To obtain better materials for electrodes and optimize the electrochemical performance of lithium batteries, researchers have made many attempts and polymer nanocomposites (PNCs) are commonly used. Graphene and graphene sheets (GNSs) are two-dimensional thin sheets with one or more atomic thicknesses. They are composed of sp^2 carbon atoms arranged in a honeycomb structure. Because of their special electronic conductivity, they are considered excellent conductive additives in nanocomposites [12]. The graphene and GNS hold a high specific surface area (theoretical value $2630\,m^2\,g^{-1}$), which improves the interface contact between active materials and the substrate. Guo et al. intended to enhance the electrochemical performance of nitroxide free radical polymer cathode by using graphene in the dispersion–deposition process. However, the loading rate of active materials in the composite electrode is only 10%, while that of graphene is as high as 60% [13]. Taking PAQS and polyimide (PI) as examples, polymer–graphene nanocomposites with high dispersion graphene were introduced by Song et al. via one-pot synthesis (Figure 15.5a) [16]. Functionalized graphene sheets (FGSs)

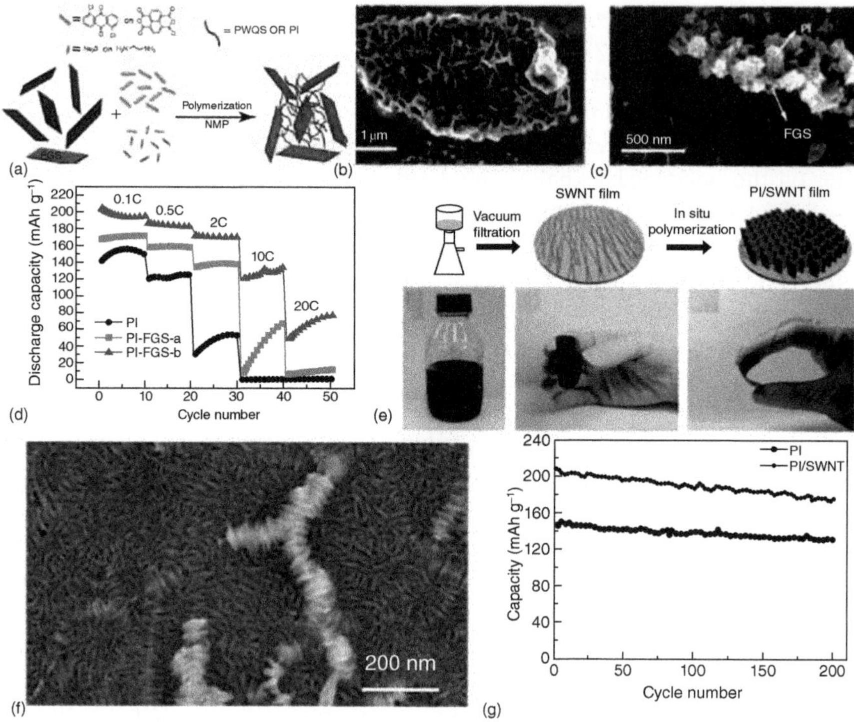

Figure 15.5 (a) In situ polymerization process of PAQS-FGS or PI-FGS nanocomposite, (b) SEM image of PAQS-FGS-a, (c) SEM image of PI-FGS-b, and (d) cycling performance of PI and its composites at different C-rates. Source: Modified from Song et al. [14]. (e) Schematic of the preparation process of PI/SWNT film; photographs of SWNT aqueous dispersion, SWNT film, and PI/SWNT film, (f) SEM images of PI/SWNT film after reaction for six hours, and (g) cycling performance of PI and PI/SWNT at a rate of 0.5 C. Source: Wu et al. [15]/John Wiley & Sons.

were prepared by the graphene thermal expansion method because of their higher electrical conductivity than chemically reduced graphene oxide (RGO). FGS was first dispersed in 1-methyl-2-pyrrolidinone (NMP) by sonication. Then, 1,5-dichloroanthraquinone (DCAQ) or 1,4,5,8-naphthalene triformicanhydride (NCTDA) were added as precursors of PAQS or PI and completely dissolved in NMP [17]. Polymer chains are formed by condensation polymerization at high temperatures after the addition of a condensate (anhydrous sodium sulfide for PAQS or ethylenediamine for PI). In the final product, FGSs are wrapped and uniformly embedded in the polymer matrix.

To find the morphology of the polymer–graphene nanocomposites, the samples were characterized by a scanning electron microscope (SEM). The SEM image in Figure 15.5b indicates that each FGS is several microns in size, with a uniform and thick PAQS coating on both sides of the FGS. For PI-FGS nanocomposites, the SEM image in Figure 15.5c showed that the FGS was uniformly coated with flower-like PI polymer particles, which are smaller than those found in pure PI. The electrochemical performance of PI samples was improved. Although the theoretical capacity of

PI based on the four-electron redox process is 367 mAh g^{-1}, only two electrons can normally be reversibly transferred in the actual charging and discharging process, reaching about half of the theoretical capacity. As shown in Figure 15.5d, the discharge capacity of pure PI at 0.1 C is 156 mAh g^{-1}, and the utilization rate is only 42% compared with the theoretical capacity of 367 mAh g^{-1}. With the addition of 6% graphene, the capacity of PI-FGS is increased to 172 mAh g^{-1} with a corresponding polymer utilization rate of 49%. In other words, graphene can increase the number of electrons per unit of PI from 1.7 to 2. Similar performance improvement was also observed for the PAQS-FGS samples [18].

Besides better electrochemical performance, PNCs are also explored for the multifunctionality of batteries. Flexible and wearable electronic products have attracted extensive attention from academia and industry due to their broad application prospects in medical monitoring, portable military equipment, intelligent textiles, etc. Thus, flexible energy storage devices are in demand to power those electronic devices. Lithium batteries are potential flexible power sources with high volumetric energy density and long cycle life [19]. However, owing to their inherent fragility, inorganic electrode materials cannot provide enough flexibility even they were loaded on curvilinear surfaces such as carbon nanotubes (CNTs), carbon nanofiber, graphene paper, and graphene foam [14]. Thus, polymers were considered because of their inherent flexibility, including conducting polymers such as polyacetylene, radical polymers such as nitroxides, and so on. Among all the organic materials considered, PI is esteemed to be a potential material for the electrodes because of its high theoretical capacity [15]. Wu et al. successfully built an electrode with single-wall carbon nanotube (SWCNT) film as current collector and PI as active materials. As shown in Figure 15.5e, the SWNT film was fabricated through vacuum filtration of SWNT aqueous dispersion. SWNT film was fixed in a polytetrafluoroethylene (PTFE) container and then PI was polymerized in situ on one side of SWNT film. The film in Figure 15.5e held good flexibility, and the SEM image in Figure 15.5f indicated that SWNT bundles were covered by PI nanoflakes fully and evenly. The PI/SWNT also obtained better electrochemical performance besides better flexibility. Using Swagelok cells, it was testified that PI/SWNT exhibited higher capacity (206 mAh g^{-1}) compared with pure PI and it held better cycling stability which retained 85% (175 mAh g^{-1}) after 200 cycles (Figure 15.5g).

Polymer–graphene/CNTs are often used to satisfy the flexibility requirement of devices. Liu et al. prepared highly flexible, bendable, and conductive films, which consisted of poly(3,4-ethylene dioxythiophene) (PEDOT), poly (styrene sulfonate) (PSS), and RGO. The composite was prepared by the addition of PEDOT/PSS into the RGO dispersions [20]. As shown in Figure 15.6e, RGO is covered by PEDOT/PSS, and there are some re-stackings of several layers of RGO sheets that are separated by PEDOT/PSS, leading to the increase of the porosity and surface area [21]. A device was made by a poly(vinyl alcohol) (PVA)/H_3PO_4 solid-state electrolyte between two symmetric RGO-PEDOT/PSS films (Figure 15.6f). The electrochemical performance of the supercapacitor is shown in Figure 15.6g–i. The CV curve of this device depicts a more rectangular shape and larger area than that

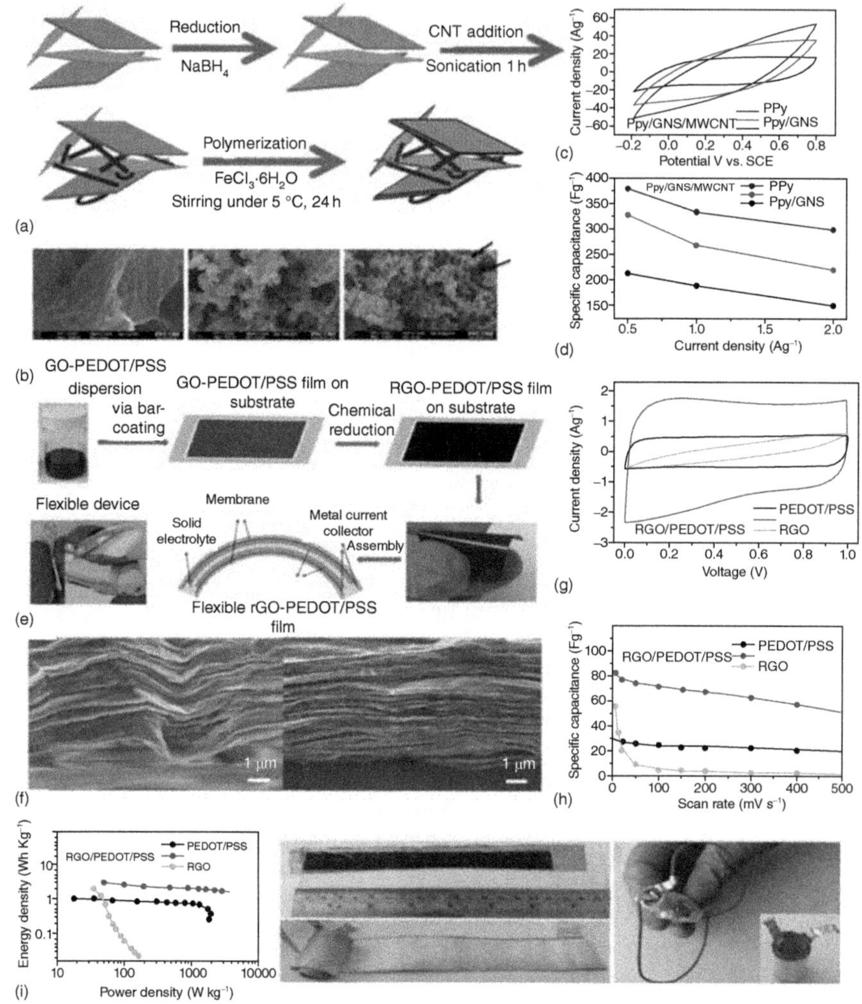

Figure 15.6 (a) Schematic diagram of the synthesis route of the PPy/GNS/MWCNT composites. PPy: polypyrrole; GNS: graphene nanosheets; MWCNT: multiwalled carbon nanotubes, (b) FESEM images of GNS, pure PPy, and PPy/GNS/MWCNT, (c) CV curves of the samples at a scan rate of 100 mV s^{-1} in 1M $NaNO_3$ electrolyte, (d) specific capacitance of PPy, PPy/GNS, and PPy/GNS/MWCNT as a function of the various current densities. Source: Kim et al. [21]/Korean Carbon Society. Copyright 2017, Korean Carbon Society. (e) SEM images of RGO and RGO-PEDOT/PSS, (f) schematic illustration of the preparation process of RGO-PEDOT/PSS films and the structure of assembled supercapacitor devices, (g) CVs of pure RGO, PEDOT/PSS, and RGO-PEDOT/PSS devices at a scan rate of 50 mV s^{-1}, (h) the specific capacitance of RGO, PEDOT/PSS, and RGO-PEDOT/PSS electrodes was calculated from CV, (i) Ragone plot of RGO, PEDOT/PSS, and RGO-PEDOT/PSS electrodes, and (j) a demo device used to power a green light-emitting diode (LED). Source: Nie et al. [22]/John Wiley & Sons.

of the PEDOT/PSS and RGO devices (Figure 15.6g). The specific capacitance of the supercapacitor is higher than either of the pure PEDOT/PSS or RGO at all scan rates of 5–500 $mV\,s^{-1}$ (Figure 15.6h). The Ragone plot in Figure 15.6i indicates the improvement in the power and energy densities of RGO-PEDOT/PSS composite electrodes. A supercapacitor was also fabricated to power a light-emitting diode (LED) to demonstrate the ability of the supercapacitor [22].

To get with the fast development of flexible electronics such as electronic skin and smart energy storage clothes, supercapacitors that are flexible and have self-healing ability as well as high capacitance are needed to power those devices. At the same time, the aforementioned requirements also become demanding on electrolytes. The fabrication of multifunctional electrolyte by polyacrylic acid (PAA) was dual crosslinked by hydrogen bonding and vinyl hybrid silica nanoparticles (VSNPs) [23]. The procedure of the fabrication is shown in Figure 15.7d. A supercapacitor was formed with VSNPs-PAA film as electrolyte and CNT paper-based PPy as symmetric electrodes [26]. The picture in Figure 15.7e shows that the cut wound on the supercapacitor recovers to restore the conductivity, which means that the supercapacitor is self-healable. To further investigate the specific self-heal ability of the supercapacitor, the electrochemical performance was tested before and after self-healing. The healing efficiency got from CV curves and GCD curves indicates the same trend that the healing efficiency is ~100% during all breaking–healing cycles (Figure 15.7f). A demo of the supercapacitor is also shown to demonstrate its ability to light up an LED after self-healing (Figure 15.7g) [27].

The fabrication of a novel structural composite supercapacitor based on CNTs fibers/polymer electrolyte interleaves embedded between carbon fiber fabrics and infused by epoxy is reported. The method is simple and compatible with industrial composite fabrication techniques. Once embedded in the composite structure, the electric double-layer capacitor (EDLC) interleaves had identical electrochemical properties as before, with Coulombic efficiency of 98% and low ESR of 35 $Ohm\,cm^2$ even when charged at 3.5 V (Figure 15.8). *In situ* electrochemical measurements during four-point bending were performed on composites with EDLC interleaves at different ply positions to study the tensile, neutral, and compressive scenarios (Figure 15.9). The results show that electrochemical properties were essentially retained under high flexural deformations and stresses (70 MPa) [29]. Changes in ESR produce 10–37% increases in energy density upon initial loading and a drop of 22% at large strains. Even after sample failure by buckling, the devices could still be charged and discharged, with 69% and 93% of the initial energy and power densities, respectively. The composites produced here stand out for their high power density (30 $W\,kg^{-1}$), 1 to 3 orders of magnitude higher than the state of the art, combined with the highest measured energy density (37.5 $mWh\,kg^{-1}$) for the structural composites. Although their interlaminar properties are expected to be substantially reduced by the presence of the soft matrix interleaves, the flexural modulus and flexural strength achieved were still 60 GPa and 153 MPa, respectively.

As a demonstration of the engineering possibilities to improve interlaminar properties offered by these materials, grid-shaped interleaves are produced. Synchrotron 3D tomography shows that this design leads to epoxy regions interconnecting CF

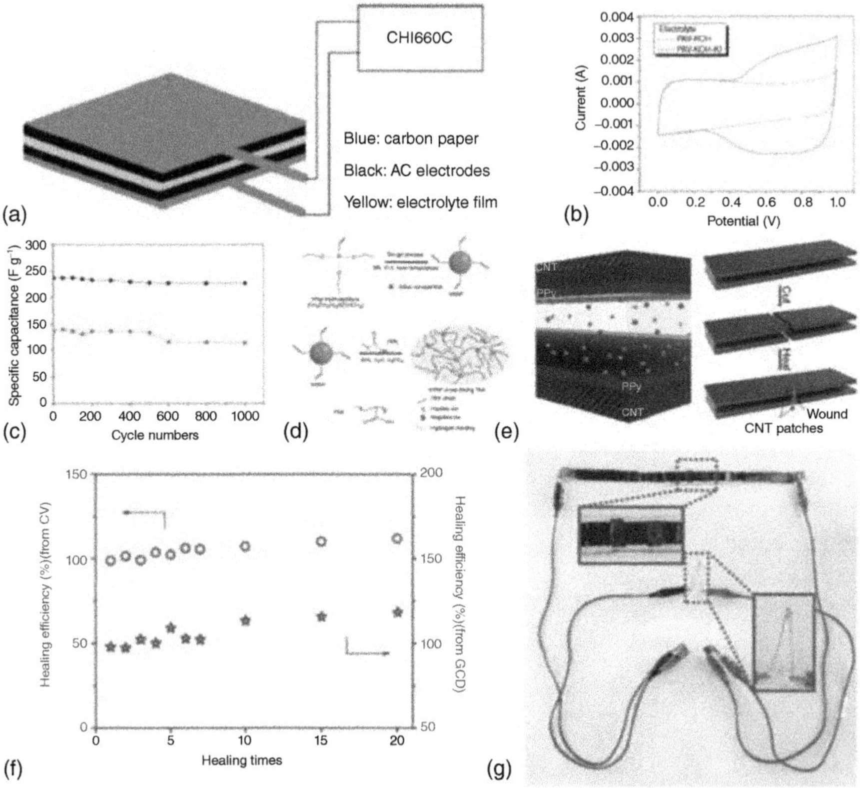

Figure 15.7 (a) Schematic representation of the supercapacitor, (b) CVs of supercapacitor with different electrolytes at a scan rate of $5\,mV\,s^{-1}$, (c) specific capacitances of the supercapacitors with PVA–KOH 'circle' and PVA–KOH–KI 'star' electrolyte during long-term cycling. Source: Yu et al. [24]/with permission of Elsevier, (d) preparation of VSNPs from vinyltriethoxysilane and preparation of the VSNPs-PAA electrolyte from VSNPs (crosslinker), acrylic acid (AA, main monomer), ammonium persulfate (APS, initiator), and phosphoric acid (pH and water content regulator), (e) schematic of the supercapacitor comprising the VSNPs-PAA polyelectrolyte and CNT paper-based PPy electrodes, (f) healing efficiency calculated from CV and GCD curves, and (g) photo of three supercapacitors connected in series, one of which has self-healed twice to power an LED bulb after self-healing (insets show enlarged profiles of the twice self-healed supercapacitor and the lit LED bulb). Source: Huang et al. [25]/Springer Nature/CC BY-4.0.

through holes in the interleaf. The ability to change the shape and size of the hole pattern in the interleave opens a wide range of design parameters to obtain desired composite properties. To remove redundant elements in the interleaves, particularly the plastic tape insulator and the Al support, reducing their thickness is important. This will avoid excessive distortion of the adjacent CSF and thus enable a deeper study into the mechanics of these novel structures, both experimentally and *in silico* [28]. Current collector-free architectures would also enable a wider design envelope and lend themselves to compression molding, stamping, performing, and other processing steps envisaged for these multifunctional composites. As pointers for

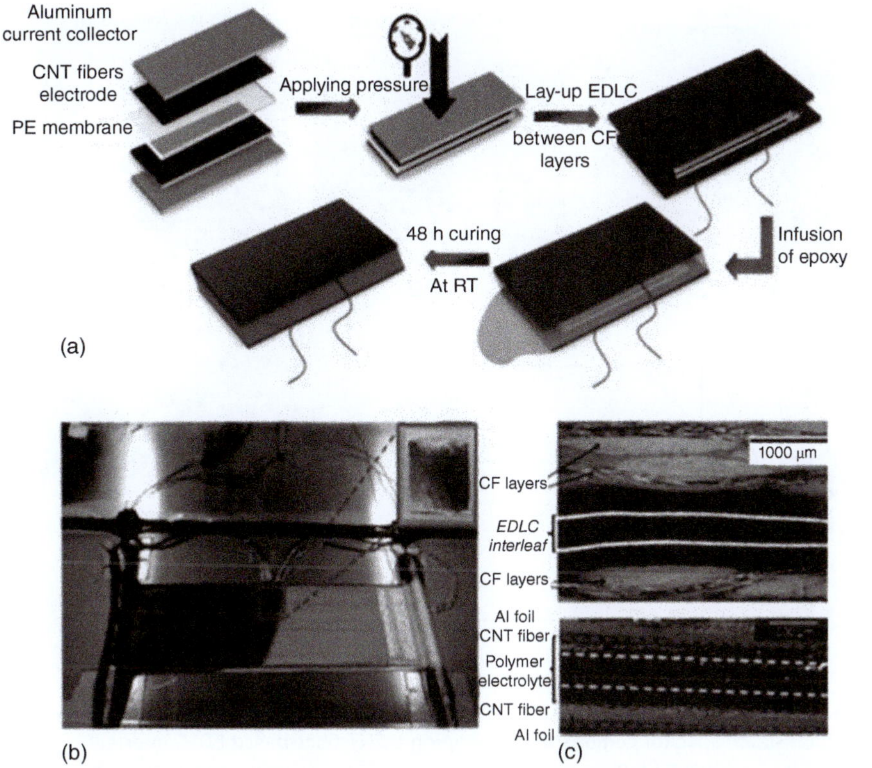

Figure 15.8 Alternative structural composite supercapacitor architectures using both CNT fibers and CF as active material, without the current collector, separator, or insulating layers. (a) Scheme of different EDLC configurations in the composite, (b) charge–discharge profiles comparing CNT fiber/CNT fiber and CNT fiber/CF device configurations, and (c) photographs of the structural supercapacitor powering a red LED in the two configurations. Source: Senokos et al. [28]/with permission from Springer Nature/CC BY 4.0.

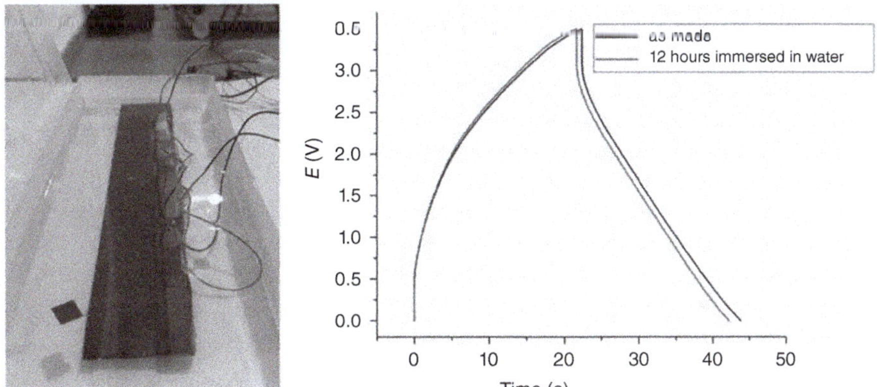

Figure 15.9 Encapsulation of the EDLC interleaf by epoxy. (a) Photo of structural supercapacitor composite immersed in water, and (b) comparison of CD profiles before and 12 hours after immersion in water show nearly identical electrochemical properties. Source: Senokos et al. [28]/with permission from Springer Nature/CC BY 4.0.

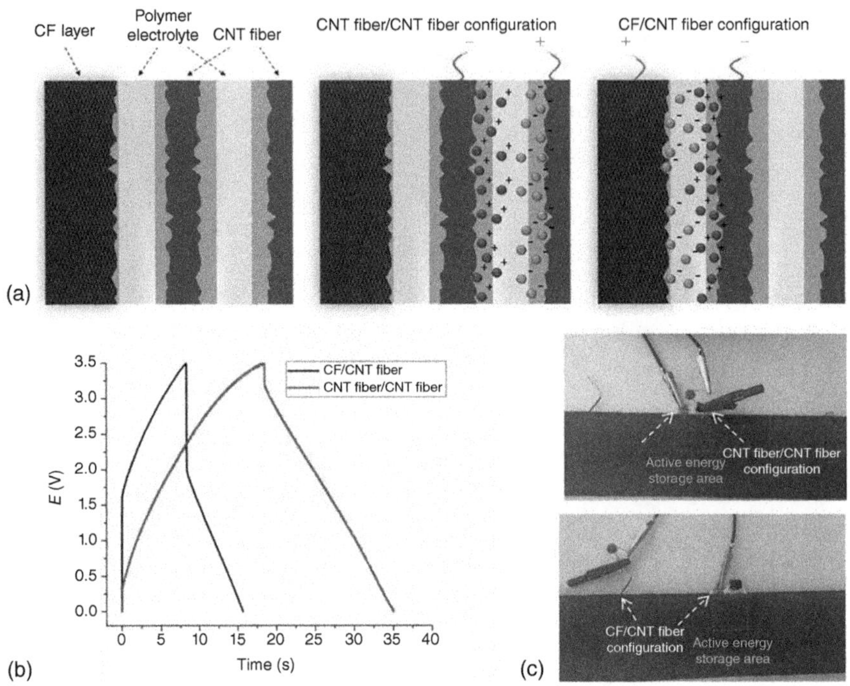

Figure 15.10 Infusion of epoxy into CF/EDLC/CF lay-up. (a) Scheme of the fabrication of structural supercapacitor composite by stamping a CNT fiber-based EDLC interleaf, embedding it between CF plies and infusion/curing of epoxy resin, (b) photographs of a CF/EDLC/CF lay-up during epoxy infusion and the 4 cm^2 EDLC interleaf embedded in it (inset), and (c) optical micrograph of composite cross section (top) showing successful integration of EDLC/CF/epoxy in the laminate and scanning electron micrograph (bottom) of integrated EDLC interleaf. Source: Senokos et al. [28]/with permission from Springer Nature/CC BY 4.0.

further improvements in energy and power density, and highlight the possibility to use both CF and CNT fibers as electrodes, which reduces weight by eliminating electrical insulators while increasing the fraction of composite material used for energy storage. The energy storage properties can be also enhanced by the introduction of materials producing Faradaic processes (Figure 15.10).

Researchers proposed a porous nanocomposite fabric (PNF) by incorporating nano-Al_2O_3 fillers into cellulose acetateporous networks to develop wearable and green triboelectric nanogenerators (TENGs). This PNF was fabricated using a dry casting method, and it was considered an effective tribopositive material. When a 10 wt% solution concentration and 10 wt% nano-Al_2O_3 fillers were incorporated in the PNF, the PNFTENG could deliver the maximum electrical performance, with a VOC of 448 V and a power density of 2.5 mW cm^{-2}. The considerable outputs were attributed to the synergistic effect between the proper pore content, tribo-active ingredient, and uniformly distributed nano-Al_2O_3 nanoparticles [30]. Furthermore, the device also exhibited good mechanical strength, deformability, flexibility, and washability, which provides a guarantee for real-world applications (Figure 15.11).

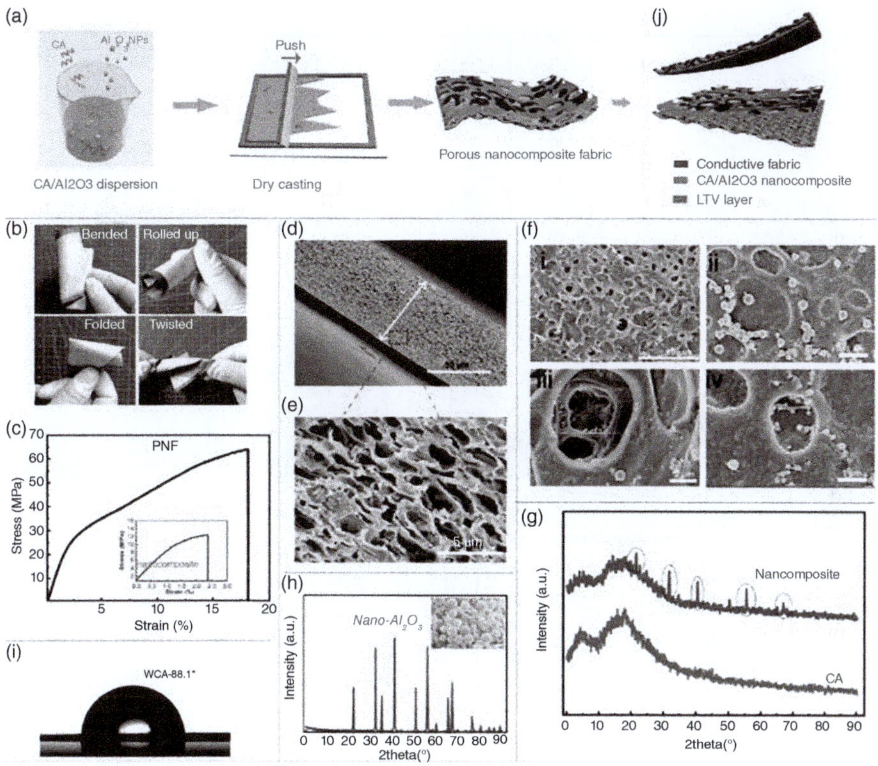

Figure 15.11 Preparation, characterization of the porous nanocomposite fabric, and structure of the PNF-TENG. (a) Schematic illustration of the fabrication for the PNF, (b) digital images of the PNF, showing the softness and deformability, (c) stress–strain curves of the PNF and porous nanocomposite, (d, e) cross-sectional SEM images of the optimal porous nanocomposite, (f) SEM images of the surface morphology at different amplitudes and locations on the surface of the nanocomposite with 10 wt% nano-Al_2O_3 and 10 wt% CA concentrations, (g, h) XRD spectra of Al_2O_3, CA, and the nanocomposite, the inset in (h) shows the rough surface of Al_2O_3 nanoparticles, (i) water contact angle of the PNF, and (j) structural design of the PNF-TENG. Source: Bai et al. [30]/with permission from American Chemical Society.

Based on these advantages, the PNF-TENG was effectively capable of charging commercial capacitors, powering wristwatches, and LED arrays and exhibited great potential as a wearable power source. Also, a PNF-TENG-based smart elbow supporter and a grip ball were designed and further employed for monitoring human motion states in real-time during sports and health training, demonstrating the feasibility of the device as a promising self-powered sensor [31].

Further, to explore the potential of the PNF-TENG as a wearable self-powered sensor, it was used to provide real-time and auxiliary monitoring of human movements in sports and fitness training. A kind of PNF-TENG-based smart sports elbow supporter was designed to detect the swing amplitude of an arm during a dumbbell exercise. As shown in Figure 15.12g, the magnitude of current signals from the sports elbow supporter at a large swing amplitude is different from that at a small swing

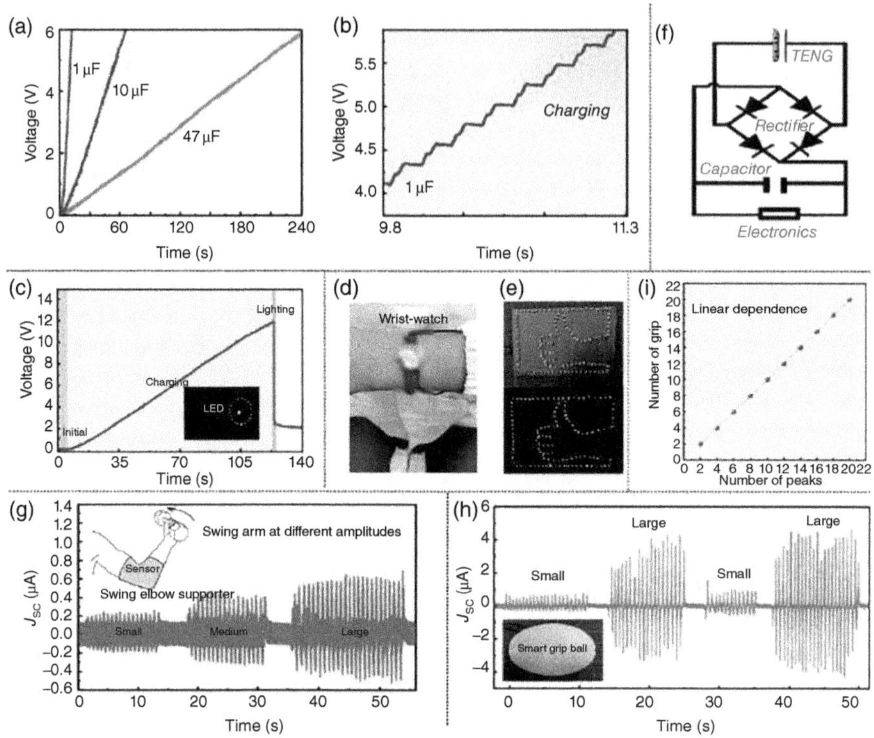

Figure 15.12 Demonstration of the PNF-TENG as a wearable power source and self-powered sensor. (a) Voltage charging curves of capacitors with different capacitances (1, 10, 47 μF), (b) step voltage rise of the capacitor with a capacitance of 1 μF, (c) charging and discharging curves of a capacitor (4.7 μF), with an inset showing lit LED powered by the capacitor, demonstration of (d) driving a wristwatch and (e) the LED arrays with a "thumb" sign powered by the PNF-TENG, (f) simplified circuit diagram showing electronics charged by the PNF-TENG with the aid of abridging rectifier, (g) detected current output from the PNF-TENG-based sports elbow supporter, (h) detected signals from a PNF-TENG-based homemade grip ball during human grip strength testing, and (i) linear relationship between the number of output peaks and the number of grips. Source: Bai et al. [30]/with permission from American Chemical Society.

amplitude. When the swing amplitude of an arm, in turn, changed to small, medium, or large states, the value of detected output signals was enhanced gradually [32]. The elbow supporter delivered a current of about 0.7 μA under a large swing amplitude, higher than the output at a small swing amplitude, which is because the elbow supporter could exert larger deformation under a large swing amplitude. Thus, the exercise intensity of the body in the dumbbell training process can be determined and assessed by the amplitude of the signal detected by the smart elbow supporter, which provides positive guidance for humans. Furthermore, a PNF-TENG-based smart grip ball for monitoring grip strength in real time is designed. As displayed in Figure 15.12h, the electrical output amplitude of the grip ball exhibits a strong dependence on the grip strength. The distinct amplitudes of current signals would be generated when the grip ball experiences grips with different intensities. Also, the current signals exhibit good reproducibility under a certain grip strength, which affirms the excellent working reliability and stability of the PNF-TENG-based grip ball. Based on this, we could identify the intensity of grips according to the amplitude of the signal, which is beneficial to assessing body exercise conditions. The grip ball not only monitors the intensity of grips but also tracks and records the number of grips during the grip training process [33]. It was observed that the number of grips is linearly related to the number of current signals in Figure 15.12i. When various numbers of grips (2, 4, 6, 8,10, 12, 14, 16, 18, and 20) were applied to the grip ball in succession, the corresponding numbers of electrical signals would be detected. These application demonstrations of a smart sports elbow supporter and a grip ball in sports and health exercises imply that the PNF-TENG has a bright future in the field of human motion sensing.

$Ti_3C_2T_x$ (MXene), a thriving member of the two-dimensional (2D) materials family, has shown increasing potential in a myriad of applications, ranging from printable electronics to energy storage and separation membranes [34]. Nevertheless, the dilemma of its oxidative instability and the easy disintegration of its assemblies in contact with water have been restricting its real-life use. Also, the report shows tartaric acid, a natural source, as a non-innocent additive in the MXene composite. In water, it can, above all, inhibit oxidation of $Ti_3C_2T_x$ and hold individual components in the composite $Ti_3C_2T_x$/poly(3,4-ethylenedioxy thiophene): polystyrene sulfonate) ($Ti_3C_2T_x$/PEDOT: PSS) firmly together; equally important, it can boost fourfold the composite's electron conductivity in comparison to the additive-free equivalent [35]. To showcase its practical value, a tartaric-acid-treated, water-stable MXene/PEDOT: PSS conductive coating is made, which serves as electrodes for an ultrafast supercapacitor; among all 2D materials-based assemblies, the designed supercapacitor deliver shows high performance in an alternating current filtering application [36].

The unexpected multifunction of tartaric acid to satisfy such design requirements and allow for the preparation of a $Ti_3C_2T_x$/PEDOT: PSS composite system with its conductivity value beyond the state of the art is reported (Figure 15.13a). $Ti_3C_2T_x$ is

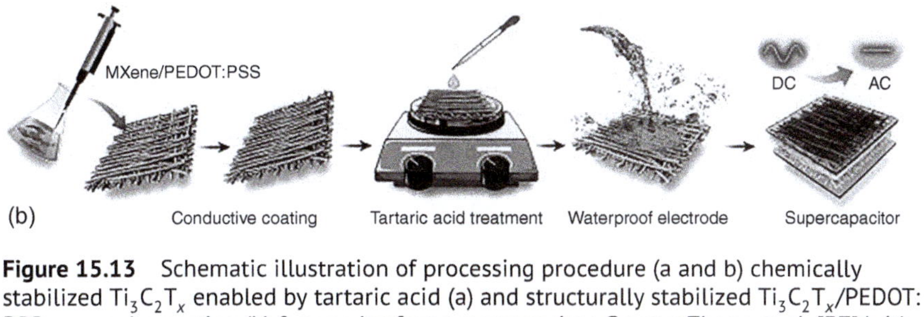

Figure 15.13 Schematic illustration of processing procedure (a and b) chemically stabilized $Ti_3C_2T_x$ enabled by tartaric acid (a) and structurally stabilized $Ti_3C_2T_x$/PEDOT: PSS composite coating (b) for an ultrafast supercapacitor. Source: Zhang et al. [37]/with permission of Elsevier.

chosen as an exemplar here to stand for MXene because of its high conductivity and its well-established synthesis and assembly chemistry (Figure 15.13b). As a proof of concept, ultrahigh rate filtering capacitor electrodes were fabricated from our design concept and delivered excellent stability in aqueous conditions, together with areal capacitance of 1 149 mF cm^{-2} and a phase angle of 80° at 120 Hz, and are the best of all 2D-materials-based assemblies ever reported [37].

References

1 Tripathi, B.P. (2018). High-performance aqueous redox flow batteries based on redox-active polymers and Janus membranes. Research & Technology Development Compendium on Materials for Energy Storage, pp. 13–17.

2 Gowthaman, S. (2018). Studies on the strength and durability of ZnO nanowire/T1000 carbon/epoxy composites for flywheel energy storage. Research & Technology Development Compendium on Materials for Energy Storage, pp. 51–53.

3 Nagarale, R.K. (2018). Indigenous polymer electrolyte membranes for energy devices: redox flow battery and reverse electrodialysis. Research & Technology Development Compendium on Materials for Energy Storage, pp. 79–80.

4 Parasuraman, A., Lim, T.M., Menictas, C., and Skyllas-Kazacos, M. (2013). Review of material research and development for vanadium redox flow battery applications. *Electrochim. Acta* 101: 27–40.

5 Ke, X., Prahl, J.M., Alexander, J.I.D. et al. (2018). Rechargeable redox flow batteries: flow fields, stacks, and design considerations. *Chem. Soc. Rev.* 47: 8721.

6 Tarascon, M. (2008). Towards sustainable and renewable systems for electrochemical energy storage. *ChemSusChem* 1: 777–779.

7 Larcher, D. and Tarascon, J.M. (2015). Towards greener and more sustainable batteries for electrical energy storage. *Nat. Chem.* 7: 19–29.

8 Okubo, M., Hosono, E., Kim, J. et al. (2007). Nanosize effect on high-rate Li-ion intercalation in $LiCoO_2$ electrode. *J. Am. Chem. Soc.* 129: 7444–7452.

9 Hosono, E., Kudo, T., Honma, I. et al. (2009). Synthesis of single-crystalline spinel $LiMn_2O_4$ nanowires for a lithium-ion battery with high power density. *Nano Lett.* 9: 1045–1051.

10 Chen, H., Armand, M., Demailly, G. et al. (2008). From biomass to a renewable $Li_xC_6O_6$ organic electrode for sustainable Li-ion batteries. *ChemSusChem* 1: 348–355.

11 Koshika, K., Sano, N., Oyaizu, K., and Nishide, H. (2009). An ultrafast chargeable polymer electrode based on the combination of nitroxide radical and aqueous electrolyte. *Chem. Commun.* 7: 836–838.

12 Guo, W., Yin, Y.X., Xin, S. et al. (2012). Superior radical polymer cathode material with a two-electron process redox reaction promoted by graphene. *Energy Environ. Sci.* 5: 5221–5225.

13 Song, Z.P., Xu, T., Gordin, M. et al. (2012). Polymer-graphene nanocomposites as ultrafast-charge and discharge cathodes for rechargeable lithium batteries. *Nano Lett.* 12: 2205–2211.

14 Song, Z.P., Zhan, H., and Zhou, Y.H. (2010). Polyimides: promising energy-storage materials. *Angew. Chem. Int. Ed.* 49: 8444–8448.

15 Wu, H.P., Shevlin, S.A., Meng, Q. et al. (2014). Flexible and binder-free organic cathode for high-performance lithium-ion batteries. *Adv. Mater.* 263338–263343.

16 Koo, M., Park, K.I., Lee, S.H. et al. (2012). Bendable inorganic thin-film battery for fully flexible electronic systems. *Nano Lett.* 12: 4810–4816.

17 Dunn, B., Kamath, H., and Tarascon, J.M. (2011). Electrical energy storage for the grid: a battery of choices. *Science* 334: 928–935.

18 Wang, K., Lu, S., Wu, Y. et al. (2013). Super-aligned carbon nanotube films as current collectors for lightweight and flexible lithium-ion batteries. *Adv. Funct. Mater.* 23: 846–853.

19 Suga, T., Konishi, H., and Nishide, H. (2007). Photocrosslinked nitroxide polymer cathode-active materials for application in an organic-based paper battery. *Chem. Commun.* 17: 1730–1732.

20 Liu, Y., Weng, B., Razal, J.M. et al. (2015). High-performance flexible all-solid-state supercapacitor from large free-standing graphene-PEDOT/PSS films. *Sci. Rep.* 5: 17045.

21 Kim, H.S., Jun, Y., and Kim, S. (2017). Capacitance behaviors of conducting polymer-coated graphene nanosheets composite electrodes containing multi-walled carbon nanotubes as additives. *Carbon Lett.* 23: 63–68.

22 Nie, N., Mengmeng, H., Liu, J. et al. (2021). The application of polymer nanocomposites in energy storage devices. In: *Polymer Nanocomposite Materials: Applications in Integrated Electronic Devices* (ed. Y. Zhou and G. Ding). Wiley-VCH.

23 Hu, L., Pasta, M., Mantia, L.F. et al. (2010). Stretchable, porous, and conductive energy textiles. *Nano Lett.* 10: 08–714.

24 Yu, H., Wu, J., Fan, L. et al. (2011). Improvement of the performance for quasi-solid-state supercapacitor by using PVA-KOH-KI polymer gel electrolyte. *Electrochim. Acta* 56: 6881–6886.

25 Huang, Y., Zhong, M., Huang, Y. et al. (2015). A self-healable and highly stretchable supercapacitor based on a dual crosslinked polyelectrolyte. *Nat. Commun.* 6: 10310.

26 Kim, D.H., Lu, N.S., Ma, R. et al. (2011). Epidermal electronics. *Science* 333: 838–843.

27 Sun, J.Y., Keplinger, C., Whitesides, G.M., and Suo, Z. (2014). Ionic skin. *Adv. Mater.* 26: 7608–7614.

28 Senokos, E., Ou, Y., Torres, J.J. et al. (2018). Energy storage in structural composites by introducing CNT fiber/polymer electrolyte interleaves. *Sci. Rep.* 8: 3407.

29 Lorca, J. et al. (2011). Multiscale modeling of composite materials: a roadmap towards virtual testing. *Adv. Mater.* 23: 5130–5147.

30 Bai, Z., Xu, Y., Li, J. et al. (2020). An eco-friendly porous nanocomposite fabric-based triboelectric nanogenerator for efficient energy harvesting and motion sensing. *ACS Appl. Mater. Interfaces* 12: 42880–42890.

31 Hu, Y. and Zheng, Z. (2019). Progress in textile-based triboelectric nanogenerators for smart fabrics. *Nano Energy* 56: 16–24.

32 Wang, Z.L. (2020). Triboelectric nanogenerator (TENG)-sparking an energy and sensor revolution. *Adv. Energy Mater.* 10: 2000137.

33 Pan, R., Xuan, W., Chen, J. et al. (2018). Fully biodegradable triboelectric nanogenerators based on electrospun polylactic acid and nanostructured gelatin films. *Nano Energy* 45: 193–202.

34 Gund, G.S., Park, J.H., Harpalsinh, R. et al. (2019). MXene/polymer hybrid materials for flexible AC-filtering electrochemical capacitors. *Joule* 3: 164–176.

35 Wu, M., Chi, F., Geng, H. et al. (2019). Arbitrary waveform AC line filtering applicable to hundreds of volts based on aqueous electrochemical capacitors. *Nat. Commun.* 10: 2855.

36 Shi, H., Zhang, P., Liu, Z. et al. (2021). Ambient-stable two-dimensional titanium carbide (MXene) enabled by iodine etching. *Angew. Chem. Int. Ed.* 60: 8689–8693.

37 Zhang, M., Heraly, F., Yi, M., and Yuan, J. (2021). Multitasking tartaric-acid-enabled, highly conductive, and stable MXene/conducting polymer composite for ultrafast supercapacitor. *Cell Rep. Phys. Sci.* 2: 100449.

16

Modeling and Simulation Techniques

16.1 Introduction

The modeling and simulation methods at the molecular level usually employ atoms, molecules, or their clusters as the basic units [1]. The most popular methods include molecular mechanics (MMs), molecular dynamics (MDs), and molecular computational (MC) simulation. Modeling of polymer nanocomposites (PNCs) at this scale is predominantly directed toward the thermodynamics and kinetics for the formation of molecular structure and their interactions. Figure 16.1 describes the equation of motion for each method and the typical properties predicted from them [3]. The two widely used molecular scale methods are MD and MC [4].

16.2 Modeling and Simulation of Polymer Nanocomposites

16.2.1 Nanocomposite thermodynamics

The formation of stable nanocomposites depends on the thermodynamics of the multicomponent mixture concerned. In the case of polymer–nanoparticle mixtures, their final structures are strongly influenced by the characteristics of nanoparticles (e.g. size, shape, and aspect ratio), polymer (e.g. molecular weight, structure, polarity, and its compatibility with the particle), and surfactant, if employed. The remarkably improved properties are usually observed from the structure in which nanoparticles are uniformly dispersed in a polymer matrix. Therefore, to achieve such PNCs and the maximal property improvement, it is very important to determine thoroughly the effects of various factors on the final structure and then isolate the thermodynamic conditions for the stable and uniform dispersion of nanoparticles. Experimentally, it is very difficult to ascertain these issues at a nanoscale.

Computer modeling and simulations have shown great success in colloid–polymer solutions. However, the colloid particles in such systems are much larger than the gyration radius of a polymer. These methods may not directly be extendable to nanoparticle–polymer systems in which particle size is equivalent to the gyration radius. Therefore, to elucidate the formation of nanoparticle–PNCs at a molecular

Polymer Nanocomposites for Energy Applications, First Edition.
T. Daniel Thangadurai, Manjubaashini Nandhakumar, Sabu Thomas, and Ange Nzihou.

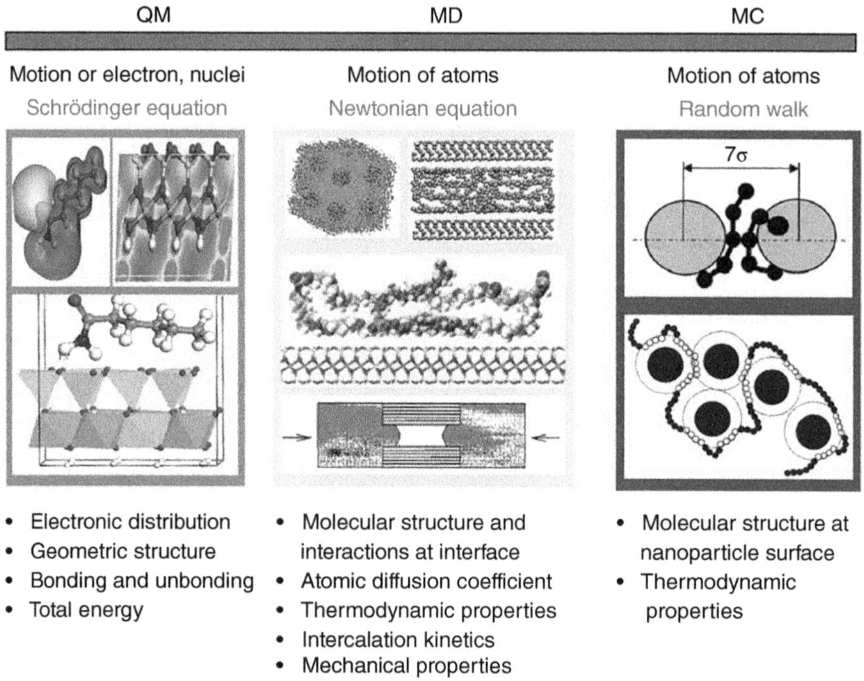

Figure 16.1 Molecular modeling and simulation methods commonly used for polymer nanocomposites. Source: Based on Heinz et al. [2].

level, some theoretical and simulation efforts have been made recently, especially on PNCs. The simulation methods used include the mean-field model [5], combined model of DFT and self-consistent field (SCF), and molecular dynamic model [6]. It is well recognized that the dispersion of layered clays or silicates in the polymer matrix may lead to three kinds of structures, i.e. conventional immiscible microstructure, intercalated nanostructure, and exfoliated/delaminated nanostructure. The formation of these equilibrium structures is believed to be determined by the interplay of entropic (i.e. intermolecular interactions), and enthalpic (i.e. configurational changes in the components) factors. Based on this consideration, Vaia and Giannelis [7] developed a modified mean-field model to predict the above three possible structures. Figure 16.2 shows the free-energy change of polymer–organoclay mixtures as a function of clay platelet separation. In their mean-field model, the free-energy change DH associated with the platelet separation and polymer incorporation is divided into two terms: one is the internal energy change DU associated with the establishment of new intermolecular interactions and the other is the combinatorial entropy change DS associated with configuration changes of the various components:

$$\Delta H = H(h) - H(h_0) = \Delta U - T\Delta S \tag{16.1}$$

where h and h_0 are the final and initial clay platelet separations and T is the temperature. Thus, $\Delta H < 0$ indicates that the intercalation process is favorable,

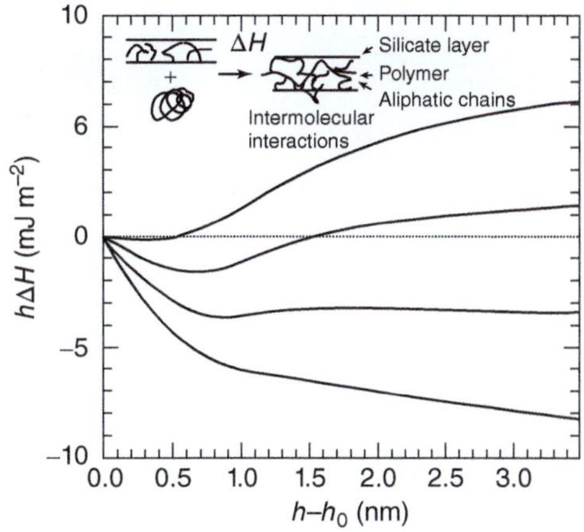

Figure 16.2 The free-energy change (DH) of polymer–organoclay mixtures as a function of clay gallery separation under different intercalation strength between polymer–organoclay: (a) $\epsilon = 0$; (b) $\epsilon = -4$; (c) $\epsilon = -8$; and (d) $\epsilon = -12\ \mathrm{mJ\ m^{-2}}$. The inset shows the intercalation process of polymer in organoclay. Source: Uribe Calderon [8]/McGill University.

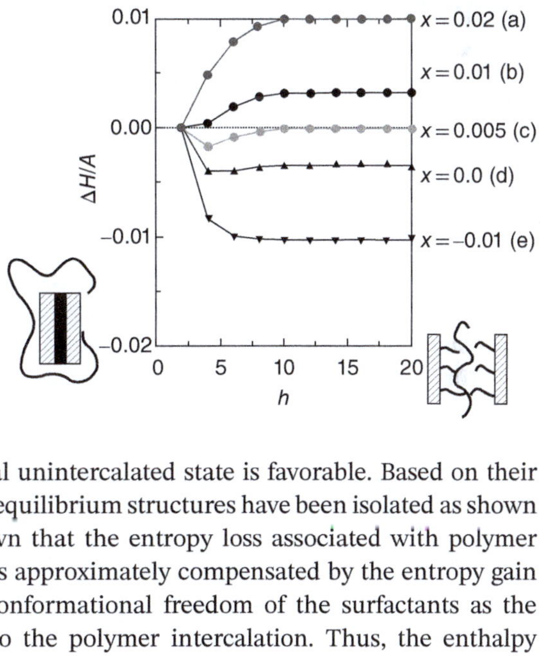

Figure 16.3 Free-energy change per unit area (DH/A) as a function of surface separation (*h*) for various interaction parameter values (*w*): the left diagram shows the reference state where the grafted chains form a melt between the surfaces and the right diagram shows the surfaces that are separated by the intercalated polymers. Source: Uribe Calderon [8]/McGill University.

while $\Delta H > 0$ indicates the initial unintercalated state is favorable. Based on their calculated results, three kinds of equilibrium structures have been isolated as shown in Figure 16.3. It has been shown that the entropy loss associated with polymer confinement in the clay gallery is approximately compensated by the entropy gain associated with the increased conformational freedom of the surfactants as the gallery distance increases due to the polymer intercalation. Thus, the enthalpy determines whether or not polymer intercalation will take place. From this point of view, it is possible to select potentially compatible polymer and organoclay systems to produce materials with desirable structures. The greatest advantage of the mean-field model is that it can explore the effect of various aspects of the polymer and organoclay on nanocomposite formation. However, the assumptions such as the separation of configurational terms and intermolecular interactions and the further separation of the entropic behavior of the components somewhat limit the usefulness of the model [9].

Ozmusul and Picu further investigated the static structure of polymer chains filled with nanoparticles through MC lattice simulation [10]. The overall conformation of the chains is expressed statistically with the subchain segments as defined in

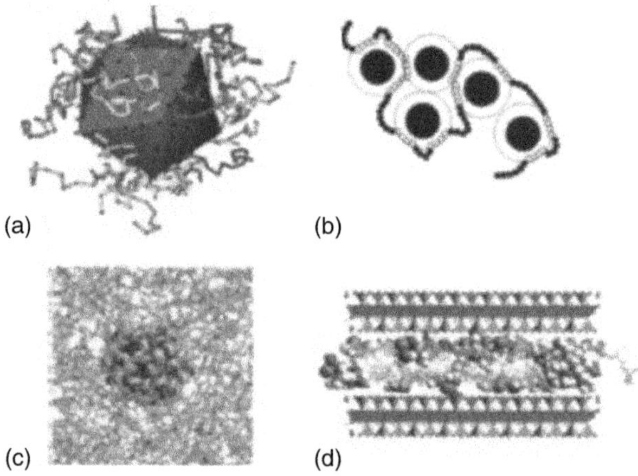

Figure 16.4 Representative models of nanoparticle-reinforced polymer systems: (a) one spherical nanoparticle in a polymer, (b) one silica nanoparticle in polyimide, (c) multiple nanoparticles in a polymer, and (d) polymer intercalated nanocomposite. Source: Heinz et al. [2]/with permission from American Chemical Society.

Figure 16.4, bridges, dangling segments (tails), loops, and trains. Their results supported the notion that the mechanical reinforcement of fillers is due to the polymer-mediated transient network created among the fillers [11]. In addition, it was shown that increasing filler loading leads to a larger fraction of bridges probably due to the decrease in mean distance between fillers.

By using the MD method, Zeng and coworkers investigated the molecular packing and chain orientation of intercalated quaternary alkyl ammoniums in organoclays in terms of atomic density distribution, chain title angle, order parameter, and

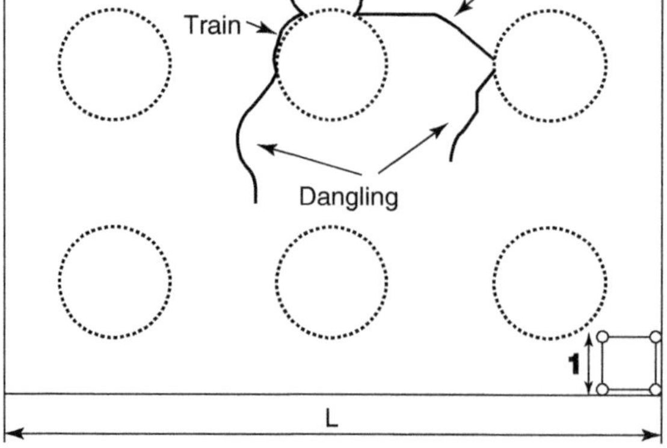

Figure 16.5 2D representation of the 3D polymer nanocomposite model with a unit cell of the lattice being sketched in the right lower corner. Source: Based on Heinz et al. [2].

conformation [12]. They observed various layered structures for the intercalated alkyl ammoniums, from monolayer, bilayer, pseudo-trilayer, to pseudo-quadrilateral structure (Figure 16.5). These layering structures are closely related to the alkyl chain length and cationic exchange capacity of clays, and the long alkyl chains, for example, octadecyl trimethyl ammoniums, do not lie flat within a single layer but interlace. Moreover, the intercalated long alkyl chains demonstrate a predominant trans-conformation, although there is also extensive gauche conformation. In the case of polyurethane nanocomposites, both the hard and soft segments demonstrate layered structure in the clay gallery [13]. More importantly, phase separation, a common phenomenon in bulk polyurethane, has not been observed in such intercalated polyurethane. The absence of phase separation is attributed to the nanoconfinement and the competitive interactions among clay surface, surfactant, and polyurethane. Recently, Greenwell et al. presented large-scale MD simulations for the systems of primary amine intercalated clays [14]. They found that the interlayer spacing depends on the number of intercalated ammonium groups, which agrees well with the results of Paul et al. [15] on organoclays modified with quaternary alkyl ammoniums. The obtained results showed that amine functional groups do not interact strongly with clay platelets, whereas ammonium groups do. In addition, it is observed that the long-wavelength undulations in the clay platelets are believed to be related to the exceptionally large supercell considered (Figure 16.6). The undulation phenomenon perhaps supports the notion of flexible

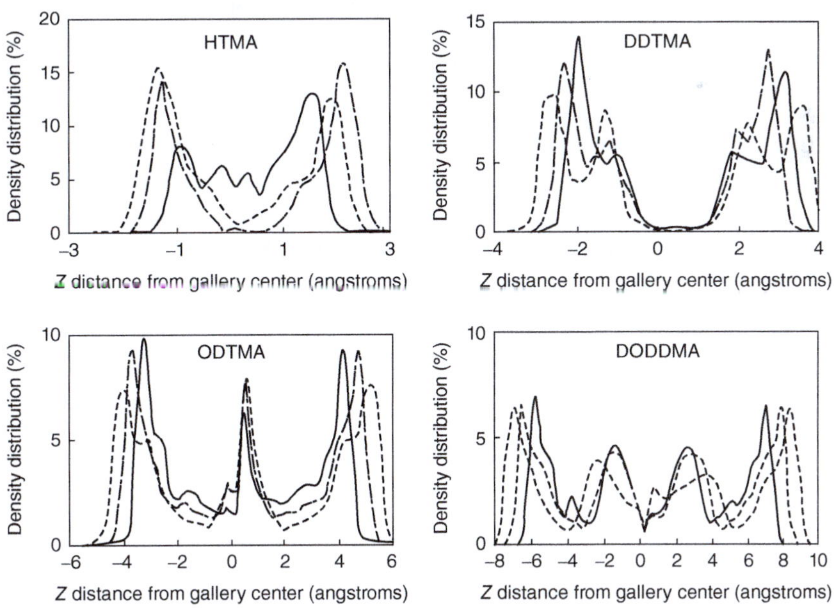

Figure 16.6 Methylene group density distributions of organoclays with different alkyl chain lengths and CEC as a function of z distance from gallery center: solid curve, CEC 85 meq 100 g^{-1}; dashed curve, CEC 102 meq 100 g^{-1}; dotted curve, CEC 119 meq 100 g^{-1}. Here, the surfactants include hexyltrimethyl (HTMA), dodecyltrimethyl (DDTMA), octadecyltrimethyl (ODTMA), and dioctadecyldimethyl (DODDMA). Source: Based on Heinz et al. [2].

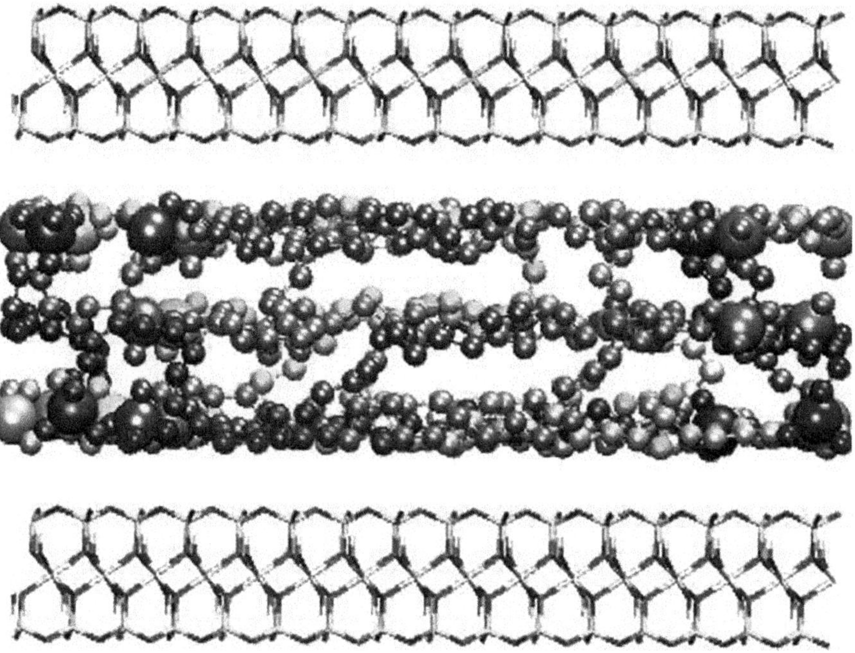

Figure 16.7 Snapshot at 1000 ps of octadecyl trimethyl–clay. Clay platelets are represented by a stick model and each surfactant chain is represented by a ball model with a different color for better visualization and includes nitrogen (large ball), united carbon (small ball) of the hydrocarbon chain, and oxygen (medium ball). Source: Based on Heinz et al. [2].

other than rigid clay platelets. Recently, Heinz et al. investigated the structure and dynamics of alkylammonium-modified montmorillonite employing MD simulation and experimental study. They examined the effects of montmorillonite cation exchangeable capacity, ammonium head groups, and chain length [16, 17] (Figures 16.7 and 16.8).

16.2.2 Atomistic MD Simulation of Graphene-Based PMMA Nanocomposites

As a model polymer, we have chosen poly(methyl methacrylate) (PMMA) whose nanocomposites with graphene sheet (GS) have been studied in detail over the years with several techniques. For example, Ramanathan et al. used sonication to break rigid nanoplatelets of expanded graphite apart into thinner platelets, which were dispersed next in a PMMA solution using high-speed shearing methods [18]. This led to an increase in the Tg by 29 °C at 0.05 wt% loading of the matrix in functionalized graphene sheet (FGS) and up to an 80% enhancement of Young's modulus at 1 wt% loading in FGS. Similar observations have been reported by Li and McKenna for GO/PMMA nanocomposites [19]. One reason for the extraordinary mechanical properties of FGS-PMMA nanocomposites is the enhanced interfacial interactions of oxygen functionalities across the surface of graphene with PMMA

Figure 16.8 Snapshot of the 350, 840-atom supercell after 0.5 ns of MD simulation showing a perspective view of the rectilinear supercell, the clay sheets exhibiting gentle undulations. The color scheme is C gray, H white, O red, N blue, Si orange, Al green, Mg magenta, and Na brown. Source: Based on Heinz et al. [2].

chains. FGS contain pendant hydroxyl groups, which may form hydrogen bonds with the carbonyl groups of PMMA. The additional enhancement comes from the nanoscale surface roughness of FGS, the defects caused during thermal exfoliation of the precursor graphite oxide and their wrinkled topology at the nanoscale due to their extremely small thickness. These can enhance mechanical interlocking with the polymer chains, which also leads to better adhesion. Atomistic MD simulations have shown strong adhesion of PMMA chains (especially of its side groups) on graphene and considerably slower segmental and chain mobility in the interfacial area [20]. According to simulation data, local mass density, segmental dynamics, and chain terminal relaxation differ from the bulk behavior up to several nanometers from the graphene surface. In the remaining of this chapter, we will focus on a methodology, initially proposed for a simpler class of systems (glassy vinyl polymers such as polypropylene and polystyrene) [21], which allows the determination of the mechanical properties of PMMA nanocomposites filled with GS (functionalized or nonfunctionalized), based on small-strain deformation experiments on the computer of microscopically detailed model structures. The procedure involves several modeling and mathematical steps and allows computing the elastic constants (Young's modulus E, bulk modulus B, shear modulus G, and Poisson's ratio ν) of a polymeric glass under the assumption that vibrational contributions of the hard degrees of freedom are not significant; as a result, estimates of the elastic constants can be obtained by computing changes only in the total potential energy of static microscopic structures subjected to simple deformation modes. For glassy atactic polypropylene for which the method was first developed and implemented by Theodorou and Suter, elastic constants were predicted within 15% of the experimentally measured values. All the simulations have been performed with an all-atom force field, allowing for a direct comparison of the computed

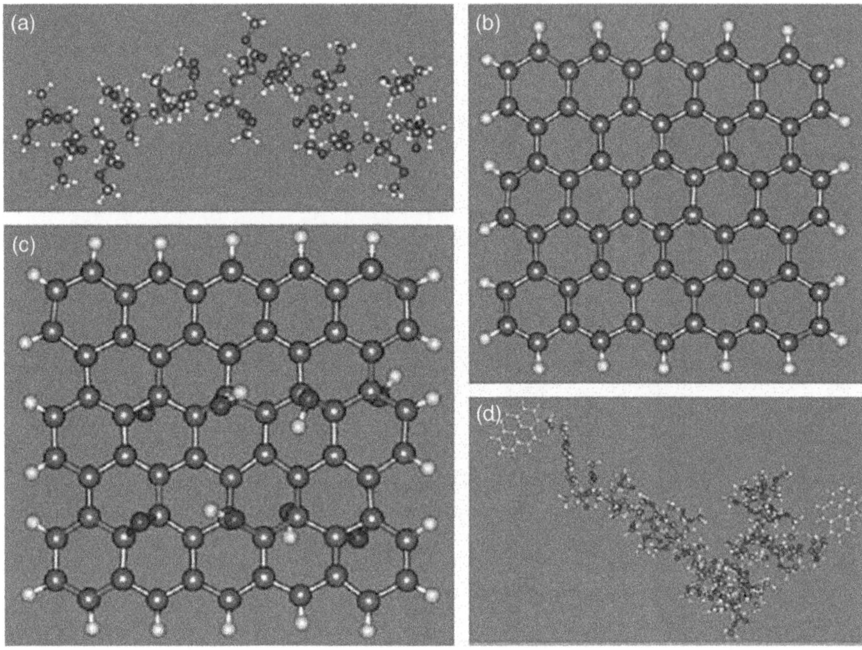

Figure 16.9 Typical atomistic structures of (a) an sPMMA chain, (b) a nonfunctionalized graphene sheet (GS), (c) a functionalized graphene sheet (FGS), and (d) a functionalized (py-sPMMA-py) chain. Source: Emmanuel et al. [22]/Walter de Gruyter.

with available experimental data. DREIDING [2] has been chosen because it combines simplicity with accuracy (for acrylic polymers). Since DREIDING does not provide information about the values of partial charges of PMMA atoms, these were borrowed by the OPLS-AA force field. Additional technical details (such as the parameter values of all bonded and non-bonded interactions describing intra- and inter-atomic contributions to potential energy) can be found in two published articles. Figure 16.9a–d provides typical atomistic structures of a PMMA chain, a GS, a GO, and a functionalized PMMA chain with pyrene groups added to its two ends.

16.3 Systems Simulated and Simulation Strategy

We focus on sPMMA, at $P = 1$ atm. The simulations were performed with strictly monodisperse samples with the model system consisting of 27 chains of a degree of polymerization $X = 15$ (corresponding to a molecular weight of 1503.75 g mol^{-1}). The nonfunctionalized and functionalized GS had lateral dimensions 12 Å × 12 Å. Three model systems were studied: (i) the neat sPMMA matrix (no GS added; it will be denoted as sPMMA in the following), (ii) its nanocomposite with three nonfunctionalized monolayer GS (it will be denoted as GS-sPMMA in the following) corresponding to 5.67 wt% concentration in GS, and (iii) its nanocomposite with three functionalized monolayer GS or GO (it will be denoted as FGS-sPMMA in the following) corresponding to 6.54 wt% concentration in GO. The surface concentration of GO in hydroxyl (–OH) and epoxy (–O–) groups in the latter system was chosen to match as closely as possible the experimentally determined concentration reported by Ramanathan et al. through elemental analysis [23]. To build initial configurations of all systems we used MAPS and to execute the MD simulations we used LAMMPS [24]. All initial configurations were subjected to static structure optimization using a MM algorithm to remove overlaps, and the resulting minimum potential energy structures were annealed to 500 K for several hundreds of nanoseconds to render them completely amorphous before quenching them down to room temperature, also to completely equilibrate them at all length scales. The rectangular parallelepiped simulation cells of initial sides 40 Å × 40 Å × 40 Å subject to full periodic boundary conditions are used. Technical details regarding the execution of the MD simulations (type of thermostat–barostat used, calculation of electrostatic interactions, calculation of LJ interactions and of the tail corrections, integration of equations of motion, time step, etc.) can be found in the two relevant publications [25, 26].

The coarse-grained (CG) MD simulation study of PNCs containing nanorods with homogeneous and patchy surface chemistry/functionalization is modeled with isotropic and directional nanorod–nanorod attraction, respectively. The PNC morphology is impacted by the nanorod design (i.e. aspect ratio, homogeneous, or patchy surface chemistry/functionalization) for nanorods with a diameter equal to the Kuhn length of the polymer in the matrix. For PNCs with 10 vol% nanorods that have an aspect ratio ≤5, percolated morphology is observed with directional nanorod–nanorod attraction and phase-separated (i.e. nanorod aggregation) morphology with isotropic nanorod–nanorod attraction [27]. In contrast, for nanorods with higher aspect ratios, both types of attractions result in aggregated nanorods morphology due to the dominance of entropic driving forces that cause long nanorods to form orientationally ordered aggregates. For most PNCs with isotropic or directional nanorod–nanorod attractions, the average matrix polymer confirmation is not perturbed by the inclusion of up to 20 vol% nanorods. The polymer chains in contact with nanorods (i.e. interfacial chains) are on average extended and statistically different from the conformations the matrix chains adopt in the pure melt state (with no nanorods). In contrast, the polymer chains far from nanorods (i.e. bulk chains) adopt the same conformations as the matrix chains adopt in the pure melt state. The effect of other parameters, such as attraction strength, nanorod volume fraction, and matrix chain length, for PNCs with isotropic or directional nanorod–nanorod attractions, were determined [28]. Collectively, these results provide valuable design rules to achieve specific PNC morphologies (i.e. dispersed, aggregated, percolated, and orientationally aligned nanorods) for various potential applications.

By interacting PNCs with nanorods via isotropic repulsion or attraction, the "isotropic model" is used. To represent the PNCs with nanorods interacting via directional attractions, the "directional model" is employed (Figure 16.10).

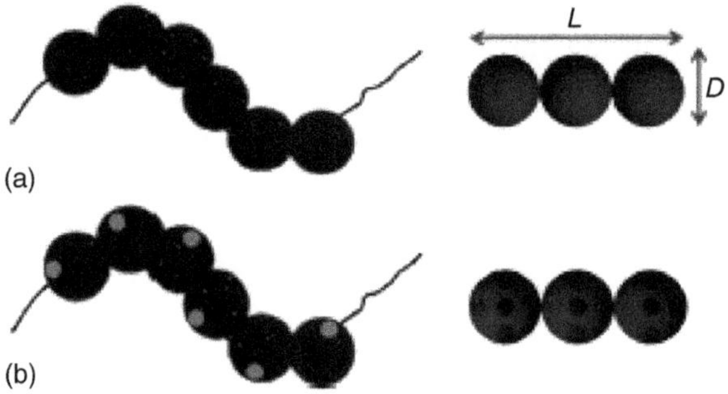

Figure 16.10 Schematic of the CG model for the: (a) matrix chains and nanorods in PNCs with isotropic nanorod–nanorod attraction and (b) matrix chains and nanorods in PNCs with directional nanorod–nanorod attraction. Embedded directional interaction sites are shown as cyan spheres in the gray polymer chain and darker spheres in the brown nanorod.

The nanorod–nanorod radial distribution functions are employed for all three nanorod lengths and all three solvent/matrix lengths of the isotropic and directional models. For PNCs with isotropic nanorod–nanorod attraction (Figure 16.11a, c, and e), the contact peak increases with L, showing the increasing contribution of entropically driven depletion effect in addition to the enthalpic effect from nanorod–nanorod attraction [30]. It is also worth noting that nanorods in $N = 1$ solutions, in general, show a higher contact peak than analogous PNCs with $N = 20$ and 80 matrix chains because the translational entropy gain of the solvent beads upon nanorod aggregation is expected to be larger than the translational entropy gain of the polymer chains. For PNCs with directional nanorod–nanorod attraction (Figure 16.11b, d, and f), the increase in contact peak with nanorod length is less significant than that for the analogous isotropic PNCs. The plots for $L = 5d$ (Figure 16.11d) and $L = 15d$ (Figure 16.11f) with directional attraction are similar, indicating that with limited possible nanorod packing configurations mainly determined by the placements of interaction sites on the nanorod surface, the local nanorod packing is similar across nanorod lengths. For PNCs, the first peak in the nanorod–nanorod radial distribution functions seen in the isotropic PNCs "splits" into two discrete and narrow peaks in directional PNCs. This is mainly due to the regularly distanced interaction sites on the nanorod surface in the model, leading to highly refined relative positions between nanorod beads. These sharp discrete peaks can only be seen in experiments if the functional groups or "patches" on the nanorod surface are as perfectly spaced out on the nanorod surface as of the model. If the patches lack this regular periodicity, then the contact peaks in the radial distribution function for the directional PNCs will be smeared out.

For short nanorods, at similar effective nanorod–nanorod attractions, the patchy nanorod surface functionalization leads to a percolated nanorod morphology within the PNC, while homogeneous nanorod surface functionalization leads to a phase-separated aggregate of nanorods [23]. For long nanorods, regardless of the type of nanorod surface functionalization, nanorods form either finite-sized or percolating aggregates/bundles with high short-range orientationally order, how the attraction strength between nanorods affects the morphology is also found to

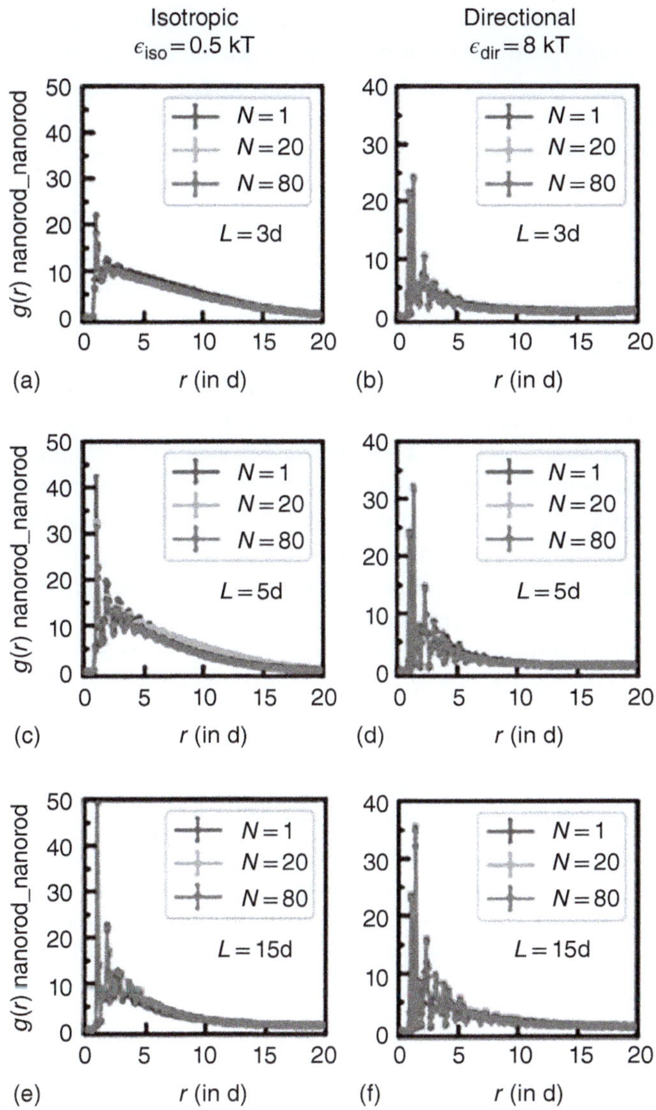

Figure 16.11 Nanorod–nanorod radial distribution functions, $g(r)_{\text{nanorod–nanorod}}$, for PNCs with $\phi_r = 0.10$ and (a) $L = 3d$ nanorods with isotropic nanorod–nanorod attraction and $\epsilon_{iso} = 0.5$ kT, (b) $L = 3d$ nanorods with directional nanorod–nanorod attraction and $\epsilon_{dir} = 8$ kT, (c) $L = 5d$ nanorods with isotropic nanorod–nanorod attraction and $\epsilon_{iso} = 0.5$ kT, (d) $L = 5d$ nanorods with directional nanorod–nanorod attraction and $\epsilon_{dir} = 8$ kT, (e) $L = 15d$ nanorods with isotropic nanorod–nanorod attraction and $\epsilon_{iso} = 0.5$ kT, and (f) $L = 15d$ nanorods with directional nanorod–nanorod attraction and $\epsilon_{dir} = 8$ kT. Error bars represent the standard deviation from 15 configurations collected from 3 simulation trials, and when not visible, error bars are smaller than the symbol size. Source: Based on Jayaraman [29].

be dependent on whether the nanorod has a homogeneous or patchy functionalization. For homogeneously functionalized nanorods, at all aspect ratios explored (i.e. $L/D = 3$, 5, and 15), as the isotropic attraction strength increases, the morphology remains qualitatively unchanged, yet the nanorod–nanorod contact peak increases and nanorod orientational order increases. For patchy functionalized nanorods and smaller aspect ratios (i.e. $L/D = 3$) as the directional attractive strength increases, the PNC morphology goes from dispersed nanorods to percolated nanorods. The $L = 5$d nanorods consistently form a percolating network structure, while $L = 15$d nanorods consistently form finite-sized or percolating, one-dimensional bundles. The impact of the addition of nanorods on the conformation of matrix chains was also obtained. At nanorod volume fractions of 0.05, 0.10, and 0.20, regardless of the nanorod surface functionalization, the conformation of all the matrix chains on average remains the same as the matrix chains in a pure melt (with no nanorods) [31]. However, this is a cumulative effect of the interfacial chains (i.e. chains in contact with the nanorods) being extended and the bulk chains (i.e. chains far from nanorods) either being compacted or remaining statistically the same as the matrix chains in the pure melt. It was noted that as nanorods and matrix chains interact via purely repulsive potential, any observed perturbation of matrix conformation originates purely from the physical presence and arrangement of nanorods. It will be interesting to study how the morphology (i.e. nanorod arrangement as well as matrix chain conformations) is altered when we consider isotropic vs. directional attractive interactions between nanorod and matrix chains.

Dielectric capacitors have been the major enabler for many applications in advanced electronic and electrical power systems because of their capability for ultrafast charging/discharging and ultrahigh power density. The low energy densities of polymer dielectrics used in these capacitors have not been able to meet the ever-increasing demands for compact, reliable, and efficient electric power systems. PNCs, in which high-dielectric constant (k) nanofillers are incorporated in the polymer matrix, have been actively pursued (Figure 16.12). Researchers

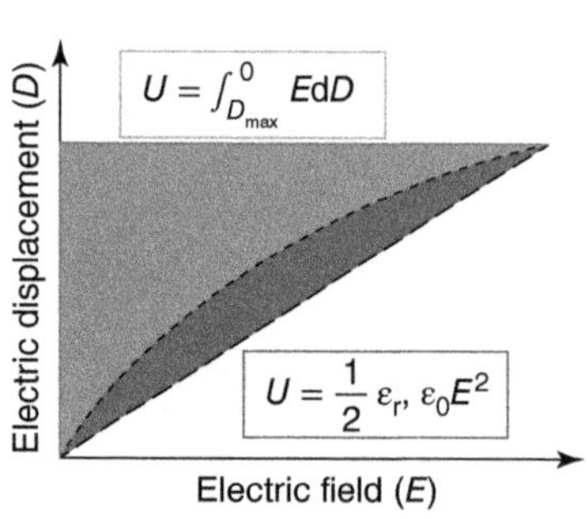

Figure 16.12 Schematic illustration of the electric displacement (D) and discharge energy density (U) with an applied electric field (E). For nonlinear dielectrics, the discharge energy density is determined from the area, whereas for linear dielectrics, the discharge energy density is determined from the triangular area. Note: $\varepsilon_0 = 8.85 \times 10^{-12}$ F m^{-1} is the vacuum permittivity, and ε_r is the relative dielectric permittivity of the dielectric. Source: Shen et al. [32]/Materials Research Society.

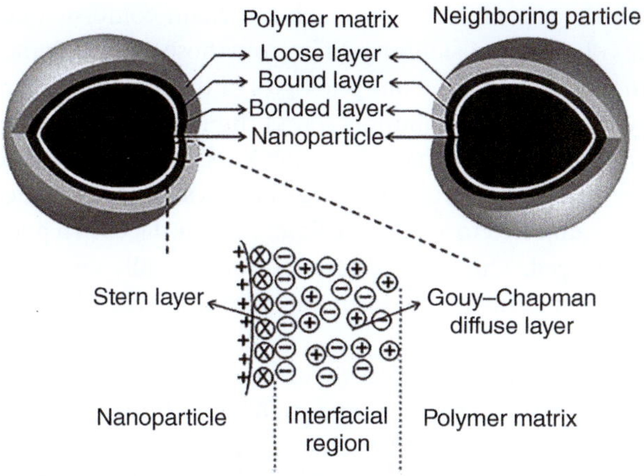

Figure 16.13 Multicore model for interfaces between inorganic nanoparticles and a polymer matrix in a polymer nanocomposite. The model consists of a bonded layer, a bound layer, and a loose layer. As the nanoparticle is positively charged, an electrical double layer (comprising the Stern and Gouy–Chapman diffusion layers) forms in the interfacial region and overlaps the three layers of the model. Source: Modified from Shen et al. [32]/Materials Research Society.

have determined two theoretical considerations for concomitantly increasing the dielectric permittivity and breakdown strength of nanocomposites: critical interfacial polarization and local electric field distribution. In the framework of these considerations, the progress toward PNCs with high energy densities based will be on two approaches: (i) core–shell-structured PNCs and (ii) dielectric anisotropy (Figure 16.13) [33].

16.4 Interface and Interfacial Polarization

In PNC dielectrics, the interface between the ceramic nanoparticles and the polymer matrix is defined as the range over which the permittivity changes from the value of the ceramic nanoparticles to that of the polymer matrix. For instance, for a 10-nm diameter nanoparticle with a 5.0-nm-thick interface, the interfacial volume fraction exceeds 300%. Given their high volume fraction in the composites, interfaces are the dominant feature of nanocomposite dielectrics. Two characteristics of interfaces are critical in describing their dielectric responses, namely how they are chemically formed and how the charges are distributed across the interfacial regions. To describe the formation of interfaces, Tanaka et al. proposed a multilayer core model for a spherical inorganic filler embedded in a polymer matrix [22]. The first layer is the bonded layer, which corresponds to a transition layer that is tightly bonded by ionic, covalent, or hydrogen bonds or van der Waals forces, to both the nanoparticle and the polymer. The second layer (the bound layer) is an interfacial region consisting of a layer of polymer chains strongly bound to or interacting with the bonded layer and the surface of the nanoparticle. The third layer is a region loosely coupled and interacting with the second layer, in which the chain conformation, chain mobility, and even free volume or crystallinity differ from those of the polymer matrix. In addition to the chemical bonds, the morphologies of the interface layers are also strongly affected by the charge distribution across the interfacial regions. Within these regions, Coulombic forces (long range), van der Waals forces (moderate range), and electrostatic forces (short range) are involved and can significantly alter the interface layers [32]. When subjected to an electric field, the interfaces in a polymer composite can become charged. The charged interface is generally described in terms of three charged layers that invariably form and determine the dielectric characteristics of the interface.

On the nanoparticle side, a double layer associated with surface states is induced by immobile charged impurities, trapped carriers, and mobile electrons and holes in the nanoparticle. After the nanoparticle becomes charged, the surrounding polymer responds by establishing a screening counter-charge confronting the charge on the nanoparticles. The mobile ions in the polymer will migrate to establish a diffuse electrical double layer (the so-called Stern–Helmholtz double layer) around the nanoparticle, which is of molecular thickness and can have a high charge density. The nanoparticle side of the Stern–Helmholtz double layer consists of adsorbed ions or dipoles, whereas the polymer side of this layer (also called the outer Helmholtz plane [OHP]) is determined by the nearby ions. Extending beyond the OHP into the polymer is the so-called Gouy–Chapman diffusion layer. In this region, the mobile charges are distributed in such a way that the counter-charges with opposite polarity are diffused outward from the OHP to the Debye shielding length, decaying exponentially with distance according to the Born approximation. Filled with mobile charge, the Gouy–Chapman diffusion layer could play a dominant role in determining the dielectric response of polymer composite dielectrics. By considering the charge density and electric potential within the interface layers, the core–shell model provides plausible explanations for the extreme enhancement in the dielectric response of polymer composites [34]. However, its basic assumption is that the electric field remains unaltered outside the core–shell-type nanoparticles, which is not the case in real polymer composites [35].

16.5 Simulation Techniques Based on Hydrogen Storage

Solid-state hydrogen storage has the potential to significantly improve upon conventional forms of technologies such as liquefaction or compression. Adsorbents with large surface areas and high hydrogen uptakes incorporated into storage tanks can, in principle, improve the volumetric density of stored hydrogen or decrease the operating pressure for a given amount of H_2 stored. As a result, numerous materials have been developed in recent years that can store significant amounts of hydrogen via physisorption.

The composites are based on a polymer of intrinsic microporosity matrix (PIM-1) containing a high surface area porous aromatic framework filler (PAF-1)

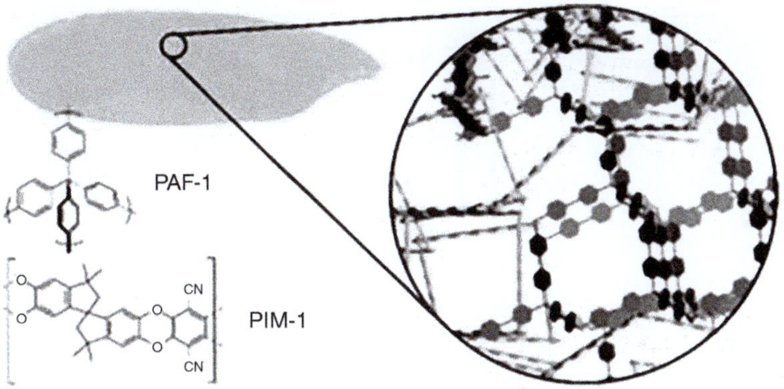

Figure 16.14 A self-standing composite membrane, with a schematic representation of the microporous structure resulting from a mixture of kinked PIM-1 (yellow) and diamondoid PAF-1 (gray), together with the chemical structures of PIM-1 and PAF-1. Source: Rochat et al. [36]/Royal Society of Chemistry/CC BY-3.0.

(Figure 16.14). The hydrogen uptake and mechanical properties of the composites are presented along with an assessment of their potential for integration into high-pressure hydrogen storage tanks, either to increase the storage capacity or to reduce the operating pressure for the same uptake. The composites are more stable and processable than systems such as finely divided physisorbed materials, and they can be made into self-standing films. In addition to retaining the processability of PIM-1, they also possess enhanced surface areas and pore volume approximately proportion incorporated into PAF-1 [37]. Hydrogen uptake measurements combined with theoretical modeling show that the composites can store up to 6.7 wt% H_2 at 77.4 K. Tensile testing and dynamic mechanical thermal analyses indicate decreasing stress and strain to failure with an increasing proportion of PAF-1, although the processability and elasticity of the compounds are maintained until the weight percentage of filler reaches 30%. These lightweight composites show promise as effective hydrogen storage materials, especially for applications where pressures up to 7.5 MPa, are required (Figure 16.15) [29].

The hydrogen fuel cell car is typically representative of the developing hydrogen industry. Safe, high-efficiency, and economical hydrogen storage technique is a key to ensuring favorable run of hydrogen energy car [38]. Among many hydrogen storage patterns including high-pressure gaseous storage, cryogenic liquid storage, and chemical hydrogen storage, high-pressure gaseous storage has become the most popular technique [39]. Composite high-pressure hydrogen storage vessel has been increasingly applied to the hydrogen fuel cell car. The design of a composite vessel involves various integrated parameters such as the progressive failure properties, the burst pressure, and fatigue lifetime [36]. The favorable combination of high reliability and practicability of the composite vessel is a challenging task from the beginning

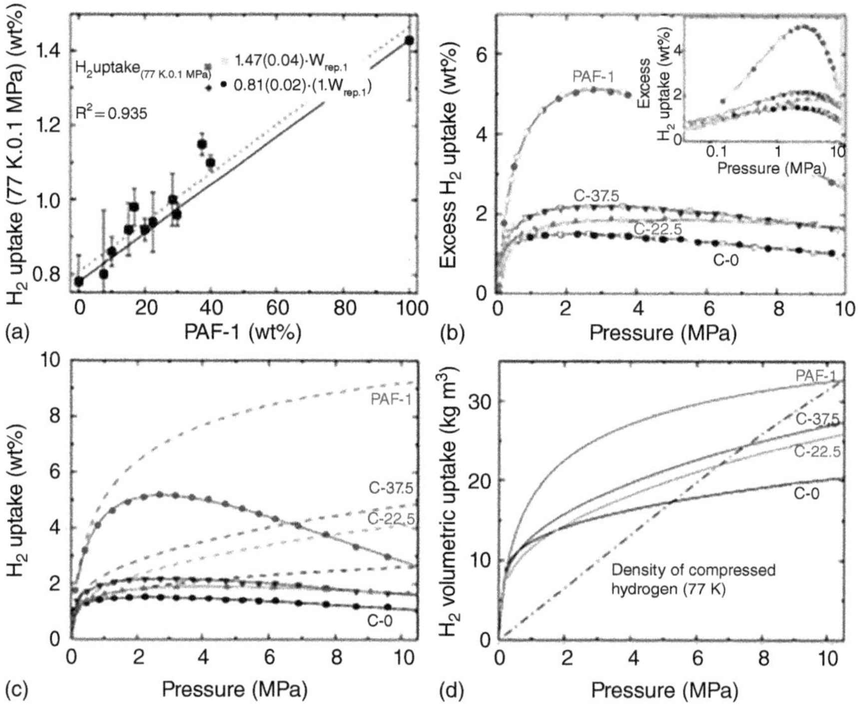

Figure 16.15 (a) Excess hydrogen uptakes (0.1 MPa, 77.4 K) for PIM-1/PAF-1 films as a function of the weight proportion of PAF-1 (average of three independent measurements). Solid and dotted lines represent theoretical and experimental "rule of mixtures," respectively, and the fitting equation is given with standard errors, (b) excess hydrogen uptake for PIM-1 (C-0), PAF-1 powder, and composites C-22.5 and C-37.5 at pressures up to 10 MPa. Filled and open symbols represent adsorption and desorption branches, respectively, (c) experimentally measured excess uptakes (symbols), fitted isotherms (solid lines), and calculated total hydrogen uptakes (dashed lines) of the same samples, and (d) total volumetric hydrogen uptakes of the same samples at 77.4 K. The dash-dotted line shows the volumetric density of compressed hydrogen gas at 77.4 K. Source: Rochat et al. [36]/Royal Society of Chemistry/CC BY-3.0.

of design (Figure 16.16). To reach the aim of large-capacity hydrogen storage, the composite vessel includes an inner aluminum layer and outer carbon fiber/epoxy composite wound layers in China [41]. In general, the composites are used to make the pressure vessel by placing them in different orientations for different layers and in a common orientation within a layer. These layers are stacked in such a way to achieve high stiffness and strength. The designable properties of the composite vessel enable it flexibly applicable to different working conditions and environments by adapting the fiber wound patterns, the dome shapes, and the layer thickness (Figure 16.17) [42].

Figure 16.16 Flow chart of finite element modeling of a composite vessel. Source: Based on Chapelle and Perreux [40].

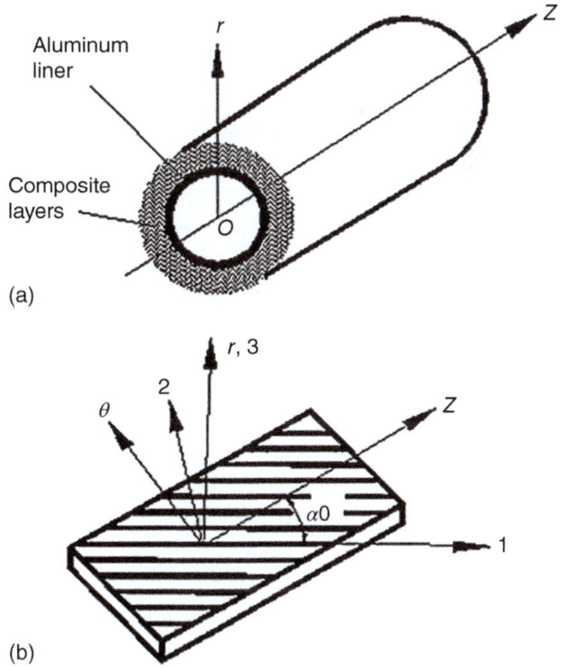

Figure 16.17 Lightweight composite high-pressure hydrogen storage vessel used in the hydrogen fuel cell vehicle. Source: Zheng et al. [42]/with permission of Elsevier.

16.6 Finite Element Modeling of a Composite Hydrogen Storagevessel

Accurate and fast modeling of the composite vessel is an important work in the design, which involves many parameters such as the wound angle and thickness. Parametric modeling by exacting these parameters as the design variables facilitates the strength prediction and optimization for the composite vessels with different sizes [43].

Design theories: In general, the composite vessel is considered a composite laminated structure. Now, there are two main design theories: grid theory and composite laminate theory.

Grid theory: The basic assumptions according to the grid theory are as follows: (i) only the longitudinal carbon fiber bears the pressure and (ii) the effects of wound patterns are neglected. The grid theory can be used to calculate the thickness of each composite layer and the longitudinal *in situ* fiber strength. The strength value is inversely solved by the burst pressure of the composite vessel. Chen [44] gave a set of design methods of the composite vessel based on the grid theory, and the total composite wound thickness h_f is expressed as:

$$\begin{cases} h_{f\alpha} = \dfrac{R_0 P_b}{2\sigma_{fb}\cos^2\alpha_0}, \quad h_{f\theta} = \dfrac{R_0 P_b}{2\sigma_{fb}}(2 - \tan^2\alpha_0) \\ h_f = h_{f\alpha} + h_{f\theta} = \dfrac{3R_0 P_b}{2\sigma_{fb}} \end{cases} \tag{16.1}$$

where $h_{f\alpha}$, and $h_{f\theta}$ are longitudinal and hoop wound thickness, P_b is the burst pressure of composite vessel, and σ_{fb} is the *in situ* strength of carbon fiber. It should be pointed out the grid theory is merely an ideal design method that largely depends on the processing parameters and the sizes of the experimental composite vessel. This shows that an efficient design theory is necessary [45].

The basic assumptions of the classical composite laminate theory (CCLT) are as follows: (i) perfect bonding appears at the interface between each layer, (ii) the mechanical properties of composite laminates are substituted by those of a middle plane, and (iii) the normal stress is neglected at the section parallel to the middle plane. Lifshitz et al. proposed a method using the CCLT to calculate the stress and strain in a nonsymmetric wound pressure vessel with thick metal liners [46]. However, the last two assumptions above do not strictly hold for a 3D composite vessel. In this case, Parnas et al. derived the elastic stress and displacement solutions for 3D cylindrical composite laminates based on the CCLT and generalized plane strain assumption [47]. As the liner is assumed isotropic and the composite layers are considered transversely isotropic, Chapelle et al. further derived the analytical solutions of the stress, strain, and displacement for 3D cylindrical composite laminates [40]. Figure 16.18a shows the cylindrical part of the composite vessel and Figure 16.18b describes a representative volume.

Figure 16.18a shows the principal direction (1, 2, 3) of a composite layer under the cylindrical coordinate (r, θ, z). If the axial strains at all layers are assumed constant

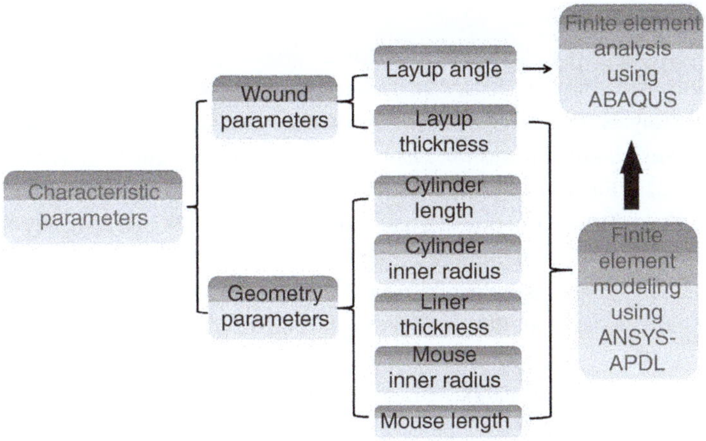

Figure 16.18 (a) Composite hydrogen storage vessel (transverse section) and (b) schematic illustration of the relationship between the on-axis coordinate (1,2,3) and the off-axis coordinate (r,θ,z). Source: Liu et al. [43]/with permission of Elsevier.

and the shear strains are independent of z, the strain–displacement relationship is given by:

$$\varepsilon_r^{(k)} = \frac{du_r^{(k)}}{dr}, \quad \varepsilon_\theta^{(k)} = \frac{u_r^{(k)}}{r}, \quad \varepsilon_z^{(k)} = \frac{du_z^{(k)}}{dz} = \varepsilon_0, \quad \gamma_{z\theta}^{(k)} = \frac{du_\theta^{(k)}}{dz} = \gamma_0 r \tag{16.2}$$

where ε_z is the axial strain, $\gamma_{z\theta}$ is the shear strain, and γ_0 is twist per unit length. The off-axis stress–strain relationship under the hygro-thermomechanical multiphysics field is expressed as:

$$\begin{bmatrix} \sigma_z \\ \sigma_\theta \\ \sigma_r \\ \tau_{z\theta} \end{bmatrix}^{(k)} = \begin{bmatrix} \overline{C}_{11} & \overline{C}_{12} & \overline{C}_{13} & \overline{C}_{16} \\ \overline{C}_{21} & \overline{C}_{22} & \overline{C}_{23} & \overline{C}_{26} \\ \overline{C}_{13} & \overline{C}_{21} & \overline{C}_{33} & \overline{C}_{36} \\ \overline{C}_{16} & \overline{C}_{26} & \overline{C}_{36} & \overline{C}_{66} \end{bmatrix}^{(k)} \begin{bmatrix} \varepsilon_z \\ \varepsilon_\theta \\ \varepsilon_r \\ \gamma_{z\theta} \end{bmatrix}^{(k)} - \begin{bmatrix} \alpha_1 \\ \alpha_2 \\ \alpha_3 \\ \alpha_6 \end{bmatrix}^{(k)} \Delta T - \begin{bmatrix} \beta_1 \\ \beta_2 \\ \beta_3 \\ \beta_6 \end{bmatrix}^{(k)} \Delta C \tag{16.3}$$

where the symbols $\acute{C}_{ij}$ ($i, j = 1, 2, 3, 6$) are off-axis elastic constants of composites. The symbols σ_z, σ_θ, σ_r, and $\sigma_{z\theta}$ are the off-axis axial, hoop, radial, and shear stresses, respectively. ε_z, ε_θ, ε_r, and $\gamma_{z\theta}$ are the corresponding strains. ΔT and ΔC are temperature and moisture concentrations, respectively. α_i ($i = 1, 2, 3, 6$) and β_i ($i = 1, 2, 3, 6$) are the corresponding thermal and moisture expansion coefficients, respectively. For the kth layer, the equilibrium equation under the cylindrical coordinate is given as:

$$\frac{1}{r}\frac{d\left(r\sigma_r^{(k)}\right)}{\partial r} - \frac{\sigma_\theta^{(k)}}{r} = 0 \tag{16.4}$$

After introducing the displacement continuity conditions, the radial stress continuity conditions as well as the axial equilibrium and zero torsion conditions between two neighboring layers, the analytical solutions for the strain, stress, and displacement can be derived by Eqs. (16.2)–(16.4).

References

1 Mishra, A., Shetti, N.P., Basu, S. et al. (2020). Recent developments in ionic liquid-based electrolytes for energy storage supercapacitors and rechargeable batteries. In: *Green Sustainable Process for Chemical and Environmental Engineering and Science*, 199–221. Elsevier.

2 Heinz, H., Vaia, R.A., Krishnamoorti, R., and Farmer, B.L. (2007). Selfassembly of alkylammonium chains on montmorillonite: effect of chain length, head group structure, and cation exchange capacity. *Chem. Mater.* 19: 59–68.

3 Wang, Y., He, Q., Ding, K. et al. (2015). Electrocatalysis in fuel cells. *J. Electrochem. Soc.* 162: F755–F763.

4 Kalowekamo, J. and Baker, E. (2009). Estimating the manufacturing cost of purely organic solar cells. *Sol. Energy* 83: 1224–1231.

5 De Wild-Scholten, M.J. and Veltkamp, A.C. (2007). Energy, environmental life cycle analysis of dye-sensitized solar devices; status and outlook. *In: 22nd European Photovoltaic Solar Energy Conference*, Milan, pp. 3–7.

6 Lin, Y.Y.H., Wu, Y.C., You, H.C. et al. (2014). Ultra-low temperature flexible dye-sensitized solar cell. In: *2014 International Symposium on Computer, Consumer and Control (IS3C)*, Taichung City, Taiwan, 470–473.

7 Smith, G.D., Bedrov, D., Li, L.W., and Byutner, O. (2002). A molecular dynamics simulation study of the viscoelastic properties of polymer nanocomposites. *J. Chem. Phys.* 117: 9478–9489.

8 Uribe Calderon, J.A. (2007). *Clay Modification for the production of Polystyrene Nanocomposites by Melt Processing*. Thesis: McGill University.

9 Vacatello, M. (2003). Predicting the molecular arrangements in polymer-based nanocomposites. *Macromol. Theory Simul.* 12: 86–91.

10 Ozmusul, M.S., Picu, C.R., Sternstein, S.S., and Kumar, S.K. (2005). Lattice Monte Carlo simulations of chain conformations in polymer nanocomposites. *Macromolecules* 38: 4495–4500.

11 Zeng, Q.H., Yu, A.B., and Lu, G.Q. (2005). Interfacial interactions and structure of polyurethane intercalated nanocomposite. *Nanotechnology* 16: 2757–2763.

12 Zeng, Q.H., Yu, A.B., and Lu, G.Q. (2008). Multiscale modeling and simulation of polymer nanocomposites. *Prog. Polym. Sci.* 33: 191–269.

13 Vaia, R.A. and Giannelis, E.P. (1997). Polymer melt intercalation in organically-modified layered silicates: model predictions and experiment. *Macromolecules* 30: 8000–8009.

14 Greenwell, H.C., Harvey, M.J., Boulet, P. et al. (2005). Interlayer structure and bonding in nonswelling primary amine intercalated clays. *Macromolecules* 38: 6189–6200.

15 Paul, D.R., Zeng, Q.H., Yu, A.B., and Lu, G.Q. (2005). The interlayer swelling and molecular packing in organoclays. *J. Colloid Interface Sci.* 292: 462–468.

16 Balazs, A.C., Singh, C., and Zhulina, E. (1998). Modeling the interactions between polymers and clay surfaces through self-consistent field theory. *Macromolecules* 31: 8370–8381.

17 Tanaka, G. and Goettler, L.A. (2002). Predicting the binding energy for nylon 6,6/clay nanocomposites by molecular modeling. *Polymer* 43: 541–553.

18 Famiglia, M., Ferrone, M., and Pricl, S. (2003). Computer simulation of nylon-6/organoclay nanocomposites: prediction of the binding energy. *Fluid Phase Equilib.* 212: 315–329.

19 Zeng, Q.H., Yu, A.B., Lu, G.Q., and Standish, R.K. (2003). Molecular dynamics simulation of organic-inorganic nanocomposites: layering behavior and interlayer structure of organoclays. *Chem. Mater.* 15: 4732–4738.

20 Zeng, Q.H., Yu, A.B., Lu, G.Q., and Standish, R.K. (2005). Interfacial interactions and structure of organic-inorganic nanohybrids. *J. Mater. Sci. Technol.* 21: 114–118.

21 Anderson, K.L., Sinsawat, A., Vaia, R.A., and Farmer, B.L. (2005). Control of silicate nanocomposite morphology in binary fluids: coarse-grained molecular dynamics simulations. *J. Polym. Sci., Part B: Polym. Phys.* 43: 1014–1024.

22 Skountzos, E.N. and Mavrantzas, V.G. (2020). 5 Molecular dynamics simulations of graphene-based polymer nanocomposites. In: *Carbon-Based Smart Materials*, vol. 115 (ed. A.C. Constantinos, P.K. Elias and A.D. Dimitrios), 152. Berlin, Boston: De Gruyter.

23 Ramanathan, T., Abdala, A.A., Stankovich, S. et al. (2008). Functionalized graphene sheets for polymer nanocomposites. *Nat. Nanotechnol.* 3: 327.

24 Li, X.G. and McKenna, G.B. (2012). Considering viscoelastic micromechanics for the reinforcement of graphene polymer nanocomposites. *ACS Macro Lett.* 3: 388.

25 Rissanou, A.N. and Harmandaris, V. (2014). Dynamics of various polymer-graphene interfacial systems through atomistic molecular dynamics simulations. *Soft Matter.* 10: 2876.

26 Theodorou, D.N. and Suter, U.W. (1986). Atomistic modeling of mechanical properties of polymeric glasses. *Macromolecules* 19: 139.

27 Jorgensen, W.L., Maxwell, D.S., and Tirado Rives, J. (1996). Development and testing of the OPLS all-atom force field on conformational energetics and properties of organic liquids. *J. Am. Chem. Soc.* 118: 45.

28 Mac Kerell, A.D., Bashford, D., Bellott, M. et al. (1998). All-atom empirical potential for molecular modeling and dynamics studies of proteins. *J. Phys. Chem. B* 102: 35–86.

29 Jayaraman, A. (2020). 100th anniversary of macromolecular science viewpoint: modeling and simulation of macromolecules with hydrogen bonds: challenges, successes, and opportunities. *ACS Macro Lett.* 9: 656–665.

30 Papadimitriou, K.D., Skountzos, E.N., Gkermpoura, S.S. et al. (2016). Molecular modeling combined with advanced chemistry for the rational design of efficient graphene dispersing agents. *ACS Macro Lett.* 5: 24.

31 Scienomics 2015. MAPS platform, version 3.4.2, France. http://scienomics.com/

32 Shen, Y., Lin, Y., and Zhang, Q.M. (2015). Polymer nanocomposites with high energy storage densities. *MRS Bull.* 40: 753–759.

33 Skountzos, E.N., Anastassiou, A., Mavrantzas, V.G., and Theodorou, D.N. (2014). Determination of the mechanical properties of a poly(methyl methacrylate)

nanocomposite with functionalized graphene sheets through detailed atomistic simulations. *Macromolecules* 47: 8072.
34 Chen, Y., Xu, Q., Jin, Y. et al. (2018). Design of end-to-end assembly of side-grafted nanorods in a homopolymer matrix. *Macromolecules* 51: 4143–4157.
35 Jouault, N., Kumar, S.K., Smalley, R.J. et al. (2018). Do very small POSS nanoparticles perturb s-PMMA chain conformations? *Macromolecules* 51: 5278–5293.
36 Rochat, S., Polak-Krasna, K., Tian, M. et al. (2017). Hydrogen storage in polymer-based processable microporous composites. *J. Mater. Chem.* A5: 18752–18761.
37 Kulshreshtha, A., Modica, K.J., and Jayaraman, A. (2019). Impact of hydrogen bonding interactions on graft-matrix wetting and structure in polymer nanocomposites. *Macromolecules* 52: 2725–2735.
38 Shizhao, L., Zijie, W., and Jayaraman, A. (2021). Molecular modeling and simulation of polymer nanocomposites with nanorod fillers. *J. Phys. Chem. B* 125: 2435–2449.
39 David, W.I.F. (2011). Effective hydrogen storage: a strategic chemistry challenge. *Faraday Discuss.* 151: 399.
40 Chapelle, D. and Perreux, D. (2006). Optimal design of a Type 3 hydrogen vessel. Analytic modeling of the cylindrical section. *Int. J. Hydrogen Energy* 31: 627–638.
41 Chen, P. and Zhu, M. (2008). Recent progress in hydrogen storage. *Mater. Today* 11: 36–43.
42 Zheng, J.Y., Liu, X.X., Xu, P. et al. (2012). *Int. J. Hydrogen Energy* 37: 1048–1057.
43 Liu, P.F., Chu, J.K., Hou, S.J. et al. (2012). Numerical simulation and optimal design for composite high-pressure hydrogen storage vessel: a review. *Renewable Sustainable Energy Rev.* 16: 1817–1827.
44 Zheng, J.Y., Bie, H.Y., Xu, P. et al. (2007). Investigation on standards of fully-wrapped composite tanks for on-board storage of high-pressure hydrogen. *Pressure Vessel Technol.* 24: 48–56.
45 Cox, B.N. and Yang, Q.D. (2006). In quest of virtual tests for structural composites. *Science* 314: 102–107.
46 Lifshitz, J.M. and Dayan, H. (1995). Filament-wound pressure vessel with thick metal liner. *Compos. Struct.* 32: 313–323.
47 Parnas, L. and Katirci, N. (2002). Design of fiber-reinforced composite pressure vessels under various loading conditions. *Compos. Struct.* 58: 83–95.

17

Life Cycle Analysis of Polymer Nanocomposites for Energy Storage

17.1 Scope of a Life Cycle Analysis

Nanocomposites comprising green polymers are currently drawing a lot of attention because the polymers are renewable and biodegradable, thus eliminating existing challenges revolving around biocompatibility and sustainability [1]. To get a better understanding of these composites to ultimately make them available for various applications in the market, it is important to assess the life cycle of such composites.

The scope of a Life Cycle Analysis (LCA) comprises the extraction of resources, the synthesis of inputs, product manufacture, waste treatment, packaging, use, and ends when the product is finally disposed of or recycled. Such an analysis is "cradle-to-grave." The purpose of the LCA should be stated clearly in this phase. The purpose could be a comparative assessment of nanoproducts vs. alternatives, the preparation of a nanoproduct using different synthesis routes, or the use of a nanoproduct for a new application [2]. This is especially important to aid in defining the appropriate functional unit for the study, to give a fair comparison. As an example, consider a life cycle environmental comparison of plastic vs. paper grocery bags. One paper bag might hold the same amount of groceries as two plastic bags. Thus, the appropriate functional unit, in this case, should be one paper bag vs. two plastic bags. Once the functional unit is fixed, the process boundary should be specified clearly to define the scope of the LCA [3]. Since data collection and analysis can be very time-consuming and expensive, it is tempting to narrow the inputs considered or to make the boundary too narrow, leading to invalid or flawed results. Also, since a process or product is connected to the rest of the economy through innumerable links, any attempt to get every input may make the study too expensive or time-consuming. For these reasons, proper care and vigilance are crucial during this stage to yield meaningful and interpretable results. Also, performing multiple studies with different boundaries to determine the effect of boundary limits can be helpful in certain situations [4]. The life cycle practitioner should always state the purpose of the study, which usually helps in defining the boundary (Figure 17.1).

A typical life cycle of a polymer nanocomposite product is shown in Figure 17.2. Several alternatives exist at each step in the complete life cycle. The selection of the polymeric resin depends on the application and hence the desired properties.

Polymer Nanocomposites for Energy Applications, First Edition.
T. Daniel Thangadurai, Manjubaashini Nandhakumar, Sabu Thomas, and Ange Nzihou.

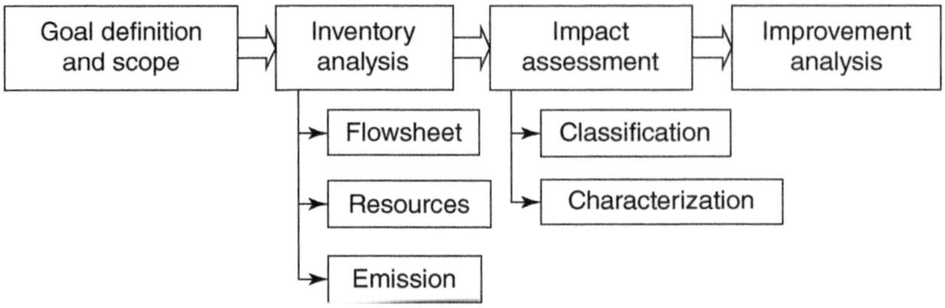

Figure 17.1 Phases of an LCA based on ISO 14040.

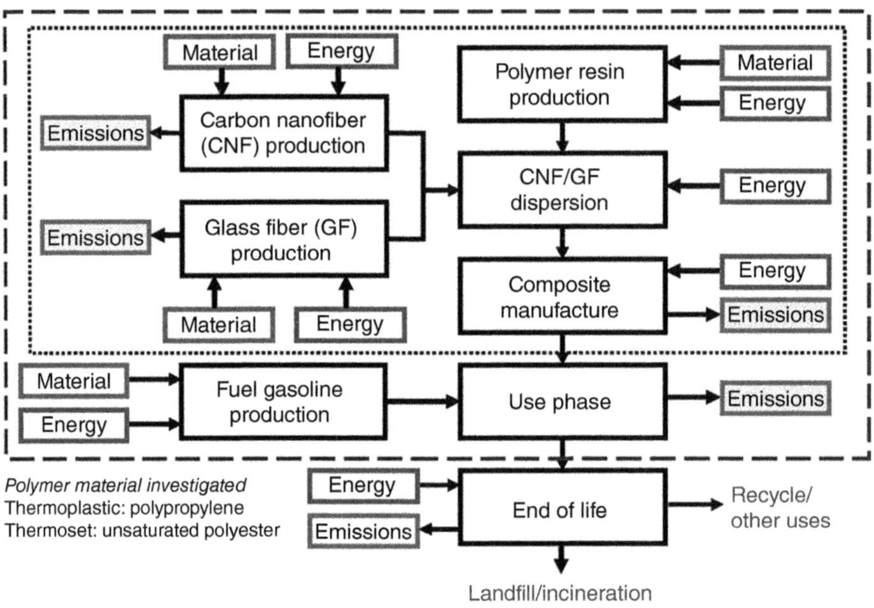

Figure 17.2 Typical life cycle of polymer nanocomposite (dotted lines indicate the system boundary for the cradle to gate comparison of PNCs with steel; dashed lines represent the boundary for the automotive body panel case study).

Polypropylene (PP) and unsaturated polyester resin (UPR) are considered as the thermoplastic and thermosetting resins, respectively, as these have been widely studied in several nanocomposite experimental studies concerning their mechanical and electrical properties. Both simple CNF and hybrid CNF-glass fiber polymer nanocomposites are evaluated [5].

17.2 Techno-Economic Evaluations of Energy Storage Systems

Each storage technology has unique characteristics and is different in terms of its appropriate application field and energy storage scale [6]. A comprehensive analysis of each storage technology needs to be performed before a decision can be made about the storage technology that is most suitable [7].

17.3 Energy and Power Density

The power density (W/kg or W/liter) is the rated output power divided by the volume of the storage device. The energy density is calculated as stored energy divided by the volume. The volume of the storage device is the volume of the whole energy storage system.

17.4 Self-Discharge

There is the portion of the energy that was initially stored and which has dissipated over a given amount of nonuse time.

17.5 Response Time

Some applications have very stringent requirements concerning the speed with which the energy has to be released or absorbed. In some applications, a few milliseconds may sometimes be the maximum response time that is acceptable [8].

17.6 Cost and Economies of Scale

The auxiliary components required by some energy storage systems determine the total system costs and are often independent of system size. For these reasons, some storage systems are only economically feasible above minimum energy content and power output [9]. To obtain the cost per output (useful) energy, the cost per unit of energy is divided by the storage efficiency. To evaluate the cost of energy storage in a frequent charge–discharge application such as load leveling, we define the per-cycle cost (the cost per unit energy divided by the cycle life).

17.7 Lifetime

The overall energy storage cost is determined by the original investment costs and its projected lifetime. The accuracy with which the lifetime can be estimated is a particularly important problem for all energy storage systems.

17.8 Storage Capacity

This is the quantity of available energy in the storage system after charging. Discharge is often incomplete. Since the discharge is often incomplete, the storage capacity is defined based on total energy stored (Wh) which is superior to that retrieved [10].

17.9 Monitoring and Control Equipment

The performance of some systems can be monitored extremely easily and cheaply, whereas in some other systems a considerable effort has to be used to monitor the available energy content and the safety of the energy storage.

17.10 Efficiency

The process of storing and drawing energy can cause considerable losses. Many auxiliary components of the energy storage system have a constant power demand, and besides, there are energy losses inherent in the storage principle [11]. These losses can be very high concerning the energy content.

17.11 Operating Constraints

The cost of providing the correct environment and operating constraints such as temperature and safety systems, etc., have to be considered for proper estimates of the lifetime expectations and overall costs [12]. Based on these criteria, the appropriateness of the energy storage system for various applications has been evaluated in the literature, such as, for flexible alternating current transmission systems, small-/medium-/large-scale applications, system efficiency, emissions control, peak shaving, and deferring facility investments in peaking generators. Many studies compare the performance of the various storage technologies in different categories based on lists of criteria from technical, economic, and environmental. Cheung et al. provided a comprehensive comparative analysis for pumped hydroelectric storage, compressed air energy storage, batteries, superconducting magnetic energy storage, flywheel, supercapacitor, and thermal energy storage [13]. Schoenung and Hassenzahl proposed a life cycle cost analysis to compare energy storage technologies [14]. Chacraetal used multiple objective optimization methods to evaluate the economic performance of energy storage technologies [15].

References

1 Sarkar, S., Gulati, K., and Poluri, K.M. (2020). Life cycle assessment and future perspectives of green polymeric nanocomposites. In: *Green Polymeric Nanocomposites*, 1–30. CRC Press.

2 Rebitzer, G., Ekvall, T., Frischknecht, R. et al. (2004). Life cycle assessment Part 1: Framework, goal and scope definition, inventory analysis, and applications. *Environ Int.* 30: 701–720.

3 Vikaskhanna, B. and Bakshi, R. (2009). Carbon nanofiber polymer composites: evaluation of life cycle energy use. *Environ. Sci. Technol.* 43: 2078–2084.

4 Khanna, V., Merugula, L., and Bakshi, B.R. (2012). Environmental life-cycle assessment of polymer nanocomposites (Chapter 2). In: *Advances in Polymer Nnanocomposites*, 33–54. Wood-head Publishing Limited.

5 Hattum, V., Leer, C., Viana, J. et al. (2006). Conductive long fiber reinforced thermoplastics by using carbon nanofibres. *Plastics, Rubber, and Composites* 35: 247–252.

6 Baker, J. (2008). New technology and possible advances in energy storage. *Energy Policy* 36: 4368–4373.

7 Makarov, Y.V., Yang, B., DeSteese, J.G., et al. (2008). Wide area energy storage and management system to balance intermittent resources in Bonneville Power Administration and California ISO control areas. Specific Northwest National Laboratory PNNL-17575.

8 Vosen, S.R. and Keller, J.O. (1999). Hybrid energy storage systems for stand-alone electric power systems optimization of system performance and cost through control strategies. *Int. J. Hydrogen Energy* 24: 1139–1156.

9 Shoenung, S.M. (2001). *Characteristics and Technologies for Long- vs Short-Term Energy Storage*. United States Department of Energy.

10 Díaz-Gonzáleza, F., Sumper, A., Bellmunt, O.G., and Robles, R.V. (2012). A review of energy storage technologies for wind power applications. *Renewable Sustainable Energy Rev.* 16: 2154–2171.

11 Tan, X., Li, Q., and Wanga, H. (2013). Advances and trends of energy storage technology in microgrid. *Electr. Power Energy Syst.* 44: 179–191.

12 Cheung, K., Cheung, S., DeSilva, R. et al. (2003). *Large Scale Energy Storage Systems*. Imperial College, London.

13 Chacra, F.A., Bastard, P., Fleury, G., and Clavreul, R. (2005). Impact of energy storage costs on economic performance in a distribution substation. *IEEE Trans. Power Syst.* 20: 684–691.

14 Schoenung, S.M. and Hassenzahl, W.V. (2003). Long- vs. short-term energy storage technologies analysis – a life cycle cost study. Sandia National Laboratories Albuquerque SAND 2783.

15 Kousksou, T., Bruel, P., Jamil, A. et al. (2014). Energy storage: Applications and challenges. *Solar Energy Mater. Solar Cells* 120: 59–80.

18

Future Research and Case Study on Energy Storage System

18.1 Introduction

Different conditions should be supported to improve the power system stability of renewable energy penetration. Mitigating power fluctuations is one of these issues. Since solar irradiation and wind speed fluctuate with time, the output power of a photovoltaic (PV) system or wind turbine also fluctuates. The durations and amplitudes of these fluctuations may be big to cause voltage and frequency variations in the power system and to damage the power quality, which is more serious in isolated power systems [1]. Therefore, energy storage systems are used to improve the voltage and frequency stability and power quality of the power system. Since the time range of power, fluctuations are usually less than a minute, and energy storage technology that is suitable for short-term energy storage applications with high ramp power rates is used such as lead-acid batteries, flow batteries, supercapacitors, flywheels, and superconducting magnetic energy storage systems [2].

Although initial studies have focused on connecting energy storage systems to DC bus of doubly fed induction generators and permanent magnet synchronous generators via a bidirectional DC–DC converter, connecting the energy storage system to the point of common coupling (PCC) has recently become popular to use the energy storage system more flexible [3]. The stochastic behavior of the renewable energy sources may affect the power system stability, especially in the case of high penetration of renewable energy sources to the power system. The energy storage systems can be used to support power system stability.

18.2 Case Study 1: Pumped Storage Hydropower (PSH) in France

Pumped storage hydropower (PSH) in France presents a unique scenario for energy storage and nuclear power due to the country's high concentration of electricity generation from nuclear power. Hydroelectric facilities in France serve

Polymer Nanocomposites for Energy Applications, First Edition.
T. Daniel Thangadurai, Manjubaashini Nandhakumar, Sabu Thomas, and Ange Nzihou.

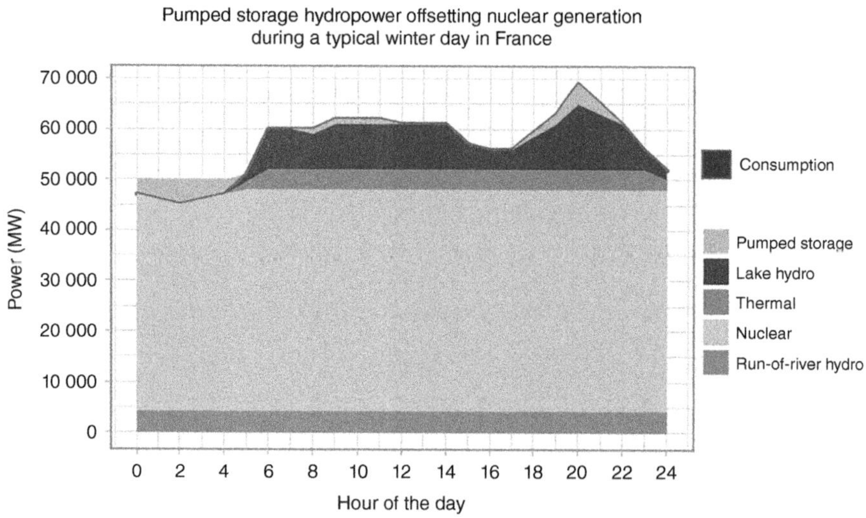

Figure 18.1 PSH powered by nuclear during times of low demand to provide additional generation capacity at a later time to meet peak demand in France.

an important secondary purpose by operating as an inexpensive and flexible form of energy storage. This energy storage capacity is critical, since the start-up time of a typical nuclear reactor can be up to 40 hours, while almost 15 000 MW of hydroelectric capacity can be brought online in a matter of minutes. Although both run-of-river- and reservoir-type hydroelectric facilities can operate as energy storage systems, the type of hydroelectric facility considered in this analysis is PSH. These facilities can be used to pump water to a higher elevation during periods of low demand so that this water can be used to generate additional electricity to meet peak load. PSH facilities make up approximately 16% of hydropower capacity in France or around 4–5 GW. On a normal day, this energy storage capacity is used to provide about four hours of additional generation during periods of high consumption, as displayed in Figure 18.1. For this case study, it was assumed that an energy storage developer was hired to address the French government's concerns and provide additional flexibility for the nation's large nuclear power fleet [4]. A timeline of one to five years was chosen because technical maturity is critical to this simulation, considering the actual energy storage deployment in France that occurred decades ago. To compare energy storage options for this case study, the weighting factors for technical maturity, economic feasibility, and environmental impact were set to 1, 0.8, and 0.5, respectively. Note that the selection of these weighting factors is mostly arbitrary and could be changed by another interested user. The resulting simple calculation yields the following energy storage options in order of recommendation: (i) molten salts, (ii) hot and cold water, (iii) compressed air energy storage (CAES), (iv) lead-acid batteries, (v) sodium-sulfur batteries, and (vi) PSH. The result of PSH being listed in sixth place is at odds with the fact that the French chose to install PSH as their primary source of energy storage.

18.3 Case Study 2: Battery Storage

The lithium-iron-phosphate battery is designed to run in any environment. The equipment has been successfully installed and run in the desert, tropical, mountainous, and coastal locations under a wide variety of temperature and humidity conditions. The power electronics used in the installation were commercially available but were custom-built for the project by Dynapower Corporation of Vermont. The actual nameplate capacity utilization for this particular project was not disclosed, but the battery system is designed for and capable of discharging to 100% depth of discharge (DoD). Nanophosphate® lithium-ion battery (LIB) technology does not have limitations on DoD or extended periods at low state of charge, unlike lead-acid battery technologies. Overall project costs were driven by equipment costs. The largest component cost for the battery itself was the lithium-ion cells. An exact percentage breakdown was not provided. Operational/variable costs are driven by energy losses that occur during electricity conversion and storage via inverters and in battery cells. Thermal management equipment consumes the most electricity as part of auxiliary power consumption. Round trip efficiency is about 80%. This includes losses due to power conversion from AC to DC and back to AC, the energy storage cells, bus bars, battery management systems, and thermal management systems. Maintenance costs are primarily made up of labor and travel costs to the Maui area. They consist of routine equipment inspections. Extended warranties provided for manufacturer parts, labor, and travel form another portion of maintenance costs. This extended warranty cost is typically a single-digit percentage of the original equipment and project development cost. The system is designed so that the customer can easily access and service it (Table 18.1). This includes replacing common parts and responding to alarms. Not all maintenance is performed by National Electrical Code (NEC) [5]. The operator may also maintain some of the system components, including the battery management systems, battery modules, and other recommended spare parts. NEC customers typically

Table 18.1 Technical information of a battery.

Round trip efficiency (AC–DC–AC)	Cost estimate (per MW)	Cell cycle life	Expected installation lifetime	Container shipping weight (2 MW/500 kWh)	Inverter shipping weight
80–85%	US$ 800 000–1.2 million	8000 cycles	20 years	59 000 lb (26 762 kg)	50 000 lb (22 680 kg)

employ their existing staff to carry out ordinary maintenance. These employees must be qualified to carry out basic electro-mechanical maintenance on industrial equipment.

18.4 Case Study 3: Solar PV Storage and Energy Shift

This project was implemented by the secretariat of the Pacific Community and funded by the European Union. It consisted of five flooded lead-acid battery installations with a total capacity of 1593 kWh. These were installed on the island archipelago of Yap combined with solar PV systems. The total solar PV capacity is 270 kWp on 5 outer islands of Yap State, encompassing 10 mini-grids. Hoppecke, a German company, manufactured the batteries. The goal was to provide access to electricity in some areas and displace diesel generation in others, estimated to cost USD 1.2 kWh^{-1}. This battery type was chosen because it was considered the most cost-effective solution for the application. It was designed to last for at least 10 years at 30 °C if DoD was limited to 20%, to maximize battery lifetime. Several containers were shipped from the factory in Germany to Yap Main Island, where a small state-owned ship then brought the batteries to individual islands. The cells were carried by hand to the battery houses. The inverter used in all the projects was manufactured by shape memory alloys, known as Sunny Island. Project cost breakdown depends on the site and application for which the battery is used (Table 18.2). Schools had a much smaller battery cost (e.g. less capacity was needed for daily load) compared to rural villages. For the same capacity of PV, villages needed much more storage to serve their evening lighting load. For decommissioning, the supply contract required a commitment to ship the used battery back to a recycling facility in Germany at the supplier's cost [6]. An alternative solution would be that the local power utility sells the lead to a nearby recycling facility.

18.5 Case Study 4: Solar and Battery Storage for Customers and Ancillary Services

Solar Grid Storage LLC provided a 500-kW AC storage and inverter combination to the headquarters of real estate developer Konterra in Maryland. The storage comprises 300 kWh of LIBs provided by All Cell Technologies. The storage and inverter

Table 18.2 Selected technical information of flooded lead-acid battery installations.

Maximum depth of discharge	Estimated calendar life	Estimated cycle life	Estimated cost (US$/kWh)	Individual cell weight
20%	15 years	8000	1340	221 kg

Table 18.3 Technical details of Power Factor 500.

Container size	Weight	Nominal battery	Battery operating voltage (V)	Peak efficiency
20 foot ISO Container	27 000 lb	235 Ah	922–996	95.3%

system was connected to solar PV panels with a capacity of 400 kW. Solar Grid provided two Power Factor 250™ installations. The intention is to provide both the customer and the grid with multiple benefits. This includes backup emergency power during grid outages and reduced system cost through additional revenue from regulation and fast-power balancing support to the local grid. Other ancillary services like spinning reserve and black start capability are possible but not easy to monetize in the present market and regulatory structure. The systems are connected to the local utility and can thus provide a range of ancillary services when required by the utility or grid operator through normal dispatch. The provision of ancillary services to the local grid is a key to Solar Grid's business model. The company's standard service is to provide the inverter and storage system (Power Factor™) at a very low cost. It pays for its installation, whereupon Solar Grid owns and operates the system for 10 years. The company recovers these costs plus a profit margin through ancillary services provided to the grid and paid for by the grid operator [7]. These business models thus utilize battery storage to capitalize on new regulatory markets for ancillary services in the United States. At the same time, it brings down the cost of the PV system and provides both customer and the grid with greater resilience (Table 18.3).

Potassium-ion batteries (PIBs) with the prominent advantages of sufficient reserves and economical cost are attractive candidates for new rechargeable batteries for large-grid electrochemical energy storage systems (EESSs) [8]. EESSs must be considered to solve the intermittent issues of clean energy, which has a great effect on storing and delivering renewable energy resources. So far, LIBs, as a representative energy storage technology, have been widely explored for portable devices, electrical vehicles, and large-grid EESSs because of their high energy density and stable cycling lifespan [9]. Unfortunately, there are some barriers for LIBs to develop sustainably, such as challenging lithium (Li) recovery, uneven distribution of Li, and increasing cost. Based on cost and resource considerations, scientists have put a lot of effort in developing a series of non-LIBs, including sodium-ion batteries (SIBs), PIBs, magnesium-ion batteries (MIBs), zinc-ion batteries (ZIBs), aluminum-ion batteries (AIBs) [10]. Since zinc, magnesium, and aluminum are less active than lithium, they could be used as anode materials for metal-ion batteries; especially, magnesium and aluminum anodes do not even form dendrites, so their corresponding ion batteries can meet the safety requirement [11]. Besides, commercial LIBs mainly use graphite with a theoretical specific capacity of 372 $mAh\,g^{-1}$ as anode material, while the capacities of zinc (820 $mAh\,g^{-1}$), magnesium (2205 $mAh\,g^{-1}$), and aluminum (2980 $mAh\,g^{-1}$) anodes are much higher than that of graphite [12].

References

1 De Haan, J., Frunt, J., and Kling, W. (2010). Mitigation of wind power fluctuations in smart grids. In: *Innovative Smart Grid Technologies Conference Europe (ISGT Europe) IEEE PES 1-8*, 11–13.

2 Qin, Z., Liserre, M., Blaabjerg, F., and Wang, H. (2013). Energy storage system by means of improved thermal performance of a 3 MW grid side wind power converter. In: *Proceedings of IEEE-IECON*, 736–742.

3 Altin, N. (2016). Energy storage systems and power system stability. In: *3rd International Smart Grid Workshop and Certificate Program (ISGWCP)*, 21–25.

4 Lubek, J. and Wakeford, S. (2015). *The Future of Hydroelectricity in France: What Does the New Energy Transition Law Mean for the Long-Delayed Renewal of Concessions?* White Plains, NY: NERA Economic Consulting.

5 Samuel, C., Johnson, F., Davidson, T. et al. (2019). Selecting favorable energy storage technologies for nuclear power, Chapter 5. In: *Storage and Hybridization of Nuclear Energy*, 119–175. Academic Press.

6 NEC Energy Solutions.

7 SPC North-REP Project: http://www.spc.int/northrep.

8 Based on Solar Grid Storage.

9 Zheng, J., Wu, Y., Sun, Y. et al. (2021). Advanced anode materials of potassium ion batteries: from zero dimension to three dimensions. *Nano-Micro Lett.* 13: 12.

10 Wang, N., Chu, C., Xu, X. et al. (2018). Comprehensive new insights and perspectives into Ti-based anodes for next-generation alkaline metal (Na^+, K^+) ion batteries. *Adv. Energy Mater.* 8: 1801888.

11 Selvakumaran, D., Pan, A., Liang, S., and Cao, G. (2019). A review on recent developments and challenges of cathode materials for rechargeable aqueous Zn-ion batteries. *J. Mater. Chem. A* 7: 18209–18236.

12 Liu, F., Liu, Y., Zhao, X. et al. (2020). Prelithiated V_2C MXene: a high-performance electrode for hybrid magnesium/lithium-ion batteries by ion co-intercalation. *Small* 16: 1906076.

Index

Polymer Nanocomposites for Energy Applications, First Edition.
T. Daniel Thangadurai, Manjubaashini Nandhakumar, Sabu Thomas, and Ange Nzihou.

d

e

q

r